Regulatory Mechanisms of Synaptic Transmission

Regulatory Mechanisms of Synaptic Transmission

Edited by

Ricardo Tapia

Department of Neurosciences
Center for Research in Cellular Physiology
Universidad Nacional Autónoma de México
Mexico City, Mexico

and

Carl W. Cotman

Department of Psychobiology
University of California
Irvine, California

Published in cooperation with
Universidad Nacional Autónoma de México

PLENUM PRESS • NEW YORK AND LONDON

Library of Congress Cataloging in Publication Data

Main entry under title:

Regulatory mechanisms of synaptic transmission.

Proceedings of a symposium, held April 14-16, 1980, in Mexico City, Mexico.
Includes bibliographical references and index.
1. Neural transmission — Regulation — Congresses. 2. Neurotransmitters — Congress-
es. I. Tapia, Ricardo. II. Cotman, Carl W. [DNLM: 1. Neuroregulators — Physiology
— Congresses. 2. Synapses — Physiology — Congresses. WL 102.8 R344 1980]
QP364.5.R43   599.01′88   81-8545
ISBN-13: 978-1-4684-3970-0   e-ISBN-13: 978-1-4684-3968-7   AACR2
DOI: 10.1007/978-1-4684-3968-7

Proceedings of a Symposium on Regulatory Mechanisms of
Synaptic Transmission, held April 14—16, 1980, In Mexico City, Mexico,
and sponsored by Universidad Nacional Autónoma de México

© 1981 Plenum Press, New York
Softcover reprint of the hardcover 1st edition 1981
A Division of Plenum Publishing Corporation
233 Spring Street, New York, N.Y. 10013

PREFACE

This book, which originated in the presentations of a symposium
sponsored by the Universidad Nacional Autónoma de México, held in
Mexico City on April 14-16, 1980, represents an attempt to analyze
some of the most relevant aspects of synaptic transmission. This topic
was chosen on the strong belief that the progress of the neurosciences
depends to a great extent on the understanding of the basic mechanisms
of synaptic function. Rather than selecting only a specific approach
or speciality, the book intends to cover this field in a multidisciplinary way. This means that neurochemical, neurophysiological
and morphological studies are mingled throughout the book, which
hopefully will help the reader to integrate the different faces of
the same problem.

Across the book the presynaptic component of synaptic
transmission and its regulation is stressed much more than the
postsynaptic phenomena. Although this might be a limitation, it has
the advantage of increasing the focus of the book on a series of
events which are gaining importance and interest every day.

The book covers several aspects of the synaptic role of three
amino acids, glutamate, γ-aminobutyrate and taurine, as well as that
of catecholamines and some peptides, emphasizing their regional
distribution and possible participation as neurotransmitters or
modulators. The mechanism of the action of calcium in neurotransmitter
release is amply revised both neurophysiologically and neuro-
chemically, including the participation of some intraterminal
proteins. Synaptic plasticity is analyzed through the study of
selective reinnervation of muscle fibers and of simple forms of
learning. Certain more integrative aspects, such as presynaptic
inhibition, facilitation by repetitive stimulation and synaptic
mechanisms in sleep, are also covered. Several of the chapters
include theoretical models based on the available experimental data.

We wish to express our gratitude to the Centro de Investigaciones
en Fisiología Celular, the Secretaría Ejecutiva del Consejo de
Estudios de Posgrado and the Dirección General de Asuntos del
Personal Académico, all of the Universidad Nacional Autónoma de

México, who made possible the organization of the symposium and the realization of this book.

The Editors

CONTENTS

CATECHOLAMINES AND ENDORPHINS AS NEUROTRANSMITTERS AND

NEUROMODULATORS

George R.Siggins

Arthur V. Davis Center for Behavioral Neurobiology

The Salk Institute, La Jolla, California 92037

INTRODUCTION

In a meeting devoted to the discussion of synaptic regulation, it seems instructive first to address the issue of what kinds of synaptic messages may be transmitted and what criteria may be developed to define the different types. This task seems particularly timely now because of the burgeoning number of substances found in brain which seem not to strictly fit our preconceived notions of a neurotransmitter. After counting all the peptides, monoamines, amino acids, nucleotides, prostaglandins and steroids that have been advanced as transmitter candidates, I come up with a number exceeding 60. It seems unlikely that such a wide variety of substances would provide only a few chemical messages. Therefore, after developing criteria for neuro-transmitters, neuromediators, and modulators, I will provide examples of substances that could satisfy the criteria for these classifications, with special emphasis on norepinephrine (NE), cyclic AMP and the opioid peptides.

Historical Perspective

The hypothesis of chemical neurotransmission was first developed for the vertebrate peripheral nervous system, which was thought to include nerve terminals that communicated to muscle tissue via release of acetylcholine or catecholamines (17,18,24,54). The chemical doctrine of Dale was later generalized to the central nervous system, although not without vigorous discussion along the way (cf. 10,19). Arising out of the controversy and research in this area was the principle that different chemicals and different neurons transmit different messages (excitation or

1

inhibition). Most of the resistance to the doctrine of central
chemical transmission seems to have arisen out of consideration for
the speed by which spinal cord synapses transmitted their messages.
Sir John Eccles, then a critic of the doctrine, felt that central
messages demanded rapid electrical transmission, although later
studies have shown that the original peripheral models for the
chemical doctrine (the autonomic nervous system) showed both
rapid and slow forms of chemical communication (12,92). The Eccles
group performed the experiments that eventually proved the
chemical doctrine for the spinal cord (16,26). However, the
narrow view that persisted until recently was that neurotransmission operated in only two modes, fast excitation via
excitatory postsynaptic potentials (EPSPs) and fast inhibition
via inhibitory postsynaptic potentials (ISPSs), both caused by the
opening of conductance channels in the post-synaptic membrane
(see below and refs. 25,44), brought about by the release of
fast-acting transmitters from large, rapidly conducting axons.

However, results from more recent electrophysiological and
ultrastructural studies on autonomically innervated smooth muscle
systems point up the fact that neurotransmission need not always
involve fast signals transferred across narrow, specialized
synaptic clefts. Features common to many neuro-effector smooth
muscle systems (see 12,13,52,76) are: 1) long junctional delays;
2) long time courses of post-junctional potential changes; 3) the
requirement for repetitive nerve activation to produce detectable,
summated responses; 4) wide junctional gaps of up to thousands of
angstroms, without synaptic specializations; 5) transmitter release
from boutons or varicosities 'en passage.'

It is now apparent that nerve cells also communicate with
one another by means of a much more complex vocabulary than a simple
rapid 'yes' or 'no'. Inklings of the idea that neuronal
transmission might not always require the rapid opening of
conductance channels was seen in the non-synaptic models of the
photoreceptor (4) and the stretch receptor (60), where 'leaky'
ion channels seemed to be closed or 'inactivated' to produce
the physiological response. The possibility that such a novel
form of response might apply also to vertebrate central neurotransmission gained credence with results of studies on mammalian
and amphibian sympathetic ganglia. Here, a new form of slow
transmission, the slow IPSP (sIPSP) and the slow EPSP (sEPSP),
occurred without an increase in membrane conductance, and co-
existed with the more usual form of rapid nicotinic depolarization
(EPSP) accompanied by a conductance increase (27,65). Although a
controversy still centers on whether the sIPSP is generated by
a decrease in ionic conductance (96) or by activation of an
electrogenic pump (65), the important point is that investigators
began to ask if such a new form of transmission might not also
apply to the central nervous system.

Shortly after the discovery of the sIPSP and the sEPSP, it became apparent that similar slow mechanisms might apply also to certain forms of synaptic transmission in brain. Thus, exogenous norepinephrine (NE) and activation of NE-containing fibers was found to evoke slow hyperpolarizations in cerebellar Purkinje cells associated with either a decrease or no change in ionic conductance (40; also see below), while muscarinic, cholinergic depolarizing responses associated with a decreased conductance were reported in spinal cord (105) and in a cortical area thought to possess cholinergic fibers (50). These responses in central neurons thus parallel those in sympathetic ganglia, where the sEPSP is thought to arise from activation of muscarinic receptors (27,47), while the sIPSP may be generated by release of a catecholamine (27, but see ref 97).

However, at about this time researchers began to anticipate the presence in brain of even more complicated forms of neuronal communication. For example, hypotheses were advanced that the post-synaptic action of certain neurotransmitters (NE, ACh) might be mediated by the generation of cyclic nucleotides (cyclic AMP and cyclic GMP; see below). In addition, with the discovery of a wide array of neuroactive peptides in brain came the likelihood that several of these might alter the responses of neurons to other neurotransmitters without having a direct action of their own. Examples of this putative form of neuronal communication were seen in iontophoresis experiments showing that TRH could alter the excitatory action of ACh (99, but see 75), and that opioid peptides could reduce synaptic efficacy and glutamate and ACh responses (2,100,102); in both cases interactions occurred without an apparent direct peptide effect on neuronal excitability.

Measures of Excitability: Transmembrane Properties

As suggested above the most conventional mechanism of hormone or transmitter action on cell excitability involves binding to a specific receptor, resulting in a change in the transmembrane permeabilities to one or more ions. The influence on membrane potential depends upon the ionic permeability changes. Each ionic species is in unequal concentration on either side of the membrane due to the relative membrane impermeability to some ions and the activity of ion pumps; thus there is a driving force, determined by the concentration gradient, for each ionic species. An equilibrium potential, E_X, can be defined for each ion (X) at which the electrical gradient is exactly equal to the chemical gradient for that ion by the Nernst equation:

$$E_X = \frac{RT}{F} \cdot \ln \frac{(X_i)}{(X_o)}$$

where R is the universal gas constant, T is absolute temperature,

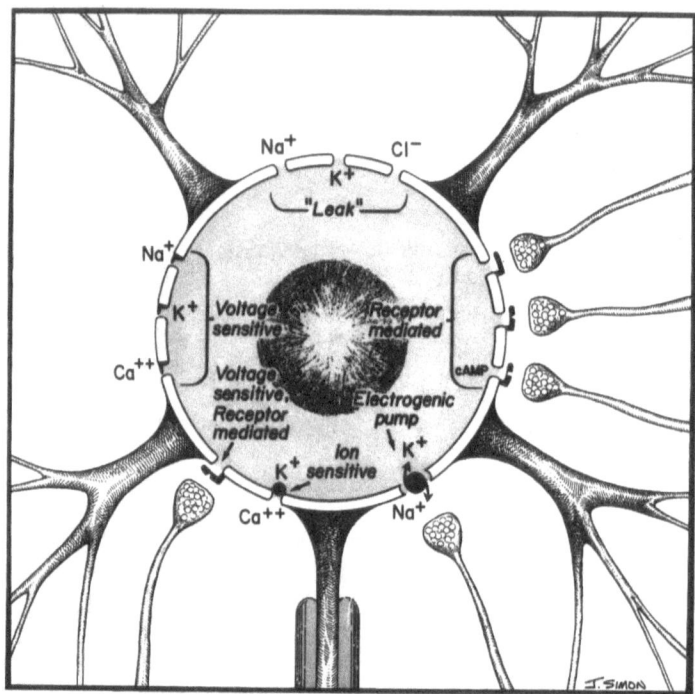

Fig. 1.   Schematic representation of several types of ion channels
and related processes thought to contribute to the electrical
activity of neurons. Most neuronal electrical activity is generated
by the flow of ions through such conductance channels. Those which
are always open (top of figure) are termed 'leak' conductances;
these, especially those for potassium and sodium ions, contribute
largely to the resting membrane potential. The receptor-mediated
channels are depicted on the right. These are activated or
inactivated by chemicals (most neurotransmitters) and are thought
to account for the conventional non-voltage-dependent or passive
responses to activation of synaptic pathways. The traditional
voltage-dependent channels (left) are those which are open only at
certain membrane potentials; these are usually thought to contribute
to the generation of an action potential, after the membrane
potential is brought to a threshold (trigger) level of
depolarization by injected current or activation of synaptic inputs.
One result of a voltage-sensitive conductance is shown at the
bottom left, wherein the entry of calcium during the action
potential triggers the efflux of potassium ions, resulting in
membrane repolarization (and sometimes the hyperpolarizing
afterpotential) at the conclusion of the action potential. Such
ion sensitive channels are opened only when a particular ion
(e.g. calcium) is present. A relatively new development is the
existence of voltage sensitive, receptor-mediated ion channels
(lower left); as with conventional synapses, such channels may be

F is the Faraday constant, and the subscripts i and o indicate inside and outside concentrations of the ion (see 25,44).

Fig. 1 is a schematic representation of the types of ionic channels likely to generate the equilibrium potentials for the major ionic species involved in determining resting transmembrane potential (RMP), action potentials and receptor mediated potential shifts in excitable cells. The approximate range of ionic equilibrium potentials involved in determining membrane potentials for most excitable cells are: Ca^{++}, +20 to +50 mV; Na^+, +20 to +40 mV; Cl^-, -10 to -30 mV; K^+, -40 to -80 mV. The resting membrane potential usually can be approximated by the sum of the contributions of Na^+, K^+ and Cl^-; thus most excitable cells display RMPs in the range of 40-100 mV, internally negative. Under resting conditions, K^+ permeability usually predominates and RMP is relatively near but slightly positive to E_{K^+}. When a hormone, neurotransmitter or second messenger increases permeability to a single ion, the potential will approach the equilibrium potential for that ion. If such an agent causes a decrease in the permeability to one ionic species, the potential will move away from the equilibrium potential for that ion toward that of the ion with the dominant permeability. Thus, an increase in K^+ permeability will generally hyperpolarize, whereas an increase in Na^+ permeability will depolarize, an excitable cell. In most brain neurons, E_{Cl^-} is more negative than RMP; here permeability increases to Cl^- are hyperpolarizing. In the vertebrate neuro-muscular junction and many synapses mediating fast excitation the transmitter opens a channel for both Na^+ and K^+, pushing the potential at the neuromuscular junction to about -15 mV, approximately midway between E_{K^+} and E_{Na^+}.

However, it is not clear what changes in ionic conductance, if any, would appear if ionic pumps were the cause of excitability changes (cf. 95). Such electrogenic pumps are thought by some to

---

opened or closed by neurotransmitters, but only at certain membrane potentials. Since many or most voltage-dependent conductances are associated with action potential mechanisms, activation of such receptors would be expected to alter properties of the spike (see text). Activation or inhibition of electrogenic ion pumps (lower right) could also contribute to receptor- or non-receptor-coupled changes in membrane potential. Generation of cyclic nucleotides, by nucleotide cyclases, possibly through activation of transmitter receptors (see text for example of cyclic AMP), could open or close ion channels directly, or alter voltage sensitive conductances or membrane pumps, thus significantly altering neuronal excitability (Taken from ref. 84).

account for certain slow synaptic potentials in sympathetic ganglia
(see 64). Moreover, electrophysiological studies of vertebrate
and invertebrate neurons suggest that some neurotransmitters can
alter excitability without directly changing membrane potential.
For example, in some neurons where E_{Cl^-} is very near RMP, transmitter-
increased permeability to Cl^- would produce no potential change,
but any other synaptic inputs would be reduced in effectiveness
because of the 'shunting' effect of the reduced resistance.
Conversely, a neurotransmitter which reduces ionic permeabilities
(increasing resistance and therefore the 'voltage drop' for a given
second synaptic current) would increase the responsiveness of
neurons to other synaptic inputs.

Another recently described phenomenon of importance is the
voltage-dependent action of some hormones and neurotransmitters.
Here, concentrations of the agonists often below those normally
producing potential changes are capable of altering ionic
permeabilities which are usually not active at the RMP. For example,
alterations of spike threshold have been seen with GABA and opioid
peptides in some cultured spinal neurons (1). Since spikes or
action potentials involve membrane conductances which are activated
by changes in membrane voltage and these actions are blocked by
specific receptor antagonists, such a novel phenomenon could be
ascribed to a receptor-mediated voltage-sensitive conductance change
(Fig. 1). Receptor-mediated changes in voltage-dependent calcium
channels are of even greater interest because calcium has so many
regulatory functions upon a variety of cellular activities (e.g.,
secretion, calcitonin activation, genome expression, cell motility).
Examples of this phenomenon are the decreases in the late (calcium)
plateau component of the spike in mammalian dorsal root cultures
produced by GABA, norepinephrine, serotonin and enkephalin (23)
and in rat sympathetic nerves by norepinephrine (42). Activation
or prolongation of voltage-sensitive Ca^{++} channels by serotonin has
been observed in aplysia neurons (45,71; see Klein, this volume).

Furthermore, Brown and Adams (11) report a muscarinic
cholinergic action (sometimes depolarizing) on neurons of
sympathetic ganglia which arises from inactivation of a voltage-
dependent K^+ channel. This latter action is similar to one action
of catecholamines on cardiac Purkinje fibers, where the slow
inactivation of a voltage-dependent K^+ conductance is responsible
in part for a depolarizing pacemaker potential. By speeding the
inactivation of this channel, norepinephrine increases the frecuency
of firing. In addition, norepinephrine (and cyclic AMP as well)
appears to enhance the slow inward current (mostly calcium) of the
cardiac potential and the outward K^+ current responsible for
repolarization; both of these effects involve voltage-sensitive
conductances.

It is against this basic and historical backdrop that an

attempt will be made to develop criteria to be used for identifying neurotransmitters, neuromodulators and mediators in nervous tissue. Where possible, I will suggest broad criteria with sufficient latitude for inclusion of possible undiscovered new forms of neuronal communication, rather than be forced to generate new terms for a process that might still be best described as, for example, neurotransmission. In this effort, the features of peripheral neurotransmission will be kept in mind as continuing models for central communication.

NEUROTRANSMITTER CRITERIA

My personalized criteria for identification of a neurotransmitter may be paraphrased and condensed from the several criteria previously suggested (98):

1. Neuronal localization of the substance and its enzymes of synthesis and degradation.

2. Release of the substance upon selective activation of a specific neuronal pathway.

3. Identical physiological response to exogenously applied transmitter and to activation of the pathway.

4. Identical action of pharmacological agents (antagonists, etc.) when tested against the effects of exogenous transmitter or of the activated pathway.

To some, these criteria may show a lack of concern for the ultrastructural bases of neurotransmission (e.g., the presence or lack of synaptic 'specializations'). However, such considerations were omitted because they are still very controversial, because they have been discussed in detail elsewhere (7,8,21), and because of continued difficulties in documenting correlations between structure and physiology. Moreover, the possibilities for a broad array of morphological forms of neurotransmission in CNS may be indicated in the peripheral nervous system: some sympathetic boutons transmit direct messages without synaptic specializations to their smooth muscle contacts in an 'en passage' fashion, but do show them at intraganglionic nerve-nerve contacts (41).

It will be noted that there is no mention in the stated criteria of the speed of transmission. Thus, as in peripheral systems, a slowly acting substance is just as much a neurotransmitter as a fast acting one. Moreover, a substance is not excluded as a neurotransmitter because it also alters (or 'modulates') the response of another neurotransmitter. If this were the case most currently-conceived neurotransmitters would be excluded because they alter membrane conductance (or resistance), and the result of this would likely be altered potentials generated by other synapses or transmitters. For example, GABA and classical IPSPs inhibit cells by increasing conductance to Cl^- and perhaps K^+ ions; such an increase in conductance would cause a decrease in the membrane such that a given synaptic current generated by another input to the cell would generate a smaller potential shift across the reduced resistance. Thus, GABA might be expected to inhibit firing by two mechanisms, hyperpolarization and reduction of other synaptic potentials. By the same reasoning, application of glutamate or activation of a fast EPSP (increasing conductance to Na^+ and reducing membrane resistance) might be expected to excite cells by depolarization, but reduction of other synaptic potentials should also be considered.

The word 'response' in criterion three above was purposely chosen as a broad term, in order to cover a wide spectrum of possible direct mechanisms of transmitter action on cell excitability. Thus, substances which hyperpolarize or depolarize by increased or decreased conductance, or by activating or inactivating electrogenic pumps, or which alter spike thresholds (with or without potential changes), perhaps via effects on voltage-dependent conductances, will all be considered neurotransmitters by these criteria because they directly affect neuronal excitability.

MEDIATOR CRITERIA

A mediator of neuronal communication might be best exemplified by the role of cyclic nucleotides as 'second messengers.' The second messenger concept as currently applied to brain has evolved from the mediator role of cyclic AMP in peripheral hormonal responses, as first suggested by Sutherland et al. (1965). Modified for neuronal transmission or local modulation (see below), this concept may be summarized as follows:

A synaptically released neurotransmitter or locally released neuroactive agent could act at certain pre- or postsynaptic receptors to activate adenylate cyclase and the synthesis of cyclic AMP. Intracellular cAMP would then initiate subsequent enzymatic or molecular events, which, among other actions (e.g. long-term trophic effects) could result in changes in membrane potential and cell discharge rate. Four major criteria may be adapted from the criteria for hormones, to establish that the action of a transmitter is mediated by a cyclic nucleotide (5,6,83).

1. Exogenous neurotransmitter substance and activation of the synaptic pathway both regulate intracellular levels of cyclic nucleotide in the postsynaptic cell.

2. The change in intracellular cyclic nucleotide content
should precede "the biological event" triggered by the transmitter
or nerve pathway.

3. Responses to the transmitter or nerve pathway should be
logically altered by drugs that specifically interact with the
nucleotide cyclase or that inhibit the appropriate phosphodiesterase.

4. Exogenous cyclic nucleotides (and analogues which activate
protein kinase) should elicit a response identical to the biological
event caused by the transmitter or nerve pathway.

Unfortunately, attempts to satisfy the second messenger
criteria for central neurons meet with considerable technical
obstacles, such as the indirect actions of systemic drugs, blood-
brain barriers to systemic agents, slow nucleotide sampling and
measurement times compared to fast synaptic events, and relative
impermeability of cyclic nucleotides into target cells. Several
of these obstacles can be partially overcome in the central nervous
system by the techniques of microiontophoresis and electrophysiology, as they have been applied to several brain areas
(see below).

In spite of these difficulties, an important new consideration
with respect to the likelihood that cyclic nucleotides mediate
the function of neurotransmitters is the idea that a neurotransmitter can generate potential and conductance changes not
by altering passive membrane properties but by altering neuronal
metabolism (or energy function) which in turn alters transmembrane
properties. This constitutes the "energy" dimension of neurotransmission as seen by Bloom (7).

MODULATOR CRITERIA

The term modulator has received increasing attention recently
as a catch-all category capable of including the action of all
substances whose actions differ from those of the classically-
conceived spinal cord transmitters. Indeed, references to
'modulators' in the literature seem to be increasing at a
frequency in direct correlation to the discovery of new neuroactive
substances; the variety of definitions of the term 'modulator'
seems almost as numerous as the number of new brain substances.
These definitions range from an emphasis on a long time-course
of action (31), to any substance (e.g., CO_2 or NH_3) released from
non-synaptic sources (28). However, a criterion common to most
definitions is the notion that a modulator should have no direct
effect of its own, but that it can alter the response of the postsynaptic neuron to other synaptic afferents (2,3). Here again,
I will apply a broad definition of modulator, using primarily

the latter criteria.   In my view, a modulator should:

   1.   Alter responses of other neurotransmitters.

   2.   Have no direct effect on spiking, membrane potential or
conductance.

   3.   Be localized in the vicinity.

   No distinction is made here between those modulators released
from neurons and those not, since such agents may be sub-clasified
as 'neuromodulators' and 'local modulators', respectively, without
loss of clarity.   Again, no consideration is given to the distance
over which the substance must diffuse (i.e., the synaptic 'gap')
or whether synaptic specializations are present (see 8); such
considerations would have assigned catecholamines and acetylcholine
of the autonomic system as 'modulators' rather than transmitters.
Nor is strict adherence paid to the criteria of Barker and Smith
(3) and Barker et al (2), who define a neuromodulator as a
substance that "alters synaptic receptor-coupled conductances
without direct activation of such conductances".   The two criteria
above are drawn in a less restrictive manner, so as to exclude
neurotransmitters which directly alter cell excitability by means
other than strict activation of conductances (such as by alteration
of electrogenic pumps, spike thresholds and inactivation of
conductances; see "Transmitter Criteria" above), and so as not
to exclude alterations of other synaptic receptors that could
operate by these other means.

   However, even with such a broadly-drawn definition of
'modulator', formidable technical difficulties are encountered in
proving their existence.   Proof of modulation of other synaptic
afferents or receptors is difficult enough, and usually relies upon
complex electrophysiological and iontophoretic techniques. Much
more difficult, however, is strict documentation of a lack of
direct effect of a substance.   This usually requires intracellular
recording.   Moreover, negative results in iontophoretic and
electrophysiological techniques are always suspect especially when
little is known of the spatial geometry on a neuron of the
receptors for a putative 'modulator'.   Thus, if the receptors are
located on dendrites some distance from the soma, where the
intracellular electrode is usually inserted potential changes
produced by diffusion of the substance to these remote sites may
be decremented to undetectable levels before they are conducted
to the soma.   Likewise, receptor-induced changes in ionic
conductance on distant dendrites are likely not to be detected by
the usual intracellular (intrasomatic) stimulation method.
Therefore, strict proof of neuromodulation requires methodologies
for tracking receptor sensitivity along the entire surface of the
neuron.

Having now applied broad criteria for classification of transmitters, modulators and mediators, and presented some caveats for experimental efforts to document the existence of these messengers in brain, we can now turn to specific examples of chemicals that appear to fulfill at least a majority of these criteria.

CENTRAL NOREPINEPHRINE AND ITS MEDIATION BY CYCLIC AMP

With the possible exceptions of acetylcholine and GABA, criteria for a central transmitter seem most completely satisfied by norepinephrine. This state of affairs arises in part because histochemical methods for localizing NE (cf. 57) are less ambigous than for other putative central transmitters, thus more easily satisfying criterion one above. Fulfillment of criterion two, release on activation, is less directly fulfilled although it is possible to detect metabolites of NE and diminished endogenous NE levels in cortex after activation of the locus coeruleus (LC), the source of NE-containing fibers to the cortex (48). Fulfillment of the last two criteria have been the subject of intensive investigations, carried out on the cerebellar Purkinje cell and the hippocampal Pyramidal cell, both targets of the LC NE-containing pathway.

Coeruleo-Cerebellar Pathway

In the cerebellar Purkinje cell, extracellular iontophoretic application of NE or stimulation of the LC was found to produce equivalent effects on the Purkinje cell, namely, inhibition of spontaneous discharge with extracellular recording (40,86) and a unique hyperpolarization associated with either no change or a decrease in membrane conductance with intracellular recording (40,88,89). This latter effect is in direct contrast to changes seen with classical spinal cord IPSP's or with iontophoresis of a putative inhibitory transmitter like GABA (88), in which the hyperpolarization is associated with increased membrane conductance. Interestingly, catecholamines also induce hyperpolarization of spinal cord motorneurons in association with a decrease in conductance, probably to sodium ions (56).

In contrast again to classical spinal cord IPSP's, the inhibitory response of Purkinje cells to single LC stimuli showed long latencies (including conduction time and synaptic delay) averaging about 125 msec and responses lasting 0.5 sec or more. Trains of pulses to LC at 10/sec, which is nearly the optimal frequency for both the LC pathway and for peripheral sympathetic nerves (see 15,29), could evoke inhibition outlasting the stimulus for up to a minute or more (40). Similar responses with long latencies and durations of action had been previously reported

for both sympathetic target tissues (e.g., smooth muscle; see
"Introduction") and for the sIPSP of sympathetic ganglia (53).
Since the response to exogenous NE has similar long latencies and
durations, and is accompanied by an unusual conductance decrease,
it would appear that transmitter criterion three is satisfied.

Unfortunately, the existence of long latencies in response
to LC stimulation confounds attempts to prove that the pathway
responsible for inhibition is monosynaptic. However, results of
iontophoretic experiments with local application of NE antagonists
provide indirect support for a monosynaptic inhibitory projection
from LC to the Purkinje cell, and in addition help to satisfy
transmitter criterion four. Thus, both effects of NE
iontophoresis and LC stimulations are comparably antagonized by
phenothiazines (32,90), by cobalt or lead (33), by lithium
Siggins and Henriksen, in preparation) as well as by prostaglandins
and beta-adrenergic antagonists (40). Furthermore, the effects
of LC stimulation are also abolished when the synthesis and storage
of NE are blocked pharmacologically and when the pathway is
destroyed with 6-OHDA (40). These pharmacological manipulations
were selective in that other pathways or neurotransmitters were
not effected.

Coeruleo-Hippocampal Projection

More recently, the properties of the LC-cerebellar circuit
have been generalized to other LC target areas. Stimulation of
LC in the awake rat produces a two-fold rise in hippocampal cyclic
AMP with no change in cyclic GMP (Segal and Guidotti, personal
communication). Physiological studies indicate that the
hippocampal NE projection from LC produces cellular effects
virtually identical to those of the locus coeruleus on cerebellar
Purkinje cells: in hippocampus, both LC and NE slow pyramidal
cell discharge with long latencies and long durations of actions,
both responses are also blocked with beta-blockers and by
prostaglandins of the E series, by lithium and by phenothiazines
(5,77,79). Moreover, the action of the LC pathway is blocked by
chronic pretreatment with 6-OHDA or acute pretreatment with
reserpine and alpha methyl tyrosine or with inhibitors of dopamine
beta hydroxylase (79).

Septum

Segal (78) has recently reported studies on neurons in the
medial septal nuclei and in the diagonal band of Broca, which
also receive a projection from the locus coeruleus. LC stimulation
was found to produce, as in the other areas already described,
an inhibition of long latency (30-100 msec) and long duration
(100-300 msec). These inhibitory effects were blocked by depletion
of NE stores or by pretreatment with 6-OHDA.

Cyclic AMP as a Mediator of Central NE Effects

With respect to second messenger mediation by cyclic nucleotides the central catecholamine-containing pathways merit investigation because they satisfy three practical considerations: 1) catecholamines meet most or all of the criteria above for a synaptic transmitter 2) biochemically, catecholamines are known to activate adenylate cyclase or elevate cyclic AMP levels in many peripheral systems and in various discrete regions of the central nervous system by definable receptors; and 3) as shown above, the source neurons and target neurons of the central catecholamine pathways have been sufficiently characterized so that their effects can be determined and related to the effects of cyclic nucleotides and related substances.

From the electrophysiologists point of view the cerebellar Purkinje cell (P-cell) is still the best candidate for a target neuron that receives a noradrenergic input capable of generating cyclic AMP postsynaptically. The data reinforcing this notion have been reviewed in detail elsewhere (5,82,83) but may be summarized (in order of mediator criteria) as follows: 1) catecholamines elevate cyclic AMP levels and increase adenylate cyclase activity in cerebellum in vitro, and exogenous NE and stimulation of the LC increase cyclic AMP histochemical immuno-reactivity in P-cells in vivo; 2) the increase in cyclic AMP immuno-reactivity is detectable at least by the time the electrophysiological effects of LC stimulation are apparent; 3) the inhibitory effects on P-cells of LC stimulation or of NE iontophoresis are potentiated by several phosphodiesterase (PDE) inhibitors and antagonized by agents (e.g., PGE_1 and E_2, MJ-1999, fluphenazine, lanthanum) known to block NE-elevated cyclic AMP levels in vitro; 4) with extra- and intra-cellular recording, responses to iontophoresis of cyclic AMP and several more potent synthetic analogues generally mimic the inhibitory hyperpolarizing action of iontophoretic NE and LC stimulation, with no change or an increase in membrane resistance.

Thus, the criteria for cyclic AMP mediation appear to be largely satisfied for the inhibitory NE input to Purkinje cells, except for the technical inability to detect postsynaptic increases in cyclic AMP at a time prior to the "biological event" triggered by NE. Although disagreement exists as to the exact percentage of P-cells inhibited by iontophoretic cyclic AMP (37,51), this discrepancy is explained by poor cell penetrability and other technical considerations (9,81). Moreover, iontophoresis of derivatives of cyclic AMP (e.g., 8-p Chlorophenyl-cyclic AMP) known to have a greater action on the protein kinase enzyme (the intracellular "receptor" for cyclic AMP) than the parent compound, can depress the activity of up to 90% of P-cells (85). In addition, the strong correlation between percentage of P-cells depressed and the potency of several derivatives in activating

protein kinase argues for an involvement of cyclic AMP-dependent
protein kinase in the depressant responses.

In studies of cultured Purkinje neurons, Gähwlier (36) has
reported potentiation of NE and cyclic AMP-induced depressions with
phosphodiesterase inhibitors. He also observed that the thresholds
for inhibitory responses to NE were 100-1000 times lower than for
cyclic AMP applied by superfusion, in keeping with predictions from
several peripheral tissues where second messenger mediation has
been proven (see 5,74).

The suggestion that actions of extracellularly-applied cyclic
AMP are mediated by conversion to adenosine and activation of an
adenosine receptor is disproven by the observations in cerebellum
and the cerebral cortex that methylxanthine-type phosphodiesterase
inhibitors potentiate NE and cyclic AMP, yet block the effects of
adenosine or 5'AMP (40,87,93). At any rate, since the effects of
adenosine are also inhibitory and proposed by the Phillis group to
be mediated by cyclic AMP (49,73), the implication is that the
physiological action of cyclic AMP is also inhibitory, thus
confirming the original findings with iontophoresis of cyclic
AMP (86-88).

Evidence similar to that found in the cerebellum exists for
cyclic AMP mediation of the inhibitory norepinephrine input to
hippocampal and cerebral cortical pyramidal cells (79,93). In brief,
exogenous catecholamine elevates cyclic AMP in vitro, the
inhibitory effects of LC activation or NE iontophoresis are affected
in a predictable way by drugs which interact with the cyclic AMP
system, and iontophoresis of cyclic AMP generally mimics the
inhibitory action of LC stimulation and iontophoretic NE. However,
immunohistochemical methods for in vivo localization of cyclic AMP
to the pyramidal cells have not yet been applied to the hippocampus
or cortex, nor has it been possible to detect cyclic AMP in these
structures prior to the NE-induced biological event. Research
suggesting that the inhibitory responses of caudate neurons to
dopamine are also mediated by cyclic AMP is reviewed elsewhere
(82,83).

Other more indirect data also support the hypothesis of
mediation of NE-induced inhibition by cyclic AMP. Thus, lithium
and lanthanum, each of which have been shown to block NE-induced
elevations of cyclic AMP in vitro (5,61,62), also both antagonize
NE-evoked inhibitions of spontaneous activity in cerebellar
Purkinje cells and hippocampal pyramidal cells (62,77). Lithium
also antagonizes LC-induced depression in hippocampus (77) and
cerebellum (Siggins and Henriksen, in preparation).

NOREPINEPHRINE AS A NEUROMODULATOR

In spite of the strong evidence favoring a transmitter role
for NE, there is data suggesting that NE could also be a modulator.
Indeed, those predisposed to define modulators on the basis of speed
or duration of action, on the basis of lack of induced increase
in ionic conductance, or on the basis of mediation through an
energetic event, have already applied the modulator label to NE
(see ref. 21). However, to my mind there is only one compelling
piece of evidence to suggest that NE has the major quality of a
modulator: it can clearly influence the responsiveness of a given
target cell to its other afferents. Studies in the awake squirrel
monkey auditory cortex (30), in the cerebellum of the anesthetized
rat (34,35), and in the hippocampus of the awake rat (80) suggest
a heterosynaptic influence of the LC in addition to the classical
mode of synaptic operation termed inhibition. For example, in the
cerebellar Purkinje cell (34,35) and the hippocampal pyramidal cell
(80), conditioning stimuli in the LC or iontophoretic application
of NE will potentiate the effects of nonadrenergic inhibitory
inputs for considerable periods. Moreover, in the cerebellum the
excitatory afferents of parallel fibers and climbing fibers are
also potentiated by NE and LC activation. Such potentiated
responses may be similar to the enhanced evoked responses seen
following LC conditioning stimuli in lateral geniculate (59) and
to the increased ratio of evoked responses to spontaneous activity
seen during iontophoresis of NE to acoustically reactive units of
the squirrel monkey auditory cortex (30).

As stated above ('Neurotransmitter Criteria') one might expect
that a substance which decreases conductance (or increases membrane
resistance) might potentiate other inputs, by virtue of its
ability to enlarge a potential produced by a given synaptic current
across an increased resistance. In verification of this, climbing
fibers EPSP's are increased in size by NE and LC activation
(88,89). Although it is suggested that the potentiating effects
of NE outlast the direct inhibitory action on extracellularly
recorded spontaneous activity (34,35), intracellular recording
is needed to verify that the membrane potential and conductance
is still not altered when the spike frequency (seen extracellularly)
has returned to normal.

Norepinephrine Effects on Voltage-Dependent Conductances

It is becoming clear that norepinephrine may have effects
on neuronal membrane properties distinct from the passive changes
in RMP and resistance, and changes in other synaptic inputs,
described above. Thus, recent intracellular investigations on
rat sympathetic neurons (42) and neurons cultured from chick
dorsal root ganglia (23) suggest that concentrations of norepinephrine below those required to evoke changes in RMP can alter

voltage-dependent properties of the neurons. In the chick dorsal
root ganglia neurons, the result is a shortening of the action
potential duration probably via a decrease in the voltage-dependent
calcium conductance (23). In the sympathetic neurons pharmacologic
investigations indicate that the voltage-sensitive inward calcium
current and calcium-dependent potassium current of the spike are
reduced by NE (42). The result of the latter actions is to reduce
the 'calcium spike,' and the late 'shoulder' and hyperpolarizing
afterpotential of normal spikes; these actions are reversed by an
alpha adrenergic antagonist. While these actions may prove to be
very significant for calcium-mediated functions such as transmitter
release, direct knowledge of the influence on neuronal
excitability must await further research.

OPIOID PEPTIDES: NEUROTRANSMITTERS OR NEUROMODULATORS?

The discovery of endorphins, the endogenous peptide ligands
for the stereospecific, naloxone-sensitive opiate receptor has
stimulated a flurry of research activity attempting to delineate
the physiological action and possible function of these peptides.
As with the catecholamines, research to date has suggested a role
for these peptides more complex than can be conceptualized either
as solely a neurotransmitter or a neuromodulator. Electrophysiologic research on opiates and opioid peptides has been
primarily through extracellular single unit recordings directed
at CNS areas with high density opiate receptors or involved with
nociception.

In such extracellular recordings, most of the stereospecific,
naloxone-antagonizable actions of opioid peptides are inhibitions
of single unit discharge (spontaneous, glutamate or ACh-evoked
activity) which are qualitatively similar throughout the mammalian
central and peripheral nervous system (see review by
Zieglgänsberger and Fry, ref. 103). However, some major exceptions
exist: naloxone-reversible excitatory responses are seen in
pyramidal cells in the hippocampus (38,63), the Renshaw cell in
the spinal cord (see 20,22) and some less well-identified cells in
various parts of the CNS (see 63). However, recent studies using
GABA blockade by bicuculline and blockade of transmitter release
by Mg^{++} ions (and intracellular recording in vitro: see below)
indicate that the excitatory responses of hippocampal pyramidal
neurons may now be viewed as a primary inhibitory effect on
neighboring inhibitory interneurons, resulting in excitation of
pyramidal cells by disinhibition (91,101).

However, intracellular recordings are required to measure
synaptically- and chemically-induced changes in membrane potential
and membrane conductance for elucidation of mechanisms of opiate
action, and recently a few such studies have been completed.

Early studies in the spinal cord showed that intravenous administration of morphine agonists depressed polysynaptic EPSPs, an effect antagonized by opiate antagonists (43). More recent studies employing intracellular recording of spinal cord neurons with simultaneous extracellular microiontophoretic application revealed that morphine and opioid peptides do not appear to change membrane potential or resting membrane resistance, but decrease the rate of rise of the EPSPs (100,102). Opiates and opioid peptides also depress the glutamate-induced depolarization in addition to synaptic activation. With respect to cat spinal neurones, microiontophoretically-applied glutamate is considered to increase the postsynaptic membrane conductance to sodium ions (104). Zieglgansberger and co-workers (100,102,103) therefore postulate that the opiates interfere with the chemically excitable sodium channels comparable to those also opened by synaptically released excitatory transmitters. The fact that the depolarizing responses to glutamate are post-synaptic and antagonized by opiate agonists suggests that the opiate receptors involved in this effect are also located on the post-synaptic membrane.

The anti-glutamate action of the opioid peptides in vivo have recently been confirmed with mouse spinal neurons grown in culture (2). Kinetic analysis indicates that the inhibitory action of the opiate peptides is brought about by a non-competitive mechanism on the postsynaptically located sodium-ionophore. However, these postsynaptic modulator-like actions of the opiates and opioid peptides are just a few of the many types described. For example, several reports of pre-synaptic influence have appeared, in which it would appear that the opiates depress transmitter release (14, 39,55), perhaps by involving mechanisms similar to the modulation of post-synaptic ionic processes. Furthermore, the study of Barker et al (2) shows a multitude of presumably post-synaptic actions of enkephalin on the cultured mouse spinal-neuron preparation, including direct positive and negative effects on membrane excitability associated with potential and/or conductance changes, abrupt naloxone-resistant depolarizations, enhancement of amino acid-induced actions without change in membrane properties, as well as the depression of glutamate actions described above. However, the possibility should be considered that some of these diverse opiate actions may arise as a pharmacological 'curiosity' in cells that do not ordinarily ever receive endorphinergic influence in vivo.

With regard to multiple actions of opioid peptides, the case of myenteric plexus neurons is an interesting one. Here, the analgesic potency of opiates correlates with their depressant effect upon the electrically-induced twitch of the guinea-pig ileum and is accurately reflected in single unit studies of myenteric plexus neurons (66,67). The stereospecific depressant effect of the opiates here can be seen in Ca^{++}-free/high Mg^{++} solution, indicating a postsynaptic effect (69,70). Interestingly, with

intracellular recording, many myenteric neurones show no change in
membrane potential or conductance in response to superfusion or
iontophoresis of the peptides. However, some other cells are seen
which show hyperpolarizing actions with increased membrane
conductance which is thought to represent conduction of hyperpolarizing responses from nearby terminals of the same cell (68).
Since this effect seems at variance with the data obtained in
central neurons cited above, a different ionic mechanism may be
involved in these neurons.

More recent studies underline the apparent diversity of opioid
action. Intracellular recordings of locus coeruleus neurons in the
in vitro brain stem slice preparation of Pepper and Henderson (72)
indicate clear hyperpolarizing responses to the opioid peptides
associated with an increased conductance, suggestive of a
transmitter-like action. By contrast, pyramidal cells of the
hippocampal slice in vitro show no direct membrane effects of
morphine or opioid peptides in concentrations up to 20 µM
(Siggins and Zieglgänsberger, in preparation; Haas and Ryall, in
preparation; R. Nicoll, personal communication); however, much
lower concentrations (0.5-1 µM) are capable of altering synaptic
responses to excitatory and inhibitory afferent stimulation (Fig. 2).
While these latter data are consistent with a modulatory action of the
peptides, intracellular recording of interneurons will be needed
to assure that the observed effects are not due to direct
hyperpolarization of inhibitory interneurons (see 101).

Opioid peptides also appear to alter voltage-dependent
conductances, as indicated by the recent report of Mudge et al (58)
showing a decrease in the duration of the action potential of
cultured dorsal root ganglia neurons by concentrations of peptide
that have no direct effect on resting membrane properties. This
shortening of the spike correlates with a decrease in substance P
release and is thought to result from an antagonism of the
inward voltage-dependent calcium current. Such an action is taken
to explain the presynaptic inhibitory actions of the opioid peptides
described above. However, the opiate effects on the spike are
recorded in the soma and the release effects involve remote
terminals. It is not clear that the calcium channels are identical
in the two locations; hence, one cannot verify a direct causal
relationship between opiate effects on the somal spike and on
transmitter release. It will also be of some interest to learn if
such actions extend to dorsal root ganglia neurons in situ.

SUMMARY

An effort has been made here to devise criteria allowing
discrimination between neurotransmitters, modulators and mediators.
However, after consideration of several technical pitfalls in

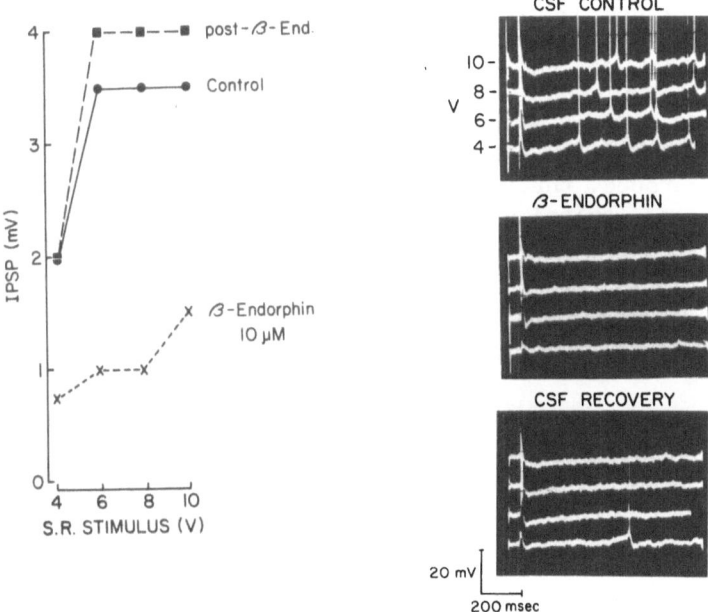

Fig. 2.   Intracellular recording of the antagonism of inhibitory
post-synaptic potentials (IPSP) in a CA1 hippocampal pyramidal
neuron (in vitro) by perfusion of 10μM beta-endorphin.   Input-output
curve on left shows the peak amplitude of the IPSPs measured after
stimulating the stratum radiatum (S.R.) Schaeffer collateral system
from CA 3 with four different voltage strengths (stimulating
electrode=bipolar concentric electrode, 150 μm tip diameter).   In
this cell, each stimulus pulse (0.1 msec duration delivered every
3 seconds) triggers an initial EPSP or spike followed by a
hyperpolarization (IPSP) due to activation of inhibitory neurons
via the recurrent collaterals of pyramidal cells.   The right
column shows the oscillographic tracings of these responses to S.R.
stimulation at the 4 different voltage strengths indicated to the
left of the top panel; these records show the dramatic reduction
in the IPSPs after 8-10 minutes of perfusion with beta-endorphin
(middle record), compared to those recorded before (top, perfused
only with artificial CSF) or after 20 minutes of wash-out with
CSF (bottom record).   This finding is in accord with the
disinhibition hypothesis of pyramidal cell excitation (91,101).
Note also the reduction in spontaneous spiking sometimes produced
by endorphin in the hippocampal slice; however, neither membrane
properties nor spike size and shape appear to be altered by this
concentration of endorphin.   The apparent resting potential of
this cell was 58 mV, recorded over a period of three hours.   From
Siggins and Zieglgänsberger (unpublished).

studies of these criteria, and examination of the properties of
two examples of neuroactive agents (norepinephrine and endorphins)
often referred to as 'modulators', it is still difficult to
conclusively classify these agents in all cases. Thus, in most
central targets where NE-fibers are known to terminate, the
synaptic actions of NE appear to have properties of both a neuromodulator and a neurotransmitter. Although much more research
needs to be pursued, the opioid peptides might be neuromodulators
for some neurons (spinal cord and hippocampal neurons) and neurotransmitters for others (myenteric plexus, locus coeruleus and
some spinal cord neurons). It may be that classification of such
peptide agonists will need to be done on a cell-by-cell basis,
with the endogenous peptides subserving a multi-faceted role in
central and peripheral neuronal communication.

As more and more endogenous ligands and transmitter-like
substances are extracted from brain, it begins to appear that the
language of neuronal communication is much richer than originally
imagined from responses of spinal neurons to the fast-acting
classical neurotransmitters. Indeed, it may evolve that the
'deviant' forms of communication or transmission are more the
rule than the exception. In the final analysis, each neurotransmitter may possess its own 'fingerprint' of holistic actions
attesting to the unique individuality of neuron types and their
neurotransmitters. Such individualities might be expected to
accomplish more sophisticated integrative operations, and hence
behaviors, than could simple rapid 'yes' or 'no' messages.

Acknowledgements. I am indebted to Drs. Donna Gruol, Quentin
Pittman, Floyd Bloom, Roy Wise, Leonard Koda, John Williams, and
Walter Zieglgansberger for their critical evaluation of ideas in
this discussion, and to Ms. Nancy Callahan for her secretarial
skills.

REFERENCES

1. Barker, J.L., Gruol, D.L., MacDonald, J.F., and Smith, T.G., Jr.
    Peptide receptor functions on cultured spinal neurons. In M.
    Trabucchi and E. Costa (Eds.), Regulation and Function of Neural
    Peptides, Raven Press, 1980 (in press).
2. Barker, J.L., Neale, J.H., Smith, T.G., Jr. and Macdonald, R.L.
    Opioid peptide modulation of amino-acid responses suggest novel
    form of neuronal communication. Science 199 (1978) 1451-1453.
3. Barker, J.L. and Smith, T.G., Jr. Peptides as neurohormones.
    Neurosci. Symp. 2 (1977) 340-373.
4. Baylor, D.A. and Fuortes, M.G.F. Electrical responses of
    single cones in the retina of the turtle. J. Physiol. (Lond.).
    207 (1970) 77-92.

5. Bloom, F.E. The role of cyclic nucleotides in central synaptic function. Rev. Physiol. Biochem. Pharmacol. 74 (1975) 1-103.

6. Bloom, F.E. The role of cyclic nucleotides in central synaptic function. In Advances in Biochemical Pharmacology, Vol. 15, Raven Press, New York, 1976, pp. 273-282.

7. Bloom, F.E. Contrasting principles of synaptic physiology: peptidergic and non-peptidergic neurons. In K. Fuxe, T. Hökfelt, and R. Luft (Eds.), The Peptidergic Neuron, Pergamon Press, New York, 1980, in press.

8. Bloom, F.E. Central noradrenergic neurons: structure-function considerations. In E. Usdin (ed.) Catecholamines: Basic and Clinical Frontiers, Pergamon Press, New York, 1979, pp. 609-618.

9. Bloom, F.E., Siggins, G.R. and Hoffer, B.J. Interpreting the failures to confirm the depression of cerebellar Purkinje cells by cyclic AMP. Science 185 (1974) 627-629.

10. Brooks, C. and Eccles, J.C. An electrical hypothesis of central inhibition. Nature 159 (1947) 760-764.

11. Brown, B.D.A. and Adams, P.R. Muscarinic suppression of a novel voltage-sensitive K^+ current in a vertebrate neurone. Nature 283 (1980) 673-676.

12. Burnstock, G., Prosser, C.L., and Holman, M.E. Electrophysiology of smooth muscle. Physiol. Rev. (1963) 482-527.

13. Caesar, R., Edwards, G. and Ruska, H. Architecture and nerve supply of mammalian smooth muscle tissue. J. Biophys. Biochem. Cytol. 3 (1957) 867-877.

14. Calvillo, O., Henry, J.L. and Neuman, R.S. Effects of morphine and naloxone on dorsal horn neurons in the cat. Canad. J. Physiol. Pharmacol. 52 (1974) 1207-1211.

15. Carpenter, F.G. and Tankersley, J.C. Response of a parasympathetic neuroeffector system to motor nerve stimulation. Am. J. Physiol. 196 (1959) 1185-1188.

16. Coombs, J.S., Eccles, J.C. and Fatt, P. The specific ionic conductances and the ionic movements across the motoneuronal membrane that produce the inhibitory post-synaptic potential. J. Physiol. 130 (1955) 326-373.

17. Dale, H.H. The action of certain esters and ethers of choline, and their relation to muscarine. J. Pharmacol. Exp. Therap. 6 (1906) 163-206.

18. Dale, H.H. Acetylcholine as a chemical transmitter of the effects of nerve impulses. J. Physiol. 48 (1914) 3iii-3iv.

19. Dale, H.H. Chemical transmission of the effects of nerve impulses. Brit. Med. J. (1934) 1-20.

20. Davies, J. and Dray, A. Substance P and opiate receptors. Nature 268 (1977) 351-352.

21. Dismukes, K. New look at the aminergic nervous system. Nature 269 (1977) 557-558.

22. Duggan, A.W., Davies, J. and Hall, J.G. Effects of opiate agonists and antagonists on central neurons of the cat. J. Pharmacol. Exp. Ther. 196 (1976) 107-120.

23. Dunlap, K. and Fischbach, G.D. Neurotransmitters decrease the

calcium component of sensory neurone action potentials. Nature
276 (1978) 837–839.

24. Elliott, T.R. The action of adrenaline. J. Physiol. (Lond.) 32
(1904) 401–467.

25. Eccles, J.C. The Physiology of Synapses, Springer-Verlag, Berlin,
1964.

26. Eccles, J.C., Fatt, P. and Koketsu, K. Cholinergic and inhibitory
synapses in a pathway from motor axon collaterals to motoneurones.
J. Physiol. 216 (1954) 524–562.

27. Eccles, R.M. and Libet, B. Origin and blockade of the synaptic
responses of curarized sympathetic ganglia. J. Physiol. (Lond.)
157 (1961) 484–503.

28. Florey, E. Neurotransmitters and modulators in the animal kingdom. Fed. Proc. 26 (1967) 1164–1178.

29. Folkow, B. Impulse frequency in sympathetic vasomotor fibres
correlated to the release and elimination of the transmitter.
Acta Physiol. Scand. 25 (1952) 49–76.

30. Foote, S.L., Freedman, R. and Oliver, A.P. Effects of putative
neurotransmitters on neuronal activity in monkey auditory cortex.
Brain Res. 86 (1975) 229–242.

31. Frederickson, R.C.A. Enkephalin pentapeptides—a review of
current evidence for a physiological role in vertebrate
neurotransmission. Life Sci. 21 (1977) 23–40.

32. Freedman, R. and Hoffer, B.J. Phenothiazine antagonism of the
noradrenergic inhibition of cerebellar Purkinje neurons. J.
Neurobiol. 6 (1975) 277–288.

33. Freedman, R. and Hoffer, B.J. A quantitative microiontophoretic
analysis of the responses of central neurones to noradrenaline:
Interactions with cobalt, manganese, verapamil and
dichloroisoprenaline. J. Neurobiol. 6 (1975) 529–539.

34. Freedman, R., Hoffer, B.J., Puro, D. and Woodward, D.J.
Noradrenaline modulation of the responses of cerebellar Purkinje
cells to afferent synaptic activity. Brit. J. Pharmac. 57 (1976)
603–605.

35. Freedman, R. and Hoffer, B.J. Interaction of norepinephrine
with cerebellar activity evoked by mossy and climbing fibers.
Exp. Neurol. 55 (1977) 269–288.

36. Gahwiler, B.H. Inhibitory action of noradrenaline and cyclic
adenosine monophosphate in explants of rat cerebellum. Nature
259 (1976) 483–484.

37. Godfraind, J.M. and Pumain, R. Cyclic adenosine monophosphate
and norepinephrine: effect on Purkinje cells in rat cerebellar
cortex. Science 174 (1971) 1257.

38. Hill, R.G., Mitchell, J.F. and Pepper, C.M. The excitation and
depression of hippocampal neurons by iontophoretically applied
enkephalins. J. Physiol. 272 (1977) 50–51P.

39. Hiller, J.M., Simon, E.J., Crain, S.H., Peterson, E.R. Opiate
receptors in cultures of fetal mouse dorsal root ganglia (DRG)
and spinal cord: predominance in DRG neurites. Brain Res.
145 (1978) 396–400.

40. Hoffer, B.J., Siggins, G.R., Oliver, A.P. and Bloom, F.E.
    Activation of the pathway from locus coeruleus to rat cerebellar
    Purkinje neurons: pharmacological evidence of noradrenergic
    central inhibition. J. Pharmacol. Exp. Ther. 184 (1973) 553-569.
41. Hökfelt, T. In vitro studies on central and peripheral monoamine
    neurons at the ultrastructural level. Zeit f. Zellforsch. 91
    (1968) 1-74.
42. Horn, J.P. and McAfee, D.A. Norepinephrine inhibits calcium-
    dependent potentials in rat sympathetic neurons. Science 204
    (1979) 1233-1235.
43. Jurna, I. Grossman, W., and Theres, C. Inhibition by morphine
    of repetitive activations of cat spinal motoneurones. Neuro-
    pharmacol. 12 (1973) 983-993.
44. Katz, B. Nerve Muscle and Synapse, McGraw-Hill, Inc., New York,
    1966, 193 pp.
45. Klein, M. and Kandel, E.R. Presynaptic modulation of voltage-
    dependent current: mechanisms for behavioral sensitization in
    Aplysia californica. Proc. Natl. Acad. Sci. U.S.A. 75 (1978)
    3512-3516.
46. Kobayashi, H. and Libet, B. Generation of slow postsynaptic
    potentials without increases in ionic conductance. Proc. Natl.
    Acad. Sci. U.S.A. 60 (1968) 1304-1311.
47. Koketsu, K. Cholinergic synaptic potentials and the underlying
    ionic mechanisms. Fed. Proc. 28 (1969) 101-112.
48. Korf, J., Roth, R.H. and Aghajanian, G.K. Alterations in
    turnover and endogenous levels of norepinephrine in cerebral
    cortex following electrical stimulation and acute axotomy of
    cerebral noradrenergic pathways. Eur. J. Pharmacol. 23 (1973)
    276-282.
49. Kostopolous, G.K., Limacher, J.J. and Phillis, J.W. Action of
    various adenine derivatives on cerebellar Purkinje cells. Brain
    Res. 88 (1975) 162-165.
50. Krnjevic, K., Pumain, R., and Renaud, L. The mechanisms of
    excitation of acetylcholine in the cerebral cortex. J. Physiol.
    215 (1971) 247-268.
51. Lake, N. and Jordan, L.M. Failure to confirm cyclic AMP as
    second messenger for norepinephrine in rat cerebellum. Science
    183 (1974) 663-664.
52. Lever, J.D., Graham, J.D.P. and Spriggs, T.L.B. Electron
    microscopy of nerves in relation to the arteriolar wall. Bibl.
    Anat. 8 (1967) 51-55.
53. Libet, B. Long latent periods and further analysis of slow
    synaptic responses in sympathetic ganglia. J. Neurophysiol. 30
    (1967) 494-514.
54. Loewe, O. Uber humorale ubertragbarkeit der herzennervenwirkung.
    I. Mitteilung. Pflugers Arch. 189 (1962) 238-242.
55. Macdonald, R.L. and Nelson, P.G. Specific opiate-induced
    depression of transmitter release from dorsal root ganglion
    cells in culture. Science 199 (1978) 1449-1451.
56. Marshall, K.C. and Engberg, I. Reversal potential for

noradrenaline-induced hyperpolarization of spinal motoneurons. Science 205 (1979) 422-424.

57. Moore, R.Y. and Bloom, F.E. Central catecholamine neuron systems: anatomy and physiology of the norepinephrine and epinephrine systems. Ann. Rev. Neurosci. 2 (1979) 113-168.

58. Mudge, A.W., Leeman, S.E. and Fischbach, G.D. Enkephalin inhibits release of substance P from sensory neurons in culture and decreases action potential duration. Proc. Natl. Acad. Sci. U.S.A. 76 (1979) 526-530.

59. Nakai, Y. and Takaoir, S. Influence of norepinephrine-containing neurons derived from the locus coeruleus on lateral geniculate neuronal activities of cats. Brain Res. 71 (1974) 47-60.

60. Nakajima, S. and Takahashi, K. Post-tetanic hyperpolarization and electrogenic Na pump in stretch receptor neurone of crayfish. J. Physiol. 187 (1967) 105-127.

61. Nathanson, J. Cyclic nucleotides and nervous system function. Physiol. Rev. 57 (1977) 157-256.

62. Nathanson, J., Freedman, R. and Hoffer, B.J. Lanthanum inhibits brain adenylate cyclase and blocks noradrenergic depression of Purkinje cell discharge independent of calcium. Nature 261 (1976) 330-331.

63. Nicoll, R.A., Siggins, G.R., Ling, N., Bloom, F.E. and Guillemin, R. Neuronal actions of endorphins and enkephalins among brain region: A comparative microiontophoretic study. Proc. Natl. Acad. Sci. U.S.A. 74 (1977) 2584-2588.

64. Nishi, S. and Koketsu, K. Origin of ganglionic inhibitory postsynaptic potentials. Life. Sci. 6 (1967) 2049-2055.

65. Nishi, S. and Koketsu, K. Early and late discharges of amphibian sympathetic ganglion cells. J. Neurophysiol. 31 (1968) 717-728.

66. North, R.A. Effects of morphine on myenteric plexus neurones. Neuropharmacol. 15 (1976) 1-9.

67. North, R.A. and Henderson, G. Action of morphine on guinea pig myenteric plexus and mouse vas deferens studied by intracellular recording. Life Sci. 17 (1975) 63-66.

68. North, R.A., Katayama, K., and Williams, J.T. On the mechanism and site of action of enkephalin on single myenteric neurons. Brain Res. 165 (1979) 67-77.

69. North, R.A. and Williams, J.T. Enkephalin inhibits firing of myenteric neurones. Nature (London) 264 (1976) 460-461.

70. North, R.A. and Williams, J.T. Actions of enkephalin on myenteric neurons. Fed. Proc. 36 (1977) 965.

71. Pellmar, T.C. and Carpenter, D.O. Voltage-dependent calcium current induced by serotonin. Nature 277 (1979) 483-484.

72. Pepper, C.M. and Henderson, G. Opiates and opioid peptides hyperpolarize locus coeruleus neurons in vitro. Science 209 (1980) 394-396.

73. Phillis, J.W., Kostopoulos, G.K. and Limacher, J.J. Depression of corticospinal cells by various purines and pyrimidines. Can. J. Physiol. Pharmacol. 52 (1974) 1227-1229.

74. Rall, T.W. Role of adenosine 3',5'-monophosphate (cyclic AMP) in actions of catecholamines. <u>Pharmacol. Rev.</u> 24 (1972) 399-409.
75. Renaud, L.P., Blume, H.W., Pittman, Q.J., Lamour, Y. and Tan, A.T. Thyrotropin-releasing hormone selectivity depresses glutamate excitation of cerebral cortical neurons. <u>Science</u> 205 (1979) 1275-1277.
76. Richardson, K.C. Electronmicroscopic observation of Auerbach's plexus in the rabbit, with special reference to the problem of smooth muscle innervation. <u>Am. J. Anat.</u> 103 (1958) 99-135.
77. Segal, M. Lithium and the monoamine neurotransmitters in the rat hippocampus. <u>Nature</u> 250 (1974) 71-73.
78. Segal, M. Brain stem afferents to the rat medial septum. <u>J. Physiol.</u> 261 (1976) 617-631.
79. Segal, M. and Bloom, F.E. The action of norepinephrine in the rat hippocampus. II. Activation of the input pathway. <u>Brain Res.</u> 72 (1974) 99-114.
80. Segal, M. and Bloom, F.E. The action of norepinephrine in the rat hippocampus. IV. The effects of locus coeruleus stimulation on evoked hippocampal unit activity. <u>Brain Res.</u> 107 (1976) 513-525.
81. Shoemaker, W.J., Balentin, L.T., Siggins, G.R., Hoffer, B.J., Henriksen, S.J. and Bloom, F.E. Characteristics of the release of cyclic adenosine 3',5'-monophosphate from micropipetts by microiontophoresis. <u>J. Cyclic Nucleotide Res.</u> 1 (1975) 97-106.
82. Siggins, G.R. The electrophysiological effects of cyclic nucleotides on excitable tissue. In H. Cramer and J. Schultz (Eds), <u>Cyclic Nucleotides: Mechanisms of Action</u>, John Wiley and Sons, Ltd., London-New York, 1977.
83. Siggins, G.R. Electrophysiological assessment of mono-nucleotides and nucleosides as first and second messengers in the nervous system. In A. Karlin, V.M. Tennyson and H.J. Vogel (Eds.), <u>Neuronal Information Transfer</u>, Academic Press, New York, 1978, p. 339.
84. Siggins, G.R. Cyclic nucleotides: regulation of cellular excitability. In J. Nathanson and J. Kebabian (Eds.), <u>Handbook of Experimental Pharmacology</u>, Vol. Cyclic Nucleotides, Springer-Verlag, Berlin, in press.
85. Siggins, G.R. and Henriksen, S.J. Analogues of cyclic adenosine monophosphate: correlation of inhibition of purkinje neurons with protein kinase activation. <u>Science</u> 189 (1975) 559-561.
86. Siggins, G.R., Hoffer, B.J. and Bloom, F.E. Cyclic adenosine monophosphate: possible mediator for norepinephrine effects on cerebellar purkinje cells. <u>Science</u> 165 (1969) 1018-1020.
87. Siggins, G.R., Hoffer, B.J. and Bloom, F.E. Studies on norepinephrine-containing afferents to Purkinje cells of rat cerebellum: III. Sensitivity evidence for mediation of norephrine effects by cyclic 3',5' adenosine monophosphate. <u>Brain Res.</u> 25 (1971) 535-553.
88. Siggins, G.R., Oliver A.P., Hoffer, B.J. and Bloom, F.E. Cyclic adenosine monophosphate and norepinephrine: effects on

transmembrane properties of cerebellar Purkinje cells. Science
171 (1971) 192.

89. Siggins, G.R., Hoffer, B.J., Oliver, A.P. and Bloom, F.E.
Cyclic AMP mediation of norepinephrine synaptic inhibition in
rat cerebellar cortex: A unique class of synaptic responses.
Nature 233 (1971) 481-483.

90. Siggins, G.R., Henriksen, S.J. and Landis, S.C. Electrophysiology of Purkinje neurons in the weaver mouse:
iontophoresis of neurotransmitters and cyclic nucleotides, and
stimulation of the nucleus locus coeruleus. Brain Res. 114
(1976) 53-65.

91. Siggins, G.R., Zieglgansberger, W., French, E., Ling, N. and
Bloom, F. Opiates and opioid peptides may excite hippocampal
(HPC) neurons by inhibiting adjacent inhibitory interneurons.
Neurosci. Abstr. 4 (1978) 414.

92. Speden, R. Electrical activity of single smooth muscle cells
of the mesenteric artery produced by splanchnic nerve stimulation
in the guinea pig. Nature 202 (1964) 193-194.

93. Stone, T.W. and Taylor, D.A. Microiontophoretic studies of the
effects of cyclic nucleotides on excitability of neurons in the
rat cerebral cortex. J. Physiol. 266 (1977) 523-543.

94. Sutherland, E.W., Oye, I., and Butcher, R.W. The action of
epinephrine and the role of the adenylcyclase system in
hormone action. Rec. Progr. Hormone Res. 21 (1965) 523-642.

95. Thomas, R.C. Electrogenic sodium pump in nerve and muscle cells.
Physiol. Rev. 52 (1972) 563-594.

96. Weight, F.F. and Padjen, A. Slow synaptic inhibition: evidence
for synaptic inactivation of sodium conductance in sympathetic
ganglion cells. Brain Res. 55 (1973) 219-224.

97. Weight, F.F. and Padjen, A. Acetylcholine and slow synaptic
inhibition in frog sympathetic ganglion cells. Brain Res. 55
(1973) 225-228.

98. Werman, R. Criteria for identification of a central nervous
system transmitter. Comp. Biochem. Physiol. 18 (1966) 745-766

99. Yarbrough, G.G. TRH potentiates excitatory actions of
acetylcholine on cerebral cortical neurons. Nature 263 (1976)
523-524.

100. Zieglgänsberger, W. and Bayerl, J. The mechanisms of
inhibition of neuronal activity by opiates in the spinal cord
of cat. Brain Res. 115 (1976) 111-128.

101. Zieglgänsberger, W., French, E.D., Siggins, G.R. and Bloom, F.E.
Opioid peptides may excite hippocampal pyramidal neurons by
inhibiting adjacent inhibitory interneurons. Science 205 (1979)
415-417.

102. Zieglgansberger, W., and Fry, J.P. Actions of enkephalin on
cortical and striatal neurones of naive and morphine/tolerant
dependent rats. In H.W. Kosterlitz (Ed) Opiates and Endogenous
Opioid Peptides, Elsevier/North-Holland, Biomedical Press,
Amsterdam, 1976, pp. 213-238.

103. Zieglgansberger, W. and Fry, J.P. Actions of opioids on

single neurons. In A. Herz (Ed), Development in Opiate Research,
Marcel Dekker, New York, 1978.

104.Zieglgansberger, W. and Puil, E.A. Action of glutamic acid on
spinal neurones. Exp. Brain Res. 17 (1972) 35-49.

105.Zieglgansberger, W. and Reiter, C.H. A cholnergic mechanism
in the spinal cord of cats. Neuropharmacol. 13 (1974) 519-527.

ENKEPHALIN RELEASE FROM THE GLOBUS PALLIDUS:

IN VITRO AND IN VIVO STUDIES

A. Bayón, R. Drucker-Colín, L. Lugo, W.J. Shoemaker*
R. Azad* and F.E. Bloom*

Departamento de Neurociencias, Centro de Investigaciones
en Fisiología Celular, Universidad Nacional Autónoma de
México, México 20, D.F. and *Arthur V. Davis Center for
Behavioral Neurobiology, The Salk Institute, La Jolla,
CA 92037

INTRODUCTION

The presence of the opioid peptides [5-methionine] -enkephalin
(Met-enkephalin) and [5-leucine] -enkephalin (Leu-enkephalin) (7) in
several regions of the central nervous system (17) and the immunohistochemical evidence for their location in fibers and cell bodies
(6) first supported their consideration as putative central neurotransmitters. The search for a model system to study the central
neurobiology of the enkephalins has led to the striatum and
especially the globus pallidus because this region has the highest
content of immuno-reactive enkephalin and enkephalin-containing
fibers. The early studies in one of our laboratories (8) showed that
during in vitro perfusion of slices from rat globus pallidus,
enkephalin-like immuno-reactive material was released by 50 mM K^+ in
a Ca^{2+}-dependent manner. In subsequent work (1) we have analyzed the
composition of these enkephalin-like substances, both in fresh and
perfused tissue and in the released material, and also some aspects
of the metabolic changes of enkephalins during in vitro release
experiments. Based on these studies, we have explored the basic
characteristics of enkephalin release from pallidal tissue in
unanesthetized-freely moving animals using a push-pull cannula
perfusion technique. Our results show that enkephalins can be
released from the pallidum, not only by chemical stimulation of the
local fibers but also, in a more physiological manner, by electrical
stimulation of the caudate nucleus.

EXPERIMENTAL PROCEDURES

In Vitro Perfusion Technique

Adult male rats (Sprague-Dawley) weighing 150-200 g were
decapitated, their brains removed, and the globus pallidus from each
hemisphere was quickly dissected (approximately 15 mg each) from
coronal slices (sagittal 1.5-2.25 mm, anterior level 5.0-7.0)
according to Konig and Klippel (10). The tissue samples were
immediately chopped in two perpendicular directions at 200-μm
intervals (McIlwain tissue chopper, Brinkmann, NY). Each globus
pallidus was suspended in a plastic perfusion chamber as previously
described (8). A slightly modified Krebs-Ringer bicarbonate buffer
(127 mM NaCl/3.73 mM KCl/1.8 mM $CaCl_2$/1.18 mM KH_2PO_4/1.18 mM $MgSO_4$/20
mM $NaHCO_3$/D-glucose (2 g/liter), previously equilibrated with O_2/CO_2,
95/5, vol/vol) containing 0.1% bovine serum albumin (Sigma,
crystalline) and 30 μg of bacitracin per ml was pumped (polystaltic
pumps and Silastic tubing) through the chambers at a flow rate of
250 μl/min. A heating bath kept the temperature inside the chambers
at 37°-38°. The perfusate was collected every 3 min for 39 min over
polypropylene tubes containing 1 ml of boiling 1 M acetic acid and
20 μl of saturated Na_2-EDTA solution. Release of enkephalin was
evoked after 24 min by a 6-min exposure to a modified perfusing
solution containing 85 mM NaCl/50 mM KCl (8).

Both freshly dissected tissue and perfused tissue were boiled
for 15 min in 1.0 M acetic acid solution with 20 μl of saturated Na_2
EDTA solution per ml (2 ml per each globus pallidus) and then
homogenized (13). These tissue extracts as well as perfusate fractions
were centrifuged at 1000 X g for 30 min, the supernatants frozen
overnight, and centrifuged once more. Aliquots of the supernatants
were lyophilized and kept at 0°C until radio-immuno-assayed or run
through high pressure liquid chromatography (HPLC). A chromatograph
equipped with a Universal Injector (U6K.) and μ Bondapak-CN columns
(0.39 X 30 cm) (all from Waters Associates) were used in this work.
The eluting mixture contained 14% glass-distilled acetonitrile and
86% 0.01 M ammonium acetate (pH adjusted to 4.2 with glacial acetic
acid); this eluant was also used to dissolve the samples (100-500 μl).
The columns were calibrated with pure synthetic peptides and their
retention times were determined by in-line absorption detection at
210 nm. After injection of the sample 400-μl fractions were collected
in polypropylene tubes (Gilson Microfractionator) with a flow rate
of 2 ml/min, then lyophylized and radio immunoassayed.

In Vivo Perfusion Technique

The push-pull cannula perfusion technique used in these
experiments is based on those previously described by Myers (12).
Briefly, stainless steel cannulae that serve as stereotaxic guides

for the push-pull cannulae were chronically implanted in the skull
of adult cats (2-2.5 kg) and albino rats (150-200 g) under sodium
pentobarbital anesthesia. The lower end of the guides was positioned
several mm. above the target area and the lumen of these guides was
temporarily blocked with short mandrels. One week after surgery the
mandrels were removed and concentric push-pull cannulae were
inserted through the guides to reach the pallidal tissue (the outer-
pull cannula was made of 21 gauge thin wall steel tubing; the inner-
push cannula is a 27 gauge needle protruding 1.5 mm from the tip of
the pull cannula). Each perfusing cannula was connected through
polyethylene tubing to the infusion and withdrawal syringes
positioned in a Harvard reciprocating pump. The perfusion of the
unanesthesized-freely moving animals was started using sterile Krebs-
Rinber bicarbonate containing BSA (0.1%) and bacitracin (30 µg/ml).
The flow rate was held constant at 23 µl/min (both in the push and
in the pull systems) and 15 min fractions were collected in the
withdrawal syringe which contained ice-cold 2N acetic acid and a
stirring bar to mix the acid with the perfusate. The collected
fractions were boiled for 15 minutes, lyophilized and resuspended
in buffer for the Leu-enkephalin RIA. The position of the tip of the
cannulae was verified after the experiment by inspection under
dissection microscope of frontal sections of the brain.

Radioimmunoassay

The lyophilisates of tissue extracts, perfusate fractions or
chromatographic fractions were dissolved in small volumes of water or
RIA buffer, aliquots were taken for scintillation counting when
necessary, and the remaining sample was subjected to radioimmunoassay
at two dilutions in duplicate. The Leu-enkephalin assay was as
described by Rossier et al. (14). The minimum detectable amount of
Leu-enkephalin is approximately 1 pg. Met-enkephalin crossreacts in
this assay to the extent of 3% on a weight basis. The Met-enkephalin
radioimmunoassay was performed as described by Gros et al (4) with
minor modifications; the same group provided the Met-enkephalin
antiserum. In our hands the minimum detectable amount of Met-enkephalin was 20 pg and the crossreactivity of Leu-enkephalin was 16%
(7% in the Gros et al. radioimmunoassay). Both assays were performed
with purified ^{125}I-labeled enkephalins.

The purified nonlabeled peptides used as standards in the
radioimmunoassays and HPLC were prepared by solid-phase synthesis
(11). [Tyrosyl-3,5-^{3}H] -enkephalin-(5-L-methionine)(33 Ci/mmol) and
[tyrosyl-3,5-^{3}H]-enkephalin-(5-L-leucine)(25.5 Ci/mmol) were
purchased from Amersham (England) and New England Nuclear (Boston, MA),
respectively, and purified through HPLC prior to use. L-[2,6-^{3}H] -
Tyrosine and carrier free ^{125}I were obtained from Amersham.

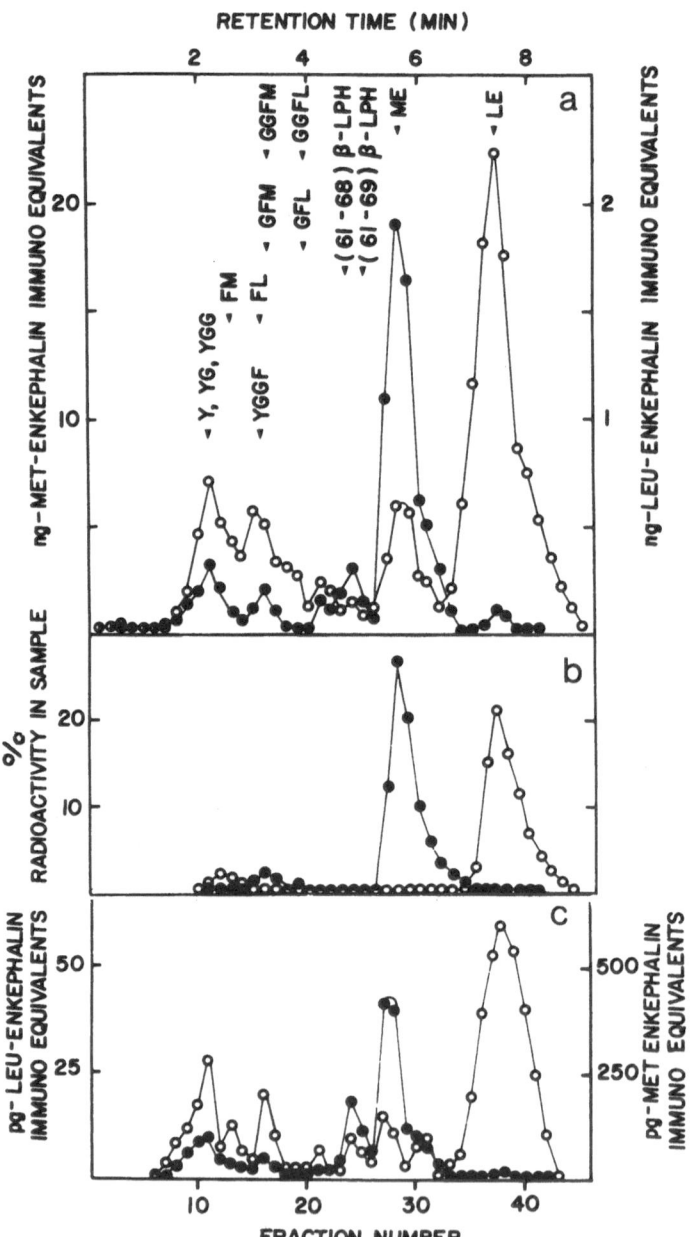

EXPERIMENTAL OBSERVATIONS AND DISCUSSION

    Early in the course of this work (1) it became clear that Met-
and Leu-enkephalin were not the only enkephalin-like substances
present in the pallidal tissue. Other compounds accounted for about
one third of the total immuno-reactivity observed both in tissue
extracts and in vitro perfusates. Fig. 1a shows the HPLC fractionation
of an extract of freshly dissected globus pallidus tissue. Each
enkephalin was identified by its retention time on the basis of
synthetic peptides and ^{3}H-labeled (Fig. 1b) internal standards. The
endogenous enkephalins were monitored by means of two radioimmunoassays using antisera raised against Met-enkephalin and Leu-
enkephalin. The identity of the peaks of immunoreactivity showing
the retention times of Met-enkephalin and Leu-enkephalin was further
assessed by their relative immunoreactivity with the Met-enkephalin
and Leu-enkephalin antisera. The immunoreactive components other than
Met-enkephalin and Leu-enkephalin showed immunoreactivity ratios,
from the two radioimmunoassays, that were intermediate between those

----

Fig. 1. HPLC fractionation of extracts from freshly dissected
pallidal tissue (a and b) and high K^+ in vitro perfusate (c). The
μ-Bondapak-CN column was calibrated with unlabeled synthetic
standards. ME stands for Met-enkephalin and LE for Leu-enkephalin.
The other standards are named by the single-letter amino acid
nomenclature (Y=Tyr, G=Gly, P=Phe, M=Met, L=Leu). Their retention
times (arrows) were determined by absorption at 210 nm. The flow
rate was kept at 2 ml/min and 400-μl fractions of the effluent were
collected. a: Radioimmunoassays using both Met-enkephalin and Leu-
enkephalin antisera were performed in each fraction of the
chromatograms (at two dilutions in duplicates). The values are
expressed in terms of ng of Met-enkephalin (•) that would produce
an equivalent immunodisplacement in the Met-enkephalin assay and,
analogously, the ng of Leu-enkephalin (o) producing an equivalent
displacement in the Leu-enkephalin assay. The total immunoreactivity
is that contained in the globus pallidus of one rat (the overall
recovery being 96%). The content of Met-enkephalin and Leu-enkephalin
in their respective peaks was estimated from the two radioimmunoassays and agreed within 5% variation. b: Radioactivity profiles of
samples, similar to those in a, that were extracted in the presence
of either tritiated Met-enkephalin (•) or tritiated Leu-enkephalin
(o) (≈ 10,000 cpm per sample). The radioactivity is expressed as %
of the cpm injected into the column. The overall recovery of
radioactivity for either tracer was 90%. The radioimmunoassay
profiles of these chromatograms were comparable to those in a. c:
HPLC fractionation of the material released from globus pallidus
slices during K^+ stimulation: Met-enkephalin (•) and Leu-enkephalin
(o) radioimmunoassay profiles.

Table 1. Comparison of the Met-enkephalin and Leu-enkephalin contents in the globus pallidus tissue and in the material released during K^+ stimulation, in vitro.

	Fresh Tissue*	Perfused Tissue *+	K^+-evoked* Release ‡
Met-enkephalin (ng)	54.8 ± 6.2	14.4 ± 1.3	0.85 ± 0.05
Leu-enkephalin (ng)	9.3 ± 0.7	4.1 ± 0.3	0.33 ± 0.02
Met/Leu enkephalin (w:w)§	5.8 ± 0.2	3.4 ± 0.2	2.7 ± 0.3

Results are expressed as the mean obtained from 3-4 independent experiments ± the standard error of the mean.

* Calculated per globus pallidus (both hemispheres).
+ After 39 min. perfusion. The tissue content immediately after release (30 min) was not significantly different.
‡ Total amount recovered during the 6 minutes exposure to the high potassium medium containing 30 µg/ml of bacitracin.
§ From ratios calculated in individual experiments.

of Leu-enkephalin and Met-enkephalin. These other peaks were also present in chromatograms of extracts from perfused tissue (not shown) and in perfusate samples, although in different proportions (Fig. 1c). Dilution curves of these fractions paralleled the standard curves in the assays. Many synthetic enkephalin and endorphin fragments had retention times in HPLC that cluster around the unidentified immuno- reactive peaks (Fig. 1a). However, these synthetic fragments show low crossreactivity in the enkephalin assay.

In a preliminary attempt to explore the relationship of these nonidentified components to the enkephalins, their molecular weights were estimated using a Sephadex G-15 column (45 X 0.7 cm), eluted with 50% acetic acid and calibrated with ^{3}H-labeled enkephalins and tyrosine; (^{3}H) tyrosine was also used as an internal standard. The unidentified immunoreactive substances were eluted before tyrosine but no significant amounts of immunoreactivity eluted before either enkephalin. Although the nature of these peaks has not been assessed, they certainly do not behave as the recently identified Met- enkephalin-Arg[6]-Phe[7] or Met-enkephalin -Lys[6] (16) since these peptides are not detected by our RIAs, which read the carboxy-terminus of the enkephalins.

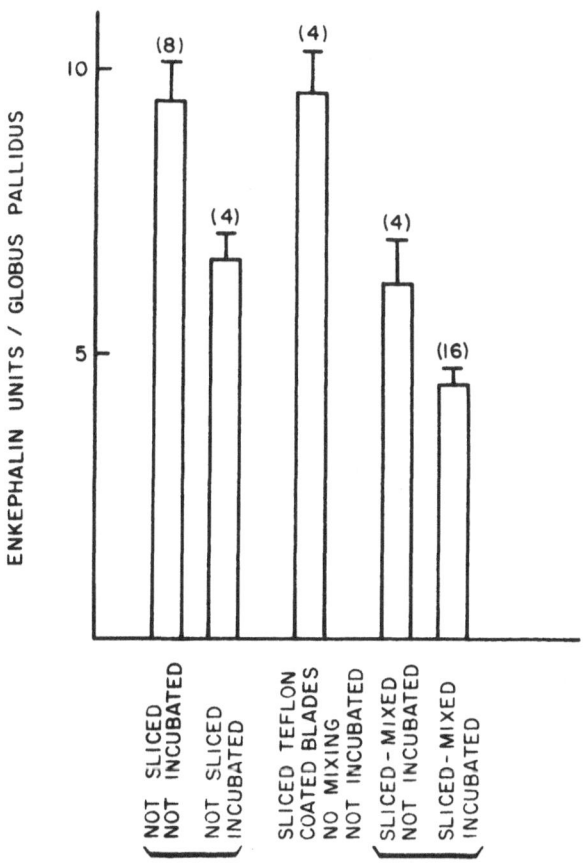

Fig. 2. Effects of slicing and incubation in the enkephalin content
of globus pallidus tissue. Enkephalin units=ng of Leu-enkephalin
that would produce an equivalent trace displacement in the Leu-
enkephalin RIA. The number of experiments is in parenthesis. Slicing
and incubation during perfusion were as described in Methods. Mixing
refers to the dispersion of the slices into the incubation chambers.

In Vitro Release Studies

During in vitro perfusion of slices from rat globus pallidus
the release of endogenous enkephalin is stimulated by a perfusing
medium containing both 50 mM K^+ and 1.8 mM Ca^{2+}. The amount of
enkephalin released in these conditions is not only 10-20 fold
higher than that measured in a low K^+ or low Ca^{2+} medium, but also
represents about 6% of the Met-enkephalin and 8% of the Leu-
enkephalin present in the tissue immediately after the stimulation
(Table 1, see also 8). The ratio of Met-enkephalin to Leu-

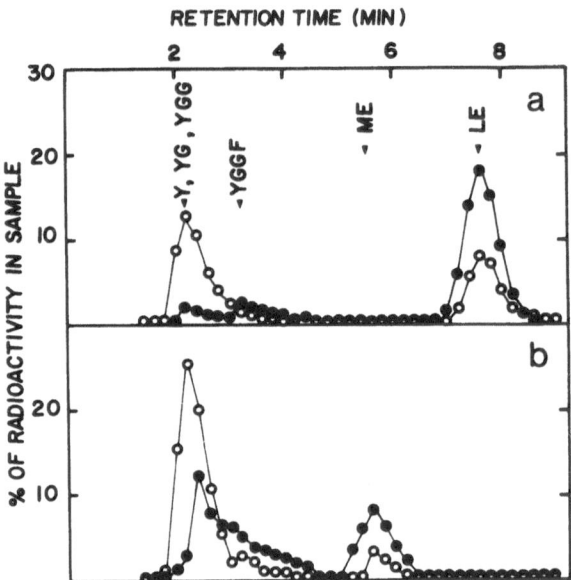

Fig. 3. HPLC (High Pressure Liquid Chromatography) of globus pallidus
perfusate obtained during K^+-stimulated enkephalin release. The
medium used in the perfusion contained trace amounts of prepurified
tritium labeled enkephalins. a: Degradation by the tissue slices
of (tyrosyl-3, 5-^{3}H) Leu-enkephalin added to the 50 mM K^+ perfusing
medium ($\simeq$ 4000 cpm in 2 ml) was 62% in the absence of bacitracin
(o) and 23% with bacitracin (30 µg/ml) (•). b: As in a, but adding
(tyrosyl-3, 5-^{3}H) Met-enkephalin ($\simeq$ 6000 cpm in 2 ml). Degradation
was 90% in the absence of bacitracin (o) and 63% with bacitracin (•).
The percentage of degradation is the mean of two experiments. Similar
percentages were found when the tracers were perfused in low K^+
media. Recovery of each tracer was nearly 100% from the perfusion
chambers and about 95% from the HPLC.

enkephalin measured in the perfusing medium during K^+ stimulation
closely resembles their ratio in the perfused tissue (Table 1, see
1 and 5). Since both Met-enkephalin and Leu-enkephalin can be
released by K^+ (in a Ca^{2+}-dependent manner) from the globus pallidus,
these chemically and pharmacologically similar pentapeptides must
be considered as performing interrelated roles, perhaps as neurotransmitters, at least in this constrained area of the nervous system.
Although this interrelationship, based on co-release, may not be
generalizable to all brain regions, our data suggest that the
correlated concentrations of Met- and Leu-enkephalin in the brain (2)
warrant extended neurobiological investigation.

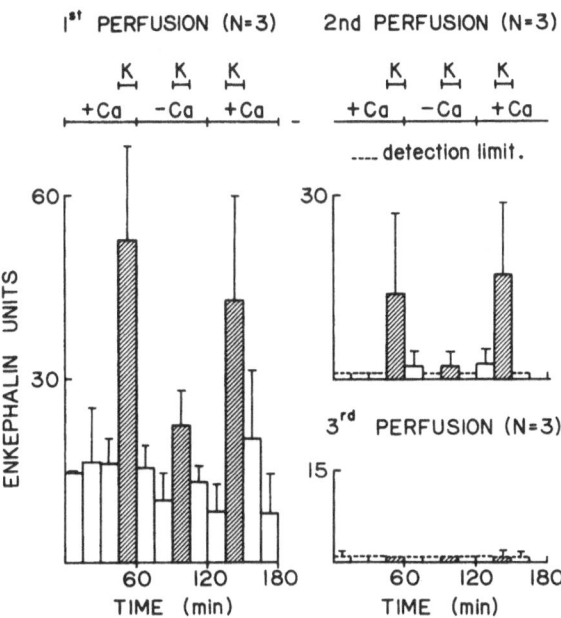

Fig. 4. In vivo release of enkephalin from the globus pallidus of the cat. Sequential push-pull cannula perfusions were performed at one week intervals after the chronical implantation of the guide-cannulae. In all perfusions the flow rate was held constant at 23 μl/min and 15 min fractions were collected on ice cold 2N acetic acid. The samples were boiled, lyophilized, redissolved in water and assayed for Leu-enkephalin immunoreactivity (1 enkephalin unit is equivalent to the trace displacement produced by 1 pg of Leu-enkephalin in the Leu-enkephalin RIA). White bars correspond to low K^+-resting release periods; 50 mM K^+ medium was perfused during the periods indicated by the hatched bars. During the second hour of perfusion Mg^{2+} was added to the medium instead of Ca^{2+}. Vertical lines represent the standard deviation of the mean; N is the number of bilaterally perfused cats. Enkephalin values lower than the minimum detectable amount are indicated by a discontinuous line on top of the bars.

It has been difficult to relate the ratio of the two enkephalins released from the pallidum to the enkephalin ratio observed in the freshly dissected tissue because a substantial and differential loss of pallidal Met-enkephalin (74%) and Leu-enkephalin (56%) (Table 1) occurs during the preparation and perfusion of the tissue. This is not a general observation when studying peptide release since, using identical experimental procedures, somatostatin does not significantly decrease in its

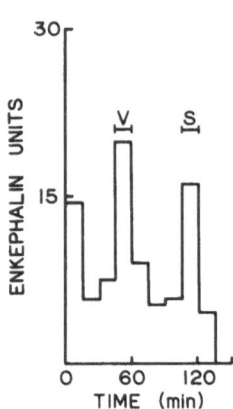

Fig. 5. In vivo release of
enkephalin from the globus pallidus
of the rat. Perfusion conditions are
indicated in fig. 4. V = low
potassium medium containing veratrine 60 µg/ml. S=electrical
stimulation of the caudo-putamen
using a coiled bipolar electrode:
Square pulses (0.5 msec duration,
applied voltage 10 V, average
courrent 100 µA) were delivered at
40 H_z, 60 H_z and 80 H_z, each during
a 5 min period. At each frequency
the trains of stimulation lasted
30 sec and were interrupted by 30
sec resting periods. Enkephalin
values are the average of release
data obtained from 3 rats.

tissue stores (9). Although part of this enkephalin loss is related
to the preparation of the slices, an important proportion of it
occurs during the incubation (Fig. 2).

It has already been reported (8) that the presence of the
protease inhibitor bacitracin in the perfusing medium does not modify
the amount of enkephalin immunoreactivity found in the tissue after
perfusion. Thus, in order to decide whether this tissue-enkephalin
loss during the perfusion is due to bacitracin-resistant degradation
inside the tissue stores, or to leakage from the tissue followed
by massive breakdown of the peptides, we studied the degradation of
each enkephalin in the perfusate by adding trace amounts of ^{3}H-
labeled Met-enkepahlin or Leu-enkephalin to the medium before
perfusing the slices (Fig. 3a and b). Degradation of (^{3}H) Met-
enkephalin in bacitracin-free medium was about 90%; bacitracin (30
µg/ml) decreased the degradation to 63%. Degradation of Leu-
enkephalin was 62% without bacitracin and 23% when it was present-
i.e. bacitracin doubled the amount of ^{3}H Leu-enkephalin recovered in
the perfusate. Since it has been reported (8) that the amount of
unlabeled endogenous Leu-enkephalin measured in the perfusate after
K^+ stimulation is also doubled by bacitracin, it seems that the
tritiated enkephalins can be used as internal standards for the
degradation of the enkephalins released from the endogenous pools.
When the total amount of enkephalins recovered in the 40 minute
perfusate (approximately 15% of that measured in the freshly
dissected tissue) is corrected for degradation, it accounts for the
tissue enkephalin lost during the incubation, which is 25-30% of the
fresh tissue content (Fig. 2). Thus, leakage of enkephalin from the
tissue and not intra-tissular degradation seems to be the major

factor to explain the tissue loss of enkephalin during perfusion.
The reason for this lability of the enkephalin tissue stores is still
unknown and is currently being studied using in vivo preparations.
However, we would like to speculate at this point that the differences
between the two enkephalins respect to their lability in tissue
slices and degradation in the perfusate could reflect metabolic and/or
physical compartmentations.

In Vivo Release Studies

Extending our studies on enkephalin release to preparations in
vivo allowed us to investigate this phenomenon in a system that
maintains its anatomical and physiological relationships with the
rest of the brain. The preliminary experiments were performed in the
cat globus pallidus mainly because this is rutinely used in our
laboratory for push-pull cannula perfusions. Following the procedure
described in Methods, a K^+-stimulated-Ca^{+2}-dependent enkephalin
release was obtained from the globus pallidus of freely moving cats
(Fig. 4). A subsequent perfusion (a week later) produced lower
amounts of enkephalin both during resting and stimulated release and
in a third perfusion (one week interval) enkephalin in the perfusate
was below the detection limit of the assay. A first perfusion can be
extended to 8 hour-a time longer than these three sessions together
and having several cannula penetrations-without a significant decrease
in the enkephalin release levels; this indicates that long term
effects of tissue damage modify the enkephalin containing system in
the globus pallidus. When the tip of the push-pull cannula was aimed
at the interphase between the internal capsule and the globus
pallidus in order to avoid damage of the pallidal tissue, the
enkephalin collected in the perfusate was in most cases below the
sensitivity of the RIA, which shows the anatomical accuracy of the
experimental procedure.

These results were also observed and extended using rats as
experimental subjects. Both K^+ (50 mM) and veratrine were shown to
stimulate the enkephalin release from the globus pallidus. Furthermore, in rats implanted with a bipolar electrode in the caudate
nucleus, electrical stimulation of this area elicited an increased
enkephalin release in the globus pallidus (Fig. 5).

These results show that not only the local perfusion of pallidal
tissue with chemical depolarizing agents can elicit a release of
enkephalin from this region (rich in enkephalin fibers) but that also
electrical stimulation of a separate area, the neighbouring caudate
nucleus [where enkephalin-containing cell bodies have been identified
(6,15)] produces a quantitatively similar effect. This observation
is consistent with the existence of a strio-pallidal enkephalin-
containing pathway postulated on the basis of lesion experiments (3).

Altogether, the results of these in vitro and in vivo studies on the release of enkephalins from the brain region richest in these opioid peptides strongly support the concept that the enkephalins must play a role in neural communication either as neuro-transmitters or as neuromodulators.

Acknowledgements: We thank Drs. F. Dray and C. Gros for the gift of Met-enkephalin antiserum (3614), Dr. N. Ling for providing the synthetic peptides and Mrs. A. Mauss for the initial RIAs. We are grateful to Dr. L. Iversen who encouraged the initiation of this work (Partially supported by NIDA 01785).

REFERENCES

1. Bayón, A., Koda, L., Battenberg, E., Azad, R., Guillemin, R., and Bloom, F.E., Regional distribution of endorphin, methionine-enkephalin and leucine-enkephalin in the pigeon brain, Neurosci. Letters, 16 (1980) 75-80.
2. Bayón, A., Rossier, J., Mauss, A., Bloom, F.E., Iversen, L.L., Ling, N. and Guillemin, R., In vitro release of methionine^5-enkephalin and leucine^5-enkephalin from the rat globus pallidus, Proc. Natl. Acad. Sci. U.S.A., 75 (1978) 3503-3506.
3. Cuello, A.C. and Paxinos, G., Evidence for a long Leu-enkephalin strio-pallidal pathway in rat brain, Nature 271 (1978) 178-180.
4. Gros, C., Pradelles, P., Rougeot, C., Bepoldin, O., Dray, F., Fournie-Zaluski, M.C., Roques, B.P., Pollard H., Llorens-Cortés, C. and Schwartz, J.C., Radioimmunoassay of methionine-and leucine-enkephalins in regions of rat brain and comparison with endorphins estimated by a radio receptor assay, J. Neurochem., 31 (1978) 29-39.
5. Henderson, G., Hughes, J. and Kosterlitz, H.W., In vitro release of Leu-and Met-enkephalin from Corpus striatum, Nature 271 (1978) 677-679.
6. Hokfelt, T., Elde, R., Johansson, O., Terenius, L. and Stein, L., The distribution of enkephalin immunoreactive cell bodies in the rat central nervous system, Neurosci. Letters, 5 (1977) 25-31.
7. Hughes, J., Smith, T.W., Kosterlitz, H.W., Fothergill, L.H. Morgan, B.A. and Morris, H., Identification of two related pentapeptides from the brain with potent opiate agonist activity, Nature, 255 (1975) 577-579.
8. Iversen, L.L., Iversen, S.D., Bloom, F.E., Vargo, T. and Guillemin, R., Release of enkephalin from rat globus pallidus in vitro, Nature 271 (1978) 679-681.
9. Iversen, L.L., Iversen S.D., Bloom, F., Douglas, C., Brown, M. and Vale, W., Calcium-dependent release of somatostatin and neurotensin from rat brain in vitro, Nature, 273 (1978) 161-163.
10.Konig, J.F.R. and Klippel, R.A., The Rat Brain, Krieger, Huntington, New York, 1967.
11.Ling, N., Solid phase synthesis of porcine α-endorphin and

γ-endorphin, two hypothalamic-pituitary peptides with opiate
activity, Biochem. Biophys. Res. Commun., 74 (1977) 248-255.
12. Myers, R.D., An improved push-pull cannula system for perfusing
an isolated region of the brain, Physiol. & Behavior, 5 (1970)
243-246.
13. Rossier, J., Bayón, A., Vargo, T.M., Ling, N., Guillemin, R. and
Bloom, F.E., Radioimmunoassay of brain peptides. Evaluation of
a methodology for the assay of β-endorphin and enkephalin, Life
Sci., 21 (1977) 847-852.
14. Rossier, J., Vargo, T.M., Minick, S., Ling, N., Bloom, F. and
Guillemin, R., Regional dissociation of β-endorphin and enkephalin
contents in rat brain and pituitary, Proc. Natl. Acad. Sci. USA,
74 (1977) 5162-5165.
15. Sar, M., Stumpf, W.E., Miller, R.J., Chang, K.-J. and Cuatrecasas,
P., Immunohistochemical localization of enkephalin in rat brain
and spinal cord, J. Comp. Neurol. 182 (1978) 17-38.
16. Stern, A.S., Lewis, R.V., Kimura, S., Rossier, J., Gerber, L.D.,
Brink, L., Stein, S. and Udenfriend, S., Isolation of the opioid
heptapeptide Met-enkephalin-Arg^6-Phe^7 from bovine adrenal
medullary granules and striatum, Proc. Natl. Acad. Sci. USA 76
(1979) 6680-6683.
17. Yang, H.-Y., Hong, J.S. and Costa, E., Regional distribution of
Leu-and Met-enkephalin in rat brain, Neuropharmacology, 16
(1977) 303-307.

ACIDIC AMINO ACIDS AS EXCITATORY TRANSMITTERS

Carl W. Cotman

University of California, Irvine

Irvine, California 92717

INTRODUCTION

Some of the most important and exciting discoveries over these
last few years have come from the identification and study of neurotransmitters. There are many new candidates, new regulatory mechanisms
and new findings on the pathways which employ known transmitters.
Curiously, though, most of the clearly identified neurotransmitters
are inhibitory. Little is known about the excitatory pathways except
that glutamate and aspartate are the major transmitter candidates
for most excitatory pathways. The possibility that these amino acids
might be transmitters was raised almost twenty five years ago.

In the late 1950's, many substances were studied for their
ability to alter the firing rate of neurons using the newly developed
technique of ionophoresis. Amino acids were identified as powerful
modulators of neuronal firing. The dicarboxylic acids glutamate and
aspartate were excitatory, γ-aminobutyric acid and glycine were
inhibitory (10,11,20). At that time acetylcholine (Ach) was the only
CNS transmitter clearly identified. A transmitter role for glutamate
and aspartate was thought to be unlikely because these acidic amino
acids excited virtually all neurons, unlike ACh which excited only
those neurons which had a cholinergic innervation. Also the D and L
enantiomorphs of glutamate were of similar potency, had the same
duration of action on cat spinal neurons and co-application of
substances known to inhibit glutamate metabolism had little if any
effect on the glutamate evoked depolarization (10). ACh was
inactivated by acetylcholinesterase, and it was assumed other
transmitters must behave similarly. We now know that the regulatory
mechanisms and characteristics of different neurotransmitters differ

widely. As a greater understanding has been gained on the properties
expected of neurotransmitters, glutamate and aspartate have faired
well as neurotransmitter candidates though it has been difficult to
rigorously prove they are transmitters.

In this chapter I will review some of our recent work which has
been aimed at evaluating the premise that glutamate and aspartate are
neurotransmitters for certain important excitatory pathways in brain,
and will describe some of their basic regulatory mechanisms. Our
approach has been to investigate in great detail one or two
promising systems with the goal of establishing the transmitter firmly
and in the process identifying the characteristics most diagnostic
of its presence. Our studies have focused on the hippocampus,
primarily on the perforant path (see 6,7,8).

The laminar organization of the hippocampus lends itself
particularly well to the analysis of specific terminal fields (Fig.
1). In the dentate gyrus of the hippocampus, the granule cells, the
major cell type, are found in a layer. The zone which contains the
granule cell dendrites and its inputs is called the molecular layer.
Granule cells receive only two major inputs: one from the entorhinal
cortex and another from hippocampal CA4 neurons. The projection from
the entorhinal cortex terminates in the outer 3/4 of the dendritic
field; the one from CA4 neurons terminates in the inner 1/4 of the
dendritic field nearest the granule cell bodies. The entorhinal input
accounts for about 60% of the total input to the granule cells, and
is known to be excitatory. The molecular layer is sufficiently large
so that it can be readily dissected from fresh slices thereby
providing tissue where the main afferent is of entorhinal origin.

PRESYNAPTIC CRITERIA FOR GLUTAMATE AS A NEUROTRANSMITTER

One of the first approaches used in an attempt to identify
excitatory amino acid pathways was to map the distribution of
endogenous glutamate and aspartate levels in the CNS and to measure
the levels after a lesion of specific nerve tracts. However, there
is no compelling evidence that glutamate even exists in greatly
enriched concentrations in glutamate neurons. Therefore, a high
endogenous level of glutamate or aspartate, or a reduction in levels
after lesioning is not sufficient evidence alone to indicate a
transmitter role. High-affinity transport has also been proposed as
a marker for glutamate neurons. On the basis of this, several authors
have suggested that the reduction in uptake which occurs after a
lesion of one nerve terminal population is indicative of a transmitter
role.

However, glutamate uptake is not a specific marker for
"glutamate-using" neurons. Glutamate and aspartate share the same

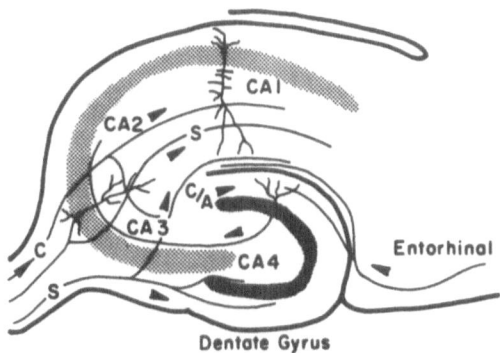

Fig. 1. Synaptic organization of the dentate gyrus. Neurons in the
entorhinal cortex form the perforant path. These fibers terminate in
the outer 2/3 of the dentate gyrus molecular layer. The dentate
granule cells give rise to the mossy fibers.

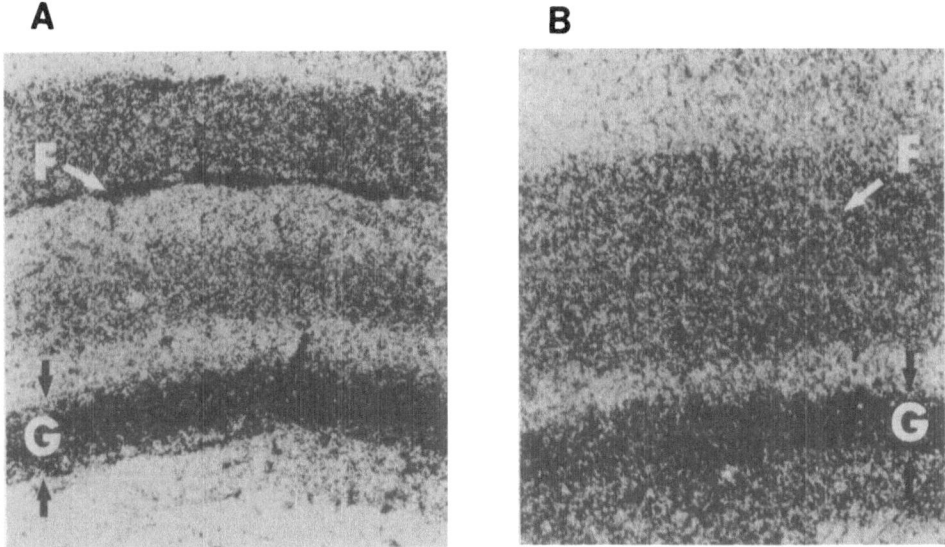

Fig. 2. Distribution of glutamate accumulated by high affinity uptake
in the dentate gyrus before (A) and after an entorhinal lesion (B).
Note the trilaminar distribution of uptake in the molecular layer with
the outer and inner thirds showing the highest concentration.

high-affinity uptake process (22), suggesting that the same uptake
sites are present on both "glutamate-using" and "aspartate-using"
neurons and glutamate uptake may exist in non-glutamate secretory
neurons as well. Also, high-affinity uptake has been demonstrated
in glial preparations (16), and autoradiography suggests that glial
uptake may predominate in vivo (18,26). Nonetheless, uptake is useful
if its limitations are realized. If uptake is not found it is
reasonable to conclude that glutamate is not a transmitter. The
perforant path zone does accumulate glutamate and this uptake is
dramatically reduced by an entorhinal lesion (29,36) (Fig. 2).

It is now well established that transmitter release in the CNS
is Ca^{2+}-dependent and stimulated by depolarization. Thus, glutamate
may serve as a transmitter in an area of the brain or spinal cord
if it is shown to be released in this manner. Such a demonstration
of Ca^{2+}-dependent release is, at present, one of the best markers
for glutamate-using pathways.

Slices of the molecular layer (or fascia dentata) or
synaptosomes isolated from the hippocampus have been employed to
analyze the characteristics of endogenous glutamate, aspartate and
GABA release. Fibers which originate from the entorhinal cortex and
project to the dentate gyrus display the appropriate release
properties. Electrical field stimulation causes glutamate release
from slices of the dentate. Release is Ca^{2+}-dependent and antagonized
by Mg^{2+} (43). Also, depolarization by 56 mM K^+ ions or veratridine
in the presence of Ca^{2+} provokes release. Release is stimulated by
depolarization alone, probably due to reversal of the carrier, but
it is 2-3 times greater in the presence of Ca^{2+}. Other divalent ions
such as Ba^{2+} substitute for Ca^{2+} (31,32). The release of newly
synthesized glutamate from glutamine or glucose shows similar
characteristics to that of endogenously loaded glutamate (14,15).

Glutamate release appears to originate from synaptic boutons
since isolated synaptosomes prepared from whole hippocampus or dentate
gyrus show an increase in Ca^{2+}-dependent release relative to less
homogenous fractions. The entry of Ca^{2+} is a necessary and sufficient
condition to evoke release since the ionophore A23187 will support
release in the absence of depolarization. In all respects GABA
release shows the same characteristics as glutamate release (7,33).

Calcium dependent glutamate release appears to come from
entorhinal boutons since it decreases after the entorhinal cortex is
removed (14,28,29). The effect is selective to glutamate since GABA
and aspartate both increase significantly. The elevation in aspartate
release is probably a result of axon sprouting by CA4 neurons (8).
The increase in GABA release is probably an adaptive response by local
interneurons to the lesion (32). Thus, presynaptic evidence strongly
favors the notion that glutamate is at least one of the transmitters

Table 1. Classes of acidic amino acid receptors in the mammalian
CNS

Agonist	D-α-AA[a]	GDEE[b]
I. N-METHYL-D-ASPARTATE	+	-
IBOTENIC ACID		
II. GLUTAMATE	-	+
AMPA[c]		
QUISQUALATE	-	+
III. KAINATE	-	-

[a]D-α-aminoadipate

[b]glutamate diethyl ester

[c](RS)-α-amino-3-hydroxy-5-methyl-4-isoxazolepropionic acid

of the perforant path.

POSTSYNAPTIC CRITERIA FOR GLUTAMATE SYNAPSES

The identification of glutamate-using pathways would be greatly
facilitated by the discovery of a specific antagonist of glutamate
responses. Only recently have compounds which selectively antagonize
amino acid responses been found. With the use of these compounds it
has become clear that several types of receptors for amino acids are
present in the mammalian CNS, probably, in fact, at least three
(Table 1).

We have tested a number of putative antagonists and it appears
that, of those tested, the glutamate analogue 2-amino-4-phosphono-
butyrate (2-APB) is the most effective. 2-APB acts as an antagonist
at the invertebrate neuromuscular junction where glutamate is the
transmitter (9). In our initial studies on hippocampal slices, we
found that D,L-2APB selectively blocked synaptic excitation of the
dentate granule cells evoked by perforant path stimulation (42).
It did not antagonize the response to mossy fiber stimulation. Our
initial data with D,L-2APB while suggestive were not, however,
without problems. The concentrations required were in the mM range and
the response was never completely suppressed. In our initial
experiments we used the racemic mixture of 2-APB. In the spinal cord
acidic amino acid receptors appear to display stereospecificity with

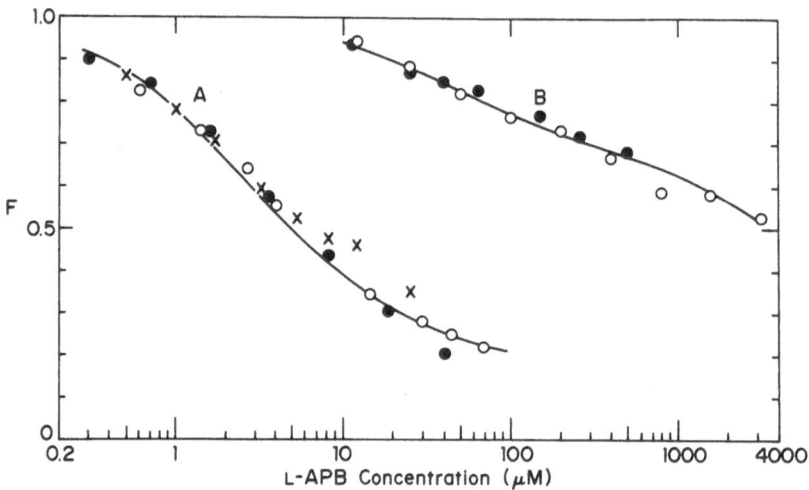

Fig. 3. Dose-response curves for the effect of L–APB on components of the perforant path. Abscissa: concentration of L–APB in the superfusing medium (µM). Ordinate: F, fraction of normal activity in the presence of drug. This is the ratio of the peak amplitude of the synaptic field potential (negativity) measured in the presence of the drug to the peak amplitude measured in the absence of the drug. <u>A</u>: Response when the recording electrode was placed in the outer one-third of the molecular layer of the dentate gyrus (data for three hippocampal slices, each from a different animal). The theoretical curve was calculated by assuming that 75% of the field potential was generated by synapses for which L–APB acts as a simple antagonist with an apparent K_D of 2.5 µM and 25% was generated by drug resistant synapses with a response like curve B. <u>B</u>: Response when the recording electrode was placed near the <u>middle</u> of the molecular layer (data for two slices, each from a different animal). The theoretical curve was derived by assuming that the response is evoked from two components for which L–APB acts as a simple antagonist. The best curve assumes 32% of a component with an apparent K_D of 45 µM and 68% with an apparent K_D of 10,000 µM. For these calculations, the relationship $F = (1 + c/K_D)^{-1}$ was assumed (F, fraction of activity; c, concentration of antagonist; K_D, its apparent dissociation constant).

the D form usually being the stronger antagonist (40). Accordingly, we have tested the D and L forms of 2–APB on the perforant path.

We found that L–APB is a powerful antagonist of synaptic transmission in the perforant path (19). Dose response relationships show that micromolar concentrations produce a maximal inhibition of about 50% and that further increasing the dose has little effect. This suggests that perhaps the perforant path consists of two

Table 2.   Relative potency of various putative antagonists on the
           lateral perforant path

Compound	Potency
L-APB	1
D-APB	1/40
D,L-APV[+]	1/100
D,A-APP[*]	1/2000

[+]D,L-2-amino-5-phosphonovaleric acid

[*]D,L-2-amino-3-phosphonopropionic acid

components, a very sensitive one and another insensitive one. Figure
3 shows the dose response curves for the lateral vs medial termination fields of the perforant path.

    By exposing a slice successively to different concentrations of
drugs and measuring the depression of the synaptic field potential
for each, graphs were obtained which depict the relative responses
of termination fields of the perforant path as a function of drug
concentration. The middle portion of the molecular layer of the
dentate gyrus requires a 500-fold higher concentration of L-APB for
50% inhibition than the outer molecular layer. Both kinds of response
originate from fibers of the perforant path, because they can both
be activated by stimulating the slices where the perforant path
traverses the subiculum. Mathematical analysis of the data suggests
that the response in the outer molecular layer includes some drug
resistant synapses. This is probably due to contamination with
synapses of the middle molecular layer, and it varies depending on
electrode placement and orientation of the plane of the slice. If
this contamination is taken into account, about 75% of the response
for the data shown can be attributed to a component for which L-APB
acts as a simple antagonist with an apparent K_D of 2.5 µM.

    The overall response from the middle of the molecular layer
spans an extremely wide range of concentrations, and no predominant
component can be discerned with a sigmoidal inhibition curve. Such
a response might be obtained from two components with apparent K_D of
about 45 µM and 10 mM for L-APB. This heterogeneity is also
demonstrated by data which shows a heterogeneous response, with the

Table 3.   Specific binding of GLU, ASP and KA to subcellular
           fractions from rat whole brain[a]

	Specific Binding Relative to Homogenate		
Fraction	GLU	ASP	KA
WHOLE PARTICULATE	1	1	1
CRUDE MITOCHONDRIAL PELLET (P_2)	1.15 ± 0.07	1.14 ± 0.57	0.73 ± 0.34
SYNAPTIC PLASMA MEMBRANES (SPM)	6.86 ± 3.07	5.20 ± 1.47	2.08 ± 0.41
SYNAPTIC JUNCTIONS (SJ)	24.35 ± 2.66	23.54 ± 8.96	22.37 ± 5.79

[a]Subcellular fractions were prepared by the method of Cotman and
   Taylor (19), and analyzed for L-^{3}H-GLU, L-^{3}H-ASP and ^{3}H-KA binding
   using a microfuge assay; at a final concentration of 100nM for
   labelled L-glutamate and L-aspartate, and 40nM for KA. Specific
   binding was determined as that which could be displaced by a 100 μM
   concentration of the respective unlabelled ligand. The values were
   taken from 3-4 separate preparations and are means $\pm$ S.E.M.

same proportions of components to β-(p-chlorphenyl)-γ-aminobutyric
acid (baclofen) (21).

    Three lines of evidence suggest that the sensitive component of
the outer molecular layer arises mainly from fibers of the lateral
entorhinal cortex: (1) it is confined to the portion of the molecular
layer known to receive projections from this region; (2) the field
potential negativity recorded during inhibition experiments follows
the relatively slow time course characteristic of the lateral
entorhinal cortex, while the more drug-resistant response recorded
from the middle molecular layer was the more rapidly-developing
negativity of medial entorhinal projections; and (3) an electrolytic
lesion to the lateral entorhinal cortex four days before the
experiment markedly reduced the most sensitive component. Thus the
perforant path, apparently homogeneous with respect to cells of
origin, shows pharmacological heterogeneity in its field of
termination. Glutamate as a transmitter of the lateral perforant path
is also consistent with the observation that the concentration of
glutamate high-affinity uptake is higher there than in the medial area

(see Fig. 2).

The response to the enantiomer D-APB and to the homologs D,L-2-amino-3-phosphonopropionic acid (DL-APP, an analogue of aspartic acid) and D,L-2-amino-5-phosphonovaleric acid (DL-APV) were tested for the outer molecular layer (Table 2). It is noteworthy that D,L-APP, with a smaller extended length which might fail to span the glutamate receptor, is the least potent while the longer homolog D,L-APV exhibits a small but readily measurable antagonist action.

These data on the relative effect of L-APB and its enantiomer and higher and lower homologs on synapses of the lateral perforant path comprise the most compelling evidence to date that a close structural analogue of L-glutamic acid can exert a highly specific inhibition of a defined mammalian neuronal pathway. Also, with an apparent K_D of 2.5 µM for the system, L-APB is now the most potent contrast strongly with those obtained for the same drugs against an N-methyl-D-aspartate-sensitive pathway of the spinal cord. In that system, D,L-APV inhibits at micromolar concentrations, D,L-APB is 100 times less potent, and its D-isomer is most active (40). Thus these data indicate the existence of a novel glutamate receptor in brain.

In support of our results Wheal and Miller (41) and Hicks and McLennan (17) have shown that ionophoretically applied GDEE will also block perforant path synaptic excitation of the dentate granule cells. D-α-AA is relatively ineffective on perforant path synapses but does block the dentate commissural response. Therefore, it appears that at least two types of excitatory amino acid receptors are present on these hippocampal cells.

Biochemical analysis would greatly aid in characterizing acidic amino acid receptors. Binding studies with acidic amino acids have revealed high-affinity receptor sites for L-glutamate kainate and NMDA in the mammalian CNS (1,3,13,23,27,34,35). The pharmacology of this binding is at best in broad agreement with electrophysiology experiments. L-Glutamate binding is displaced by the potent excitatory compounds quisqualate, D,L-homocysteate and ibotenate, but not by kainic acid (1,13,34). Kainic acid binding is displaced by glutamate and quisqualate, but this action may be indirect (24). The density of glutamate binding sites is at least one order of magnitude greater than those for kainic acid. In invertebrates many of the receptors are extrajunctional (37) and a similar situation may prevail in brain. Thus one of the first tasks is to distinguish junctional from extrajunctional receptors.

Recently, we have attempted to evaluate the relative importance of junctional and non-junctional receptors for excitatory amino acids.

Use has been made of subcellular fractionation techniques to isolate synaptic junctions from rat brain and the binding of L- $[^3H]$ - glutamate, L- $[^3H]$ -aspartate and $[^3H]$-kainate has been studied. The binding of all three ligands is enriched several fold in synaptic junctions, suggesting that acidic amino acid receptors are more concentrated there (Table 3). Greater than 70% of the glutamate, aspartate and kainate binding sites are retained in the junctional membranes. This suggests that binding sites for these compounds appear to be predominantly junctional (12).

SYNTHESIS OF NEUROTRANSMITTER GLUTAMATE

Glutamate, like any other transmitter, must be replenished. Glucose or glutamine are the two potential precursors of glutamate which are present in the highest amount in brain. Glutamate can be synthesized by hydrolysis of glutamine, a reaction catalyzed by the enzyme(s) glutaminase or it can be synthesized from glucose by oxidative metabolism and transamination of oxoglutarate. Bradford and Ward (4) showed that nerve terminals contain large quantities of glutaminase and suggested that this enzyme plays the major role in the production of transmitter glutamate. Previous studies on glutamate biosynthesis had revealed a complex compartmentation (38,39). Studies are needed on a system where glutamate is strongly expected to serve as a neurotransmitter.

Accordingly we have used dentate gyrus slices in order to evaluate the relative role of glucose and glutamine as precursors for the readily releasable glutamate pool (see 7). One way of evaluating the relative contributions of extracellular glucose and glutamine to the biosynthesis of releasable glutamate is to compare the specific radioactivity of glutamate released into the medium to that of its precursors. Glutamate that appeared in the medium during incubation with elevated K^+, Ca^{2+} and $[^{14}C]$ glutamine had a specific radioactivity 66% that of the precursor. This result suggests that about two-thirds of the glutamate released from the slices under these conditions originated from extracellular glutamine. In contrast, when $[^{14}C]$ glucose was the precursor, the specific radioactivity of glutamate released into the medium was only 16% that of the added glucose. Removal of the entorhinal cortex markedly decreased the quantity of both newly-synthesized and endogenous glutamate released. Thus more than 80% of the glutamate released from perforant path boutons under these conditions was derived from extracellular glucose and glutamine, with about two-thirds of the total originating from glutamine (14,15). Bradford et al. (5) have reported that about 80% of glutamate released from synaptosomes by elevated K^+ is derived from glutamine, in close agreement with our results. Most of the rest of the radioactive glutamate was derived from glucose. It was possible to demonstrate that most of the glutamate derived from glutamine is available for release, whereas most of the glutamate synthesized from glucose serves a different function, presumably metabolic. Thus, glutamate released by perforant path boutons appears to be derived primarily from glutamine, at least when elevated K^+ is employed as the depolarizing agent.

Glutamine appears to be synthesized primarily in glial cells, rather than in neurons (2,25,38). These data, therefore, imply a dynamic interaction between glutamate boutons and their associated glial cells. Upon release of glutamate, its biosynthesis is stimulated by an increase in both terminal uptake of glutamine and glutaminase activity. Glutamate that is released can be recaptured by the bouton or enter glial cells where it is converted to glutamine. Glutamine is then released from glia and is available to refuel the releasable glutamate stores in synaptic boutons. Thus glutamate releasing boutons and associated glial cells appear to have a symbiotic relationship. It appears that glutamine is the major precursor; however, other metabolic pathways may yet prove to play an important, if not unique, role. Hopefully one of these will prove to be present only in glutamate neurons so that a "marker" becomes available.

ACIDIC AMINO ACID ANALOGUES AS POSSIBLE TRANSMITTERS

All things considered, it seems unlikely that glutamate and aspartate are the only prominent excitatory transmitters. It is possible that amino acid analogues perhaps related to agonists already identified are transmitters. The mossy fiber system, for example, is a case in point. Mossy fiber synaptic transmission is potently suppressed by baclofen (21), a drug which appears to reduce selectively the release of acidic amino acids. However, lesion studies indicate that mossy fibers release neither glutamate nor aspartate (31).

The characteristics of the receptors suggest that the transmitter is a kainate like compound. Kainate binding sites are enriched in the CA3 region, and appear to be localized in the terminal field of the mossy fibers (12). A postsynaptic localization of these receptors could be inferred from the findings that the CA3 pyramidal cells are the most sensitive to the neurotoxic action of kainic acid (28,30).

At a subcellular level, synaptic junctions contribute the majority of the "synaptic population" of KA binding sites (see above). Specific and localized synaptic receptors imply that appropriate natural ligands exist. Our data, together with evidence that KA does not act on glutamate sites and that KA receptors are a discrete population, suggest that endogenous ligands for KA receptors exist. Such a compound would be a mediator or modulator of synaptic transmission at sites such as mossy fiber synapses.

CONCLUSION

   The evidence which supports a role of glutamate as transmitter
of the lateral perforant path is strong -- glutamate is released in a
Ca^{2+} dependent manner, the afferent possess a high-affinity uptake
system and a specific glutamate antagonist will block transmission.
It is suprising, though, that the perforant pathway which appears
to derive from a homogeneous group of cells is in fact heterogeneous,
at least with respect to postsynaptic receptor type. Most likely
many classes of receptors exist for the acidic amino acids and in
fact it is most likely that acidic amino acid analogues themselves are
the transmitters of some excitatory pathways. Our data on the mossy
fibers, for example, suggest a kainate like molecule is the
transmitter at this synapse. Analogues have not been identified but
the evidence described in this paper points toward their existence.

Acknowledgements. This work was supported by grants NS 08957 and
MH 19691. I am grateful to Mrs. D. Franks for secretarial assistance.

REFERENCES

1. Baudry, M. and Lynch, G., Two glutamate binding sites in rat
    hippocampal membranes, Eur, J. Pharmac., 57 (1979) 283-285.
2. Benjamin, A.M. and Quastel, J.H., Locations of amino acids in
    brain slices from the rat, Biochem J., 128 (1972) 631-646.
3. Biziere, K., Thompson, H. and Coyle, J.T., Characterization of
    specific, high-affinity binding sites for L- [^{3}H] glutamic acid
    in rat brain membranes, Brain Res., 183 (1980) 421-423.
4. Bradford, H.F. and Ward, H.K., On glutaminase activity in
    mammalian synaptosomes, Brain Res., 110 (1976) 115-125.
5. Bradford, H.F., Ward, H.K. and Thomas, A.J., Glutamine - A major
    substrate for nerve endings, J. Neurochem., 30 (1978) 1453-1459.
6. Cotman, C.W., Foster, A.C., and Lanthorn, T. An overview of
    glutamate as a neurotransmitter. In G. Di Chiara and G.L. Gessa
    (Eds.), Glutamate as a Neurotransmitter, Raven Press, New York,
    1980.
7. Cotman, C.W. and Hamberger, A., Glutamate as a CNS neuro-
    transmitter: Properties of release, inactivation and biosynthesis.
    In F. Fonnum (Ed.), Amino Acids as Chemical Transmitters, Plenum
    Press, New York, 1978, pp. 379-412.
8. Cotman, C.W. and Nadler, J.V., Glutamate and aspartate as
    hippocampal transmitters: Biochemical and pharmacological evidence.
    In P.J. Roberts, J. Storm-Mathisen and G. Johnston (Eds.),
    Glutamate as a Neurotransmitter, John Wiley Press, New York,
    1980, in press.
9. Cull-Candy, S.G., Donnellan, J.F., James, R.W. and Lunt, G.G.,
    2-amino-4-phosphonobutyric acid as a glutamate antagonist in
    locust muscle, Nature, 262 (1976) 408.

10. Curtis, D.R., Phillis, J.W. and Watkins, J.C., The chemical excitation of spinal neurons by certain acidic amino acids, J. Physiol. (Lond.), 150 (1960) 656-682.
11. Curtis, D.R. and Watkins, J.C., Acidic amino acids with strong excitatory actions on mammalian neurons, J. Physiol. (Lond.), 166 (1963) 1-14.
12. Foster, A.C., Mena, E.E., Monaghan, D.T. and Cotman, C.W. A synaptic localization of kainic acid binding sites, Nature, in press.
13. Foster, A.C. and Roberts, P.J., High-affinity L- [^{3}H] glutamate binding to postsynaptic receptor sites on rat cerebellar membranes, J. Neurochem., 31 (1978) 1467-1477.
14. Hamberger, A.C., Chiang, G.H., Nylen, E.S., Scheff, S.W. and Cotman, C.W., Glutamate as a CNS transmitter. I. Evaluation of glucose and glutamine as precursors for the synthesis of preferentially released glutamate, Brain Res., 168 (1979) 513-530.
15. Hamberger, A., Chiang, G.H., Sandoval, M.E. and Cotman, C.W., Glutamate as a CNS transmitter. II. Regulation of synthesis in the releasable pool, Brain Res., 168 (1979) 531-541.
16. Hertz, L., Functional interactions between neurons and astrocytes. I. Turnover and metabolism of putative amino acid transmitters, Prog. in Neurobiol., 13 (1979) 277-323.
17. Hicks, T.P. and McLennan, H., Amino acids and the synaptic pharmacology of granule cells in the dentate gyrus of the rat, Can. J. Physiol. Pharmac., 57 (1979) 973-978.
18. Kelly, J.S. and Dick, F., Differential labelling of glial cells and GABA inhibitory interneurons and nerve terminals following the micro-injection of β-^{3}H-alanine, ^{3}H-DABA and ^{3}H-GABA into single folia of the cerebellum. In Cold Spring Harbor Symposia on Quantitative Biology, Vol. XL, The Synapse, Cold Spring Harbor Laboratory, Cold Spring Harbor, New York, 1975, pp. 93-106.
19. Koerner, J.F. and Cotman, C.W. Micromolar L-2-amino-4-phosphono-butyric acid selectively inhibits perforant path synapses from lateral entorhinal cortex, submitted.
20. Krnjevic, K. and Phillis, J.W., Iontophoretic studies of neurons in the mammalian cerebral cortex, J. Physiol. (Lond.), 165 (1963) 274-304.
21. Lanthorn, T.H. and Cotman, C.W. Baclofen selectively inhibits excitatory transmission in the hippocampus, submitted.
22. Logan, W.J. and Snyder, S.H., High-affinity uptake systems for glycine, glutamate and aspartate in rat CNS synaptosomes, Brain Res., 42 (1972) 413-431.
23. London, E.D. and Coyle, J.T., Specific binding of ^{3}H-kainic acid to receptor sites in rat brain, Mol. Pharmacol., 15 (1979) 492-505.
24. London, E.D. and Coyle, J.T., Co-operative interactions at ^{3}H-kainic acid binding sites in rat and human cerebellum, Eur. J. Pharmac., 56 (1979) 287-290.

25. Martínez-Hernández, A., Bell, K.P. and Norenberg, M.D., Glutamine synthetase: Glial localization in the brain, Science, 195 (1977) 1356-1358.
26. McLennan, H., The autoradiographic localization of L- $[^3H]$ glutamate in rat brain tissue, Brain Res., 115 (1976) 139-144.
27. Michaelis, E.K., Michaelis, M.L. and Boyarsky, L.L., High-affinity glutamate binding to brain synaptic membranes, Biochem. Biophys. Acta., 367 (1974) 338-348.
28. Nadler, J.V., Perry, B.W. and Cotman, C.W., Preferential vulnerability of the hippocampus to intraventricular kainic acid, In E.G. McGeer, J.W. Olney and P.L. McGeer (Eds.), Kainic Acid as a Tool in Neurobiology, Raven Press, New York, 1978, pp. 219-238.
29. Nadler, J.V., Vaca, K.W., White, W.F., Lynch, G.S. and Cotman, C.W., Aspartate and glutamate as possible transmitters of excitatory hippocampal afferents, Nature, 260 (1976) 538-540.
30. Nadler, J.V., White, W.F., Vaca, K.W. and Cotman, C.W., Intra-ventricular kainic acid in hippocampal pyramidal cells, Nature, 271 (1978) 676-677.
31. Nadler, J.V., White, W.F., Vaca, K.W., Perry, B.W. and Cotman, C.W., Biochemical correlates of transmission mediated by glutamate and aspartate, J. Neurochem., 31 (1978) 147-155.
32. Nadler, J.V., White, W.F. and Vaca, K.W., Characterization of putative amino acid transmitter release from slices of rat dentate gyrus, J. Neurochem., 29 (1977) 279-290.
33. Sandoval, M.E., Horch, P. and Cotman, C.W., Evaluation of glutamate as a hippocampal transmitter: Glutamate uptake and release from synaptosomes, Brain Res., 142 (1978) 285-300.
34. Simon, J.R., Contrera, J.F. and Kuhar, M.J., Binding of $[^3H-]$ kainic acid, an analogue of L-glutamate, to brain membranes, J. Neurochem., 26 (1976) 141-147.
35. Snodgrass, S.R., In vitro binding studies with $[^3H]$ -N-methyl-aspartate, Abs. Soc. Neurosci., 5 (1979) 1943.
36. Storm-Mathisen, J., Glutamate and excitatory nerve endings: Reduction of glutamate uptake after axotomy, Brain Res., 120 (1977) 379-386.
37. Usherwood, P.N.R., Clark, R.B., Gration, K.A., Ozeki, M. and Patlak, J., Glutamatergic synapses and glutamate receptors in locust muscle, J. Physiol. (Paris), 75 (1979) 615-621.
38. Van den Berg, D.J. and Garfinkel, D., A simulation study of brain compartments. Metabolism of glutamate and related substances in mouse brain, Biochem. J., 123 (1971) 211-218.
39. Waelsch, H., Amino acid and protein metabolism. In K.A.C. Elliot, I.H. Page and J.H. Quasted (Eds.), Neurochemistry: The Chemistry of Brain and Nerve, Charles, C. Thomas, Springfield, Illinois, USA, 1962, pp. 288-320.
40. Watkins, J.C., Davies, J., Evans, R.H., Francis, A.A. and Jones, A.W., Pharmacology of receptors for excitatory amino acids. In G. DiChara and G.L. Gessa (Eds.), Glutamate as a Neurotransmitter, Raven Press, New York, 1980.

41. Wheal, H.V. and Miller, J.J., Pharmacological identification of acetylcholine and glutamate excitatory systems in the dentate gyrus of the rat, Brain Res., 182 (1980) 145-155.
42. White, W.F., Nadler, J.V. and Cotman, C.W., The effect of acidic amino acid antagonists on synaptic transmission in the hippocampal formation in vitro, Brain Res., 164 (1979) 177-194.
43. White, W.F., Nadler, J.V., Hamberger, A., Cotman, C.W. and Cummins, J.T., Glutamate as a transmitter of the hippocampal perforant path, Nature, 270 (1977) 356-357.

GLUTAMERGIC NEURONS: LOCALIZATION AND RELEASE OF THE TRANSMITTER

F. Fonnum, D. Malthe-Sørenssen, I. Kvale, A. Søreide,
K.K. Skrede and I. Walaas

Norwegian Defence Research Establishment
Division for Toxicology
N-2007 Kjeller, Norway

INTRODUCTION

Studies have accumulated showing that amino acids may constitute the group of neurotransmitters which dominate quantitatively in the mammalian brain (5,6,10,26,27). Glutamate is the most important candidate as an excitatory neurotransmitter. But it has several other important functions to fulfill and the problem is therefore to locate the transmitter pool of glutamate. By analogy to other amino acids and amines that function as neurotransmitters, this pool can be identified by three methods:

1) High affinity uptake process.

2) Ca^{++} dependent release on depolarization of brain slices or synaptosomes.

3) A high intraterminal localization.

By combining these 3 methods with lesion of specific brain pathways it is possible to accumulate evidence that glutamate is the neurotransmitter of a specific neuronal pathway. In the present communication we will make particular use of 1) and 3) in order to identify glutamergic fibres in the brain.

METHODS FOR IDENTIFYING GLUTAMERGIC NEURONS

High Affinity (HA) Uptake

The uptake processes into synaptosomes show a high degree of substrate specificity. Glu is taken up by the same system as Asp, but different from that of GABA, glycine and other amino acids (2). D-Asp, L-Asp and L-Glu are all taken up by the same high affinity system, whereas D-Glu is poorly taken up (7,32,54). This allows us to use D-Asp, which is metabolically stable, as a false transmitter for L-Glu. It is in this respect interesting to note that glutamate receptor agonists and antagonists such as diethylglutamate and homocysteic acid are poor inhibitors of uptake. The conformation of Glu and Asp in response to uptake or receptor function therefore differs greatly (2).

There is now strong evidence that the HA uptake of L-Glu and D-Asp is localized to specific nerve terminals. This was first demonstrated in a homogenate from the cerebellum of virus infected hamsters (61). However, these cerebella are extremely degenerated and the results should be interpreted with some caution. Subsequent work has shown, however, a specific loss in L-Glu or D-Asp uptake which parallels the degeneration of specific types of nerve terminals. Thus in lateral septum and nucleus accumbens after interruption of the fimbria/fornix (11,49,58), in striatum after hemidecortication (8,13) or in the lateral geniculate body and superior colliculus after visual cortex ablation (25,28), there is a loss in HA uptake of L-Glu or D-Asp but not in that of GABA or in any other transmitter parameter. In some of these denervated regions the loss of HA Glu uptake was more than 70%. There is no change in HA uptake of Glu in the lateral geniculate body of rat after enucleation. This shows that massive degeneration of non-glutamergic terminals in itself does not induce changes in HA uptake (28).

Also autoradiographic studies after uptake of Glu or Asp in hippocampal slices show that the highest labeling is found over terminals and axons (48). Furthermore, lesion destroying specific fibre system in the hippocampus is accompanied by a loss of labeling in a pattern expected from this degeneration (47,55). Autoradiography may therefore constitute a powerful tool in elucidating Glu-structures (53).

A complicating factor in all uptake studies is the possible interference by glial cells, particularly astroglia. Uptake of Glu as well as that of other amino acids has been demonstrated in a large series of glial cell preparations such as sensory ganglia (44), glial cells from cultures of spinal cord and medulla (25), isolated glial cell preparations (16) or primary cultures of astrocytes (17). At present the uptake of glutamate in glial cells can not be differentiated from that in neurons by substrates or inhibitors. The uptake of most glial cell preparation is, however, only a fraction of that in brain homogenate. An exception is the astroglial preparation from mouse brain where the affinity is much lower than for brain homogenates (17). It is also noteworthy that whole rat retina exhibits glial uptake characteristics towards GABA, whereas homogenates of rat retina behave similar to nerve-terminals (29). Homogenization, therefore, seems to destroy the uptake capacity of the glial cells. Further, axotomy is accompanied by gliosis in the form of microglia and astroglia depending upon the region. However, we have never been able to detect a significant increase in uptake after axotomy. In conclusion we believe that although a glial uptake of L-Glu and D-Asp is fairly well established, the quantitative contribution during uptake measurements in homogenates is small.

Levels of Endogenous Amino Acids

Glu plays an important part in the general metabolism and has many other functions beside being a neurotransmitter. All neurotransmitters, almost according to a dogma, are expected to be highly concentrated in their presynaptic elements. Thus, in the case of GABA, the concentration in nerve terminals has been estimated to be 50-150 mM (9) whereas the average concentration in brain is 2 mM. In comparison, Glu is the most abundant amino acid in brain with an average concentration of 10 mM in brain tissue. The concentration of Glu in glutamate nerve terminals is therefore expected to be much higher. Consequently, the loss of glutamergic nerve terminals should be accompanied by a reduction in the level of endogenous Glu. This is also the case as shown below. The loss of endogenous Glu is always considerably less than that of HA uptake. This only underscores the premise that the HA uptake is localized mainly to the presynaptic elements whereas Glu itself is widely distributed in the tissue. In this respect it is interesting that a decrease in endogenous Glu is not accompanied by similar changes in metabolically related amino acids, aspartate and GABA, but sometimes by an increase of glutamine. The latter is probably caused by gliosis since glial cells are responsible for glutamine synthesis.

Since HA uptake does not differentiate between glutamate and aspartate neurons, amino acid analysis is essential to discriminate between them.

LOCALIZATION OF GLUTAMERGIC FIBRES

Fornix-Fimbria Projections

The pyramidal cells in CA-3 hippocampus project bilaterally through fimbria-fornix to the lateral but not the medial part of septum (1,27,45). When the pyramidal cells were destroyed by intrahippocampal injection of kainic acid or when the fimbria-fornix was transected, HA-Glu uptake and the endogenous level of Glu fall in the lateral but not the medial septum (11). In agreement with the anatomical data the fall in HA-Glu uptake was larger after bilateral (60%) than after monolateral (40%) transection of fimbria. In addition, bilateral transection reduced the endogenous level of Glu

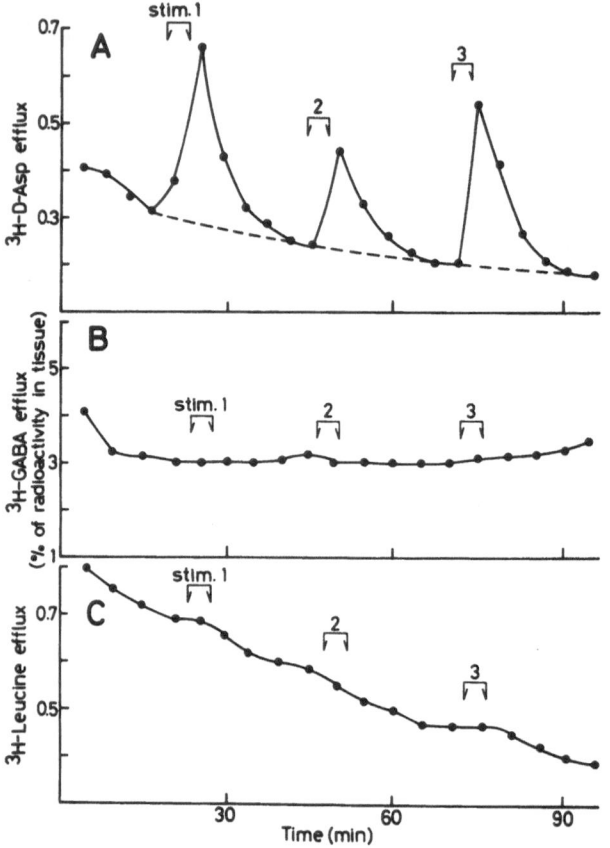

Fig. 1.   Stimulus evoked and spontaneous efflux of [3H] amino acids
from septo-hippocampal slice. Note that a stimulus evoked release
was only detected with [3H] Asp in the presence of 2mM Ca2+.

(30%). There were no significant changes in other transmitter
parameters such as HA-uptake of GABA, cholineacetyltransferase,
glutamate decarboxylase or endogenous level of aspartate. The study
therefore suggested that the hippocampal-septal fibres used Glu as
their chemical transmitter. Similar results on HA-Glu uptake were
obtained by Storm-Mathisen & Woxen-Opsahl (43).

When a slice preparation prepared from the dorso-lateral septum
with the fimbria adjoining was stimulated at the fimbria, i.e.
stimulation of the hippocampal-septal fibres (30), we were able to
demonstrate that electrical stimulation evoked the release of
exogenously added D-[3H] Asp, a false Glu transmitter (Fig. 1). The
evoked release was Ca++-dependent and specific in that neither
labeled GABA nor leucine were released under similar conditions. The

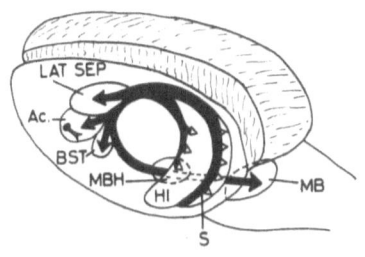

Fig. 2. Distribution of fornix/
fimbria fibres from hippocampus-
subiculum. Lat sep. lateral septum;
Ac, nucleus accumbens; NDB, nucleus
of diagonal band; MBH, mediobasal
hypothalamus; MB, mammillary body;
S, subiculum; HI, hippocampus; BST,
bed nucleus of the stria terminalis.

viability of the preparation was demonstrated by the presence of
fiber volleys and single and multiple unit changes. These studies
provided further evidence that Glu was the transmitter of the
hippocampal-septal fibre. Collaterals of these fibres go to the
stratum radiatum in hippocampus CA1 (1,27). This pathway is present
in the transverse hippocampal slice (39). We were also able to
demonstrate electrical stimulus-evoked release of exogenously added
D-[^{3}H] Asp also from this preparation (31). Similar findings have
been obtained by Wierazko & Lynch (53) who studied the release of
exogenous L-Glu. Together these experiments strongly suggest that
at least some of the fornix fibres use Glu as their chemical
transmitter.

These studies have therefore been expanded to include other
regions of the brain such as nucleus of the diagonal band, bed nucleus
of stria terminalis, nucleus accumbens, mediobasal hypothalamus and
corpus mammilare, which receive fornix fibers mainly from the
pyramidal cells in the subiculum (36,45). The results show that in
all these nuclei there was a significant decrease in HA-Glu uptake
(40-70%) and in the endogenous level of Glu (15-34%) (52). Only in
the nucleus of the diagonal band and in mammillary body was there
also a significant reduction in aspartate, but not in other amino
acids. From these studies it is clear that the subicular pyramidal
cell constitute an important center for glutamergic fibers which
distribute widely in the brain (Fig. 2). It is in this context
interesting that the mediobasal hypothalamic cells are very sensitive
to toxic doses of glutamate (35,52). This glutamate connection may
also be responsible for the endocrine functions initiated in the
hippocampal formation (24,50).

Neocortex

The distribution of glutamergic fibres in the somatic sensory
cortex of mouse was studied by autoradiography after D-Asp uptake
(47). It can be seen that a laminar pattern was readily recognized
(Fig. 3). The highest uptake was seen in layer 1, and since the
labeling was restricted to that layer, we do not believe it
represents a diffusion artifact. The subsequent layers 2 and 3 were

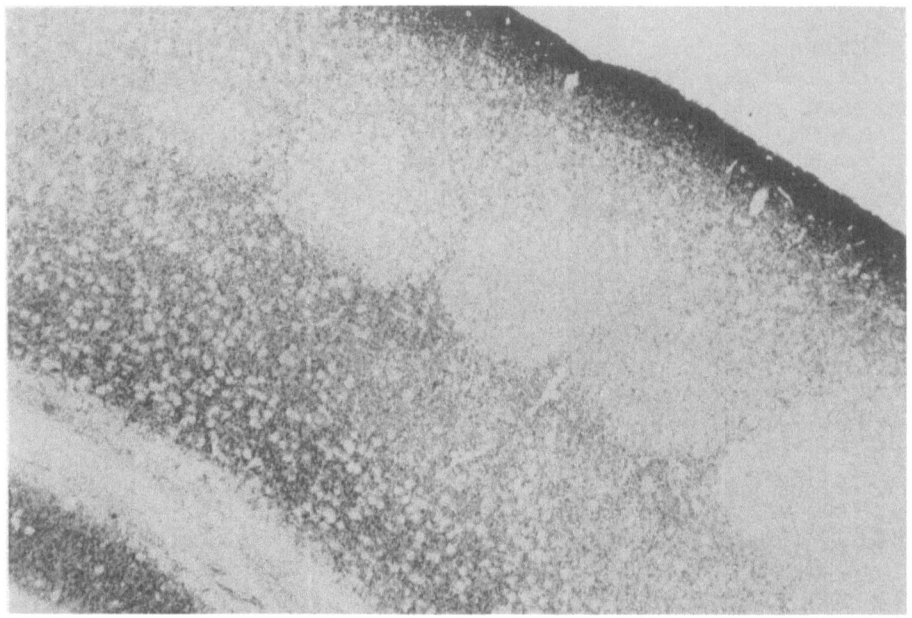

Fig. 3. Autoradiogram after uptake of labeled D-Asp into slices from somatosensory cortex of mouse. Notice the layered pattern of labeling and the presence of barrel-like structures in layer 4.

relatively pale. In layer 4 the staining conforms to the barrel pattern described by Woolsey & van der Loos (28). The darker areas correspond to the barrel sides and the lighter areas to the hollows. Layer 5 and 6 were again heavily stained. There are many candidates among efferent fibres and small local neurons which could be responsible for the heavy staining, such as the commissural fibres, thalamo-cortical fibres and small intracortical neurons. Further studies including autoradiographic staining after specific lesions may discriminate between these candidates.

Corticofugal Fibres

Several of the corticofugal fibres are probably glutamergic (Fig. 4). The ablation of the frontal or entire hemicortex was accompanied by a significant loss of HA-Glu uptake in the whole rostrocaudal direction of neostriatum (52-75%) (13). In addition, a large fall in endogenous Glu (40%) was obtained (13). A smaller change in Glu levels had been observed previously (23). A large part of the decrease in uptake in neostriatum (65-52%) could be accounted for by a lesion in frontal cortex, in agreement with anatomical studies that neostriatal fibres from this region predominate (14,22,33). In further agreement with anatomical studies, there was also a small but significant reduction in uptake (15%) on the contralateral side (4). Several biochemical investigations now support the concept that glutamate is the neurotransmitter of the cortico-striatal pathway (8,13,23,33). In agreement, the excitatory response to cortical stimulation could be suppressed by iontophoretic application of glutamate diethylester (40). Anatomists have recently started to regard nucleus accumbens as a part of ventral striatum. Although this nucleus receives most of its glutamergic fibres from fornix, it also receives a distinct input from neocortex. Decortication is therefore accompanied by a decrease in HA-Glu uptake in nucleus accumbens, particularly in the dorsal part (51).

The HA-Glu uptake is much lower in globus pallidus and substantia nigra than in neostriatum. Still, there is a significant decrease in uptake both in globus pallidus (40%) and substantia nigra (20%) after decortication; this loss is 15% and 1% of the corresponding decrease in neostriatum, respectively. The presence of cortical terminals in globus pallidus and substantia nigra have often been denied (3,37). The small loss in uptake observed in globus pallidus and substantia nigra could be due to cortical fibres transversing the regions. These fibers may give rise to axons which may in some cases have uptake properties.

The connections between thalamus and cortex in rat have been well studied. Hemidecortication was accompanied by a large reduction in D-Asp uptake (70%) in the whole of thalamus (13). Amino acid analysis after decortications showed a significant fall of Glu (38%) and to a lesser extent Asp (21%). This indicates the presence of glutamergic and perhaps some aspartergic cortico-thalamic fibres. This is in agreement with the fact that some neurons in thalamus are preferentially excited by glutamate and others by aspartate (5).

Corticofugal Fibres from Visual Cortex

The projection of visual cortex to thalamus (i.e. lateral geniculate body and superior colliculus) has been studied in more detail with regard to neurotransmitter content (25). In newborn rats the HA uptake in the lateral geniculate body is only 25% of that in the adult animal. This low activity cannot be due to corticothalamic terminals, since it did not decrease significantly after visual cortex ablation in the newborn animal. In the superior colliculus the uptake at birth was 120% (per mg protein) of the adult level. This did not decrease after visual cortex ablation either and thus may be due to local neurons or glial structures. In the adult animal, however, visual cortex ablation was accompanied by 75% reduction in HA-Glu uptake in lateral geniculate body and 50% reduction in superior colliculus (25,28).

In Figure 4 we have tried to visualize the glutamergic fibers

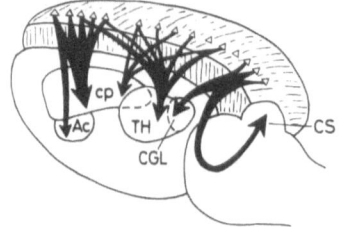

Fig. 4. Origin and distribution of glutamate fibres from neocortex. Ac, nucleus accumbens; CP, neostriatum; TH, thalamus; CGL, lateral geniculate body; CS, colliculus superior.

from neocortex to different brain regions.

When the visual cortex was removed unilaterally at birth and the animal allowed to survive for 10 weeks, the HA-Glu uptake was reduced on both sides, both in lateral geniculate body (30-40%) and in superior colliculus (40-50%). The unilateral lesion must have introduced changes on both sides. When the remaining visual cortex was removed in the adult animal, Glu uptake was only reduced in the new lesioned side of the lateral geniculate body. We did not therefore see any evidence that the remaining visual cortex had projected fibres to the opposite side. In the case of superior colliculus, we saw a further decrease in uptake on both sides. In this case therefore the results could be explained by an abarrant crossing of the visual cortex fibers induced by the neonatal lesion (Kvale & Fonnum, to be published). The plasticity of the visual cortex - superior colliculus fibres is in agreement with anatomical studies.

CONCLUSION

Glutamergic neurons can be identified by studying the changes in high affinity L-glutamate or D-aspartate uptake and the changes in endogenous glutamate tissue level after interruption of nerve fibers. Both amino acids are taken up and released in a Ca^{++} dependent manner at least from glutamergic neurons, but probably also from aspartergic neurons. Glutamate and aspartate can therefore substitute for each other and act as false and true transmitter in glutamergic and aspartergic neurons, respectively. It is therefore difficult to understand how the glutamergic neurons is capable of differentiating between the two acidic amino acids. It should be noted, however, that at the postsynaptic level at least three different receptors have been described for the acidic amino acids. There is at present a need for a better method for differentiating between aspartergic and glutamergic neurons at the presynaptic level. With these limitations in mind, we have described several cortico-fugal pathways to neostriatum, nucleus accumbens and thalamus, particularly the visual nuclei, that use glutamate as a transmitter. Several other important glutamergic pathways arise in subiculum/hippocampus and project via fornix to the lateral septum, nucleus accumbens, bed nucleus of stria terminalis, nucleus of the diagonal band, mediobasal hypothalamus and corpus mammillare.

REFERENCES

1. Andersen, P., Bland, B.H. and Dudar, J.D., Organization of the hippocampal output, Expl. Brain Res., 17 (1973) 152-168.
2. Balcar, V.J. and Johnston, G.A.R., The structural specificity of the high affinity uptake of L-glutamate and L-aspartate by rat brain slices, J. Neurochem., 19 (1972) 2657-66.
3. Bunny, B.S. and Aghajanian, G.K., The precise localization of nigral afferents in the rat as determined by a retrograde training technique, Brain Res., 117 (1976) 423-435.
4. Carman, J.B., Cowan, W.M., Powell, T.P.S. and Webster, K.E., A bilateral cortico-striate projection, J. Neurol. Neurosurg. Psychiat., 28 (1965) 71-77.
5. Curtis, D.R. and Johnston, G.A.R., Amino acid transmitters in the mammalian central nervous system, Ergebn. Physiol., 69 (1974) 94-188.
6. Curtis, D.R., Problems in the evaluation of glutamate as a central nervous system transmitter. In (Filer Jr, L.J., Garattini, S., Kane, M.K., Reynolds, W.A. and Wurtman, R.J., Eds.), Glutamic Acid. Advances in Biochemistry and Physiology, Raven Press, New York (1979), pp. 163-176.
7. Davis, L.P. and Johnston, G.A.R., Uptake and release of D- and L-aspartate by rat brain slices, J. Neurochem., 26 (1976) 1007-1014.
8. Divac, I., Fonnum, F. and Storm-Mathisen, J., High affinity uptake of glutamate in terminals of cortico-striatal axons, Nature (Lond.), 266 (1977) 377-378.
9. Fonnum, F. and Walberg, F., An estimation of the concentration of γ-aminobutyric acid and glutamate decarboxylase in the inhibitory Purkinje axon terminals in the cat, Brain Res., 54 (1973) 115-128.
10. Fonnum, F., Ed., Amino Acids as Chemical Transmitters, NATO Advances Study Series, Series A, Life Sciences, 16 (1978) Plenum Press, New York.
11. Fonnum, F. and Walaas, I., The effect of intrahippocampal kainic acid injections and surgical lesions on neurotransmitters in hippocampus and septum, J. Neurochem., 31 (1978) 1173-1181.
12. Fonnum, F., Lund Karlsen, R., Malthe-Sørenssen, D., Skrede, K.K. and Walaas, I., Localization of neurotransmitters, particularly glutamate in hippocampus, septum, nucleus accumbens and superior colliculus, Progr. Brain Res., 51 (1979) 167-191.
13. Fonnum, F., Storm-Mathisen, J. and Divac, I., Biochemical evidence for glutamate as neurotransmitter in cortico-striatal and cortico-thalamic fibres in rat brain, Brain Res., in press.
14. Goldman, P.S. and Nauta, H.J.H., An intricately patterned prefronto-caudale projection in the rhesus monkey, J. Comp.

Neurol., 171 (1977) 369-385.

15. Hattori, T., Fibiger, H.C., McGeer, P.L., Demonstration of a pallidonigral projection innervating dopaminergic neurons, J. Comp. Neurol., 62 (1975) 487-504.

16. Henn, F.A. and Hamberger, A., Glial cell function: Uptake of transmitter substances, Proc. Natl. Acad. Sci. USA, 68 (1971) 2686-2690.

17. Hertz, L., Bock, E. and Schousboe, A., Developmental Neurobiol. (1980), in press.

18. Hubbard, J. I., Mills, R.G. and Sirett, N.E., Responses in the diagonal band of Broca evoked by stimulation of the fornix in the cat, J. Physiol. (Lond.), 292 (1979) 233-249.

19. Høsli, I. and Høsli, E., Action and uptake of neurotransmitters in CNS, Tissue Culture Rev. Physiol. Biochem. Pharmacol., 81 (1978) 135-188.

20. Iversen, L.L. and Bloom, F.E., Studies of the uptake of [³H] GABA and [³H] glycine in slices and homogenates of rat brain and spinal cord by electron microscopic autoradiography, Brain Res., 41 (1972) 131-143.

21. Iversen, L.L. and Schon, F.E., The use of autoradiographic techniques for the identification and mapping of transmitter-specific neurons in CNS. In (A.M. Mandell, Ed.), New Concepts in Neurotransmitter Regulation, Plenum Press, New York (1973), pp. 153-193.

22. Kemp, J.M. and Powell, T.P.S., The cortico-striate projections in the monkey, Brain, 93 (1970) 525-546.

23. Kim, J-S., Hassler, R., Haug, P. and Paik, K-S., Effect of frontal cortex ablation on striatal glutamic acid level in rat, Brain Res. 132 (1977) 370-374.

24. Knigge, K.M. and Hayes, M., Evidence of inhibitive role of hippocampus on neural regulation of ACTH release, Proc. Soc. Exp. Biol. N.Y., 114 (1963) 67-69.

25. Kvale, I. and Fonnum, F., Development of neurotransmitter parameters in lateral geniculate body, superior colliculus and visual cortex of the rat, Neuroscience, in press.

26. Logan, W.J. and Snyder, S.H., Unique high affinity uptake systems for glycine, glutamic and aspartic acids in central nervous tissue of the rat, Nature (Lond.), 234 (1971) 297-299.

27. Lorente de No, R., Studies on the cerebral cortex-II. Continuation of the study of the Ammonic system, J. Psychol. Neurol. (Lpz), 46 (1934) 113-177.

28. Lund-Karlsen, R. and Fonnum, F., Evidence for glutamate as a neurotransmitter in the corticofugal fibres to the dorsal lateral geniculate body and superior colliculus in rats, Brain Res., 151 (1978) 457-467.

29. Lund-Karlsen, R., Neurotransmitters in the mammalian visual system. In (F. Fonnum, Ed.), Amino Acids as Chemical Transmitters, Plenum Press, New York (1978), pp. 241-256.

30. Malthe-Sørenssen, D., Skrede, K.K. and Fonnum, F., Release of D- [³H] aspartate from dorsolateral septum after electrical

stimulation of the fimbria in vitro, Neuroscience, 5 (1980) 127–
133.

31. Malthe-Sørenssen, D., Skrede, K.K. and Fonnum, F., Calcium
dependent release of D- [^{3}H]-aspartate evoked by selective
electrical stimulation of excitatory afferent fibres to
hippocampal pyramidal cells in vitro, Neuroscience, 4 (1979)
1255–1265.

32. Meibach, R.C. and Siegel, A., Efferent connections of the
hippocampal formation in the rat, Brain Res., 124 (1977) 197–224.

33. McGeer, P.L., McGeer, E.G., Scherer, V. and Singh, K., A
glutamergic cortico-striatal path, Brain Res., 128 (1977) 369–
373.

34. Nagy, J.I., Carter, D.A. and Fibiger, H.C., Anterior striatal
projections to the globus pallidus, entopduncular nucleus and
substantia nigra in the rat: The GABA connection, Brain Res.,
158 (1978) 15–29.

35. Olney, J.W., Excitotoxic amino acids: Research application and
safety implications. In (Filer Jr, L.J., S. Garattini, H.R. Kane,
W.A. Reynolds and K.J. Wurtman Eds.), Glutamic Acid. Advances in
Biochemistry and Physiology, Raven Press, New York (1979), pp. 287·
320.

36. Raisman, G., Cowan, W.M. and Powell, T.P.S., An experimental
analysis of the efferent projection of the hippocampus, Brain Res.,
89 (1966) 83–108.

37. Rinvik, E., The cortico-nigral projections in the cat, J. Comp.
Neurol., 126 (1966) 241–254.

38. Schon, F. and Kelly, J.S., Autoradiographic localization of [^{3}H]
–GABA and [^{3}H] –glutamate over satellite glial cells, Brain Res.,
66 (1974) 275–288.

39. Skrede, K.K. and Westgaard, R., The transverse hippocampal slice:
a well-defined cortical structure maintained in vitro, Brain Res.,
35 (1971) 589–593.

40. Spencer, H.J., Antagonism of cortical excitation of striatal
neurons by glutamic acid diethylester: Evidence for glutamic acid
as an excitatory transmitter in the rat striatum, Brain Res., 102
(1976) 91–101.

41. Storm-Mathisen, J., Glutamic acid and excitatory nerve endings:
reduction of glutamic acid uptake after axotomy, Brain Res., 120
(1977) 379–386.

42. Storm, Mathisen, J. and Iversen, L.L., Uptake of [^{3}H] – glutamic
acid in excitatory nerve endings. Light and electronmicroscopic
observations in the hippocampal formation of the rat, Neuroscience,
4 (1979) 1237–1253.

43. Storm-Mathisen, J. and Woxen-Opsahl, M., Aspartate and/or
glutamate may be transmitters in hippocampal efferents to septum
and hypothalamus, Neurosci. Lett., 9 (1978) 65–70.

44. Swanson, L.W., An autoradiographic analysis of the efferent
connections of the preoptic region in the rat, J. Comp. Neurol.,
167 (1976) 227–256.

45. Swanson, L.W. and Cowan, W.M., Autoradiographic studies of the

development and connections of the septal area in the rat brain. In (J. De France, Ed.), The Septal Nuclei, Plenum Press, New York (1976).

46. Swanson, L.W. and Cowan, W.M., An autoradiographic study of the organization of the efferent connections of the hippocampal formation in the rat, J. Comp. Neurol., 172 (1977) 49-84.

47. Søreide, A. and Fonnum, F., High affinity uptake of D-[^{3}H]-aspartate in the barrel subfield of the mouse somatic sensory cortex. Brain Res., in press.

48. Takagaki, G., Properties of the accumulation of D-^{14}C aspartate into rat cerebral crude synaptosomal fraction. In (F. Fonnum, Ed.), Amino Acids as Chemical Transmitters, Plenum Press, New York (1978), pp. 357-362.

49. Taxt, T. and Storm-Mathisen, J., Tentative localization of glutamergic and aspartergic nerve endings in brain, J. Physiol. (Paris), in press.

50. Velasco, M.E. and Taleisnik, S., Effect of hippocampal stimulation of the release of gonadotropin, Endrocrinology, 85 (1969) 1154-1159.

51. Walaas, I., The localization of neurons probably using amino acids as transmitters, Ph. D. Thesis, in press.

52. Walaas, I. and Fonnum, F., Biochemical evidence for glutamate as a transmitter in hippocampal efferents to the basal forebrain and hypothalamus in rat brain, Neuroscience, in press.

53. Wieraszko, A. and Lynch, G., Stimulation dependent release of possible transmitter substances from hippocampal slices studied with localized perfusion, Brain Res., 160 (1979) 372-376.

54. Woolsey, T.A. and van der Loos, H., The structural organization of layer IV in the somatic sensory region of mouse cerebral cortex: the description of a cortical field composed of discrete cytoarchitectonic units, Brain Res., 17 (1970) 205-242.

GABAERGIC SYNAPSES: DISTRIBUTION AND INTERACTION WITH OTHER

NEUROTRANSMITTER SYSTEMS IN THE BRAIN

M. Pérez de la Mora[1], K. Fuxe[2], T. Hökfelt[2], K. Andersson[2],
L. Possani[1] and R. Tapia[1]

[1]Departamento de Neurociencias, Centro de Investigaciones
en Fisiología Celular, Universidad Nacional Autónoma de
México, México 20, D.F. and [2]Department of Histology,
Karolinska Institute, S-104 01 Stockholm, Sweden

INTRODUCTION

Over the last years, a number of neuronal systems using a
different kind of neurotransmitter have been identified in the brain.
The GABA and dopamine (DA) neuronal systems, due to their biological
peculiarities, have attracted a considerable attention (for references
see 34,53,64). The DA systems are, from an anatomical point of view,
very well defined. Several ascending DA-pathways originating in the
ventral mid-brain tegmentum (areas A9 and A10 according to Dahlström
and Fuxe (10) give rise to fibers which innervate the neostriatum
(nigro-neostriatal DA pathway), subcortical limbic regions such as
nucleus accumbens and tuberculum olfactorium, cortical limbic regions
(mesolimbic DA pathways), and the frontal cortex (mesocortical DA
pathway). In addition, intrahypothalamic DA pathways such as the
tubero-infundibular pathway, which originates in the nuc. arcuatus
and the ventral part of the periventricular hypothalamic nucleus
and innervates the external layer of the median eminence, have been
described (3,18,21,63). From a functional point of view, the nigro-
neostriatal and the meso-accumbens DA pathway seem to be important
for sensorimotor integration; the tubero-infundibular pathway
participates in neuroendocrine regulation, while most of the
mesolimbic DA systems and the mesofrontal DA pathway may instead have
a role in the control of higher brain functions (for references see
34,64). The GABA system, consisting of both Golgi I and Golgi II type
neurons, includes neuronal pathways which innervate all the regions
of the brain (61). This is the major inhibitory system and might
participate in the regulation of nearly all the functions of the
central nervous system (CNS), directly and/or through its synaptic
influence on other neurotransmitter systems (for references see 53).

Anatomical, physiological and pharmacological data indicate the existence of important interactions between the GABA and DA neuronal systems. Thus, a high activity of glutamic acid decarboxylase (GAD) has been found in all the DA cell body and DA nerve terminal-rich areas of the brain (15,23,61,65). A high degree of ^{3}H-GABA uptake has also been found in some of those areas (25,26) and the existence of a GABAergic striate/pallidal pathway (13,30,40,69), which seems to control the activity of the nigro-neostriatal DA system (69), has been demonstrated. Furthermore, the administration of drugs which may act through GABAergic mechanisms, such as Lioresal, aminooxyacetic acid or γ-hydroxybutyrate, produce effects which indicate a preferential interaction between GABA neurons and the mesolimbic DA systems (5,19,23).

For several years we have been interested in the study of the GABA-DA interactions in the brain. In the course of this analysis we have studied, by means of antibodies against GAD, the distribution pattern of the GABA-nerve terminals, especially within those brain regions which contain the monoamine pathways. The dynamic aspects of the interaction have been evaluated by giving a drug acting primarily and specifically on one neurotransmitter system, and following the secondarily induced changes in the activity of the other one. GABA and DA turnover have been chosen as an index for activity of the GABA and DA systems. In this paper we summarize some of our more relevant findings concerning the interactions between the GABA and DA neuronal systems. Some results which show interactions between the GABA system and other monoamine or peptidergic systems are also considered.

DISTRIBUTION PATTERN OF GABAergic SYNAPSES WITHIN THE MONOAMINE-RICH REGIONS OF THE BRAIN.

GABA is synthesized from L-glutamic acid by the action of GAD. Since this enzyme is concentrated in GABAergic neurons (13,14,68) its presence has been used as a selective marker for the localization of GABAergic systems in the brain. In this study the GABAergic synapses were detected by the immunohistochemical visualization of GAD in the presynaptic terminals. This approach was first introduced by Saito, Barber, Wu, Matsuda, Roberts and Vaughn in their studies, at the light microscopic level, of the GABA system in the cerebellum (54) and by Roberts and collaborators in other regions (for references, see 6). The GAD antibodies used in this study were obtained in rabbits by several injections of a partially purified GAD preparation, obtained by a simple technique which involves the utilization of an affinity chromatographic column.

Antigen Purification

In general, 40 adult albino mice (Mexican local strain) were used for each preparation. Brains, without cerebellum, were homogenized in water (1:9 w/v) and the homogenate adjusted to a final concentration of 20 mM K-phosphate buffer, pH 7.2, 0.1 mM pyridoxal-5'-phosphate (PLP), 0.1 mM 2-aminoethylisothiouronium bromide (AET) and centrifuged at 100,000 g for 60 min. The whole preparation was done at 0-4°C. The supernatant was precipitated by addition of crystals of ammonium sulphate up to 55% saturation at constant pH, and centrifuged at 18,000 x g for 15 min. The precipitate was recovered in 10-15 ml of the K-phosphate buffer containing PLP and AET at the concentrations indicated above and recentrifuged in the same conditions. Ten to 15 ml of the clear supernatant was applied to a Sepharose-4B column equilibrated and eluted with 20 mM K-phosphate buffer pH 7.2 at a flow rate of 36 ml per h. Fractions of 2 ml were collected. Two main peaks of enzyme activity were recovered, one containing a high molecular weight GAD species which is excluded from the column and the other containing a low molecular weight GAD which is eluted from the included volume of the column. Fractions with high GAD activity corresponding to the included peak, were pooled and applied directly onto a Sepharose 4B-diaminobutyrate-pyridoxal-5'-phosphate column (1.2 x 5 cm) synthesized as previously described by Possani, Bayón and Tapia (48). After loading, the column was successively washed with 35 vol of 20 mM K-phosphate buffer pH 7.2, followed by 2 vol of 1 M K-phosphate buffer pH 7.2 and 4 vol of 20 mM K-phosphate buffer pH 7.2 at a flow rate of 10 to 15 ml per h. Part of the enzyme was recovered from the affinity column by eluting with 4 vol of 500 mM K-phosphate buffer, 100 mM PLP and 100 mM L-glutamic acid pH 7.0. The solubilized enzyme was concentrated by ammonium sulphate precipitation (65% saturation) and the excess PLP and L-glutamic acid removed by dialysis against 10 mM K-phosphate pH 7.2 (5 changes over a period of 10-15 hrs). This enzyme preparation constitutes the so-called "soluble GAD" and corresponds to approximately 1% of the total initial enzyme activity, taking the whole homogenate as 100%.

The remaining bound enzyme was washed with 7 vol of 20 mM K-phosphate buffer pH 7.2 and an aliquot of the sedimented Sepharose beads was taken out for enzyme activity determination. This preparation was called "Sepharose-bound GAD" as opposed to the "soluble GAD". It is not possible to report a reliable figure for the degree of purification of the enzyme preparations that we have used as antigens, since GAD is a very labile enzyme, at least in our hands. As a criterion for purity and homogeneity of the antigens, we present in Fig. 1 the results of sodium dodecyl sulphate (SDS) gel electrophoresis of both enzyme preparations obtained. The results are compared with those obtained by Matsuda et al. (39) using a GAD enzyme purified according to a much more laborious purification scheme (71).

Thus Fig. 1 indicates that the antigens that we have obtained contain at least two main protein components, with apparent molecular

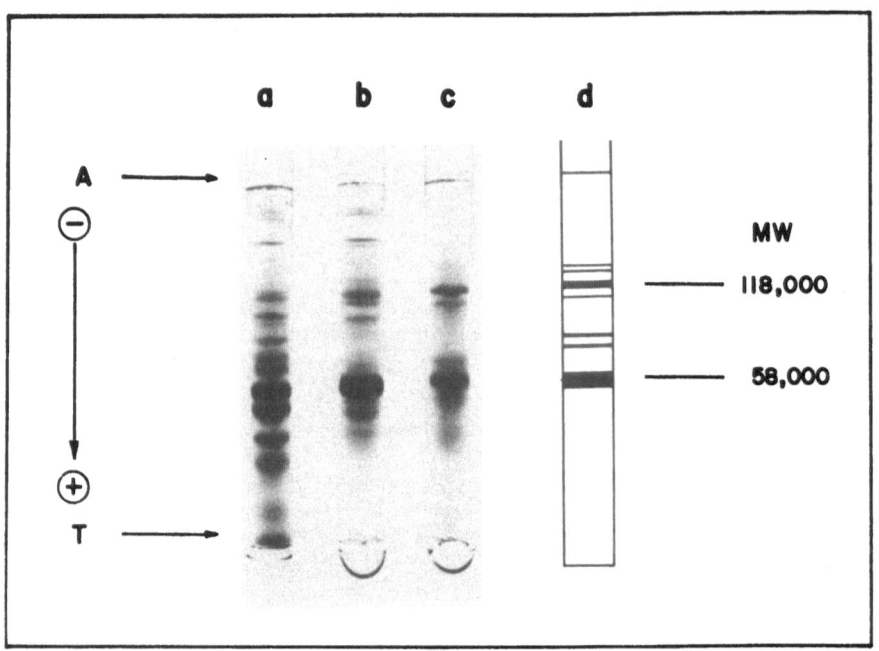

Fig. 1. Sodium dodecyl sulphate gel electrophoresis of the antigen
preparations. Polyacrylamide gel electrophoresis was run in the
presence of detergent under the same conditions as described by
Matsuda et al. (39). Lane a shows the electrophoretic pattern of
50 μg of protein taken from the supernatant of the 100,000 x g fraction (crude supernatant). Lane b shows the electrophoretic pattern
of 50 μg of protein recovered from the Sepharose-bound GAD by sodium
dodecyl sulphate and heat treatment. Lane c shows the electrophoretic
pattern of the soluble GAD (70 μg of protein). Lane d shows a diagram
drawn after the data of Matsuda et al. (39) for comparative purposes.
A means application point and T means tracking dye. - and + indicate
polarity of the electrophoresis.

weights of 57,000 ± 1,060 (means ± SEM, five experiments) and 117,500
± 570 (four experiments) respectively, corresponding to the same
molecular weights obtained by Matsuda et al. (39), for a homogeneous
GAD preparation (71). Integration of the stained gels by densitometry
shows that the 58,000 molecular weight band corresponds to 38% and
the band at 118,000 daltons to 6% of the total protein for the
Sepharose-bound GAD, and 30% and 10%, respectively, for the soluble
GAD, when analyzed under the same conditions. Since both our antigen
preparations ("soluble GAD" and "Sepharose-bound GAD") have at least
four of the seven bands reported by Matsuda et al. (39) which account
for nearly 50% of the total protein present in the SDS gels, it seems

reasonable to assume that our preparation has at least half the
specific activity (2 μmoles of GABA formed/min/g of protein) of the
GAD used by Matsuda et al. (39) for their SDS polyacrylamide gels.
Furthermore, in view of the fact that the major single protein
components present in both preparations are the same ones, it seems
also reasonable to postulate that the antibodies raised against our
antigens should have a very close specificity to that of the
antibodies obtained by Matsuda et al. (38).

Immunization Procedure

Two rabbits were immunized with the Sepharose-bound GAD, two
other rabbits with the soluble GAD, and two control rabbits were
injected with the same amounts of Sepharose-diaminobutyric pyridoxal
phosphate beads for control purposes. The GAD antigens were mixed
with an equal vol of complete Freund's adjuvant and injected
subcutaneously every two weeks, six times. The total amount of antigen
used for each of the rabbits immunized with Sepharose-bound GAD was
5.5 mg protein and for the rabbits immunized with the soluble GAD
2.5 mg protein. Two weeks after the last injection the rabbits were
bled and the serum obtained. The γ-globulin fraction was precipitated
with 40% saturation ammonium sulphate, centrifuged at low speed in
the cold and resuspended to the serum original vol in 60 mM Na-
phosphate buffer pH 7.0. The whole operation was repeated and the
final γ-globulin sediment was resuspended in a small vol of the same
buffer. The γ-globulin fractions were dialysed for 16-18 h in the
cold against several changes of 50 mM Na-phosphate buffer, pH 7.0,
centrifuged at 20,000 g to remove possible precipitates which could
be produced during the dialysis and stored in small portions at
-20°C. In order to evaluate whether or not GAD antibodies had been
produced, crude brain extracts were preincubated for various periods
of time with the γ-globulin fraction from immunized or non-immunized
rabbits. Figure 2 shows that Sepharose-bound GAD induced antibodies
inhibit GAD activity. This inhibition is progressive with time and
antibody concentration dependent. The maximal inhibition obtained
was about 80% of the respective γ-globulin treated control after
three days of preincubation. Similar results were obtained when brain
extracts were preincubated with soluble GAD, although the effects
were less apparent. No effects were found after preincubation with the
γ-globulin fraction from animals treated with Sepharose-diaminobutyric
pyridoxal phosphate. Double-immunodiffusion, immunoelectrophoresis
and counterimmunoelectrophoresis experiments were carried out to
study the specificity of the GAD antibodies. The specificity was
evaluated by looking at the number of the precipitation bands
obtained after their reaction against the components of a mouse or
rat brain crude extracts. Fig. 3 shows that by double immunodiffusion
a single precipitation band was obtained with both crude preparations.
Similar results were obtained in the electrophoretic experiments.
Both soluble and Sepharose-bound GAD gave similar results in the
immunodiffusion test and no precipitation band was obtained when the

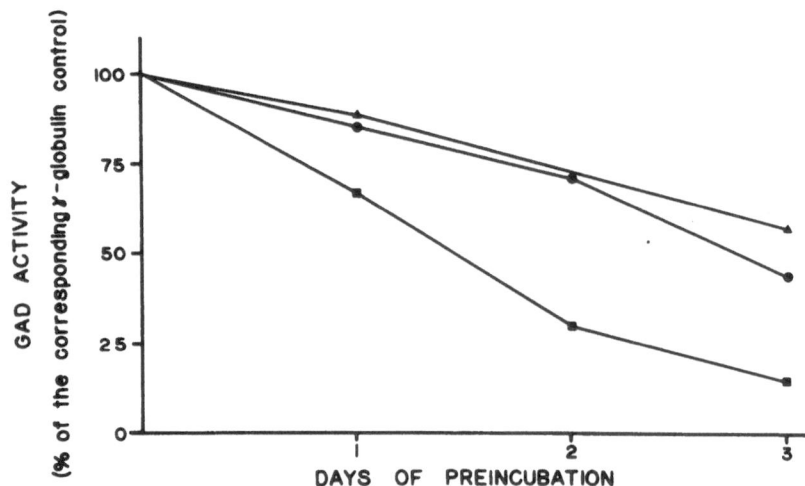

Fig. 2. Effect of GAD-antibodies on GAD activity. GAD activity was
measured in crude mouse brain extracts (1.6 mg protein) after
preincubation with either GAD antibodies or γ-globulin fraction from
non-immunized rabbits. 5.2 (■ - ■); 2.6 (▲ - ▲); and 1.3 (● - ●) mg
of γ-globulin were tested. The absolute GAD activity values without
any addition and in the presence of 5.2 mg of control γ-globulins
were, respectively (μmoles/min/g protein): 1 day, 4.0 and 2.6; 2
days, 1.6 and 1.1; 3 days, 0.8 and 0.5. With the two lowest control
γ-globulin concentrations used no effects on GAD activity were
observed. For GAD measurements see Pérez de la Mora et al. (46).

GAD antibodies were replaced by the antibodies directed against the
Sepharose carrier alone. These results indicate that the GAD
preparations used in these experiments are able to produce GAD
antibodies, since the antibodies obtained markedly inhibit GAD
activity. The characteristics of their inhibitory effects are in
close agreement with the results of Matsuda et al. (38) and Saito
et al. (55). The single band obtained in the double-immunodiffusion,
immunoelectrophoresis and counterimmunoelectrophoresis experiments
may indicate that the GAD preparation used is pure enough to
stimulate preferentially the production of specific GAD antibodies.
The above experiments also show again, in agreement with Saito et al.
(55), that GAD antibodies directed against mouse GAD can cross-react
with rat GAD, making feasible the mapping out of the GABA system
in the rat brain.

Immunohistochemical Observations

    Using those GAD antibodies it was possible by means of immuno-
histochemical methods to demonstrate GAD immunoreactive dots and
fiber like structures in all parts of the brain (16,17,18,20,24). No

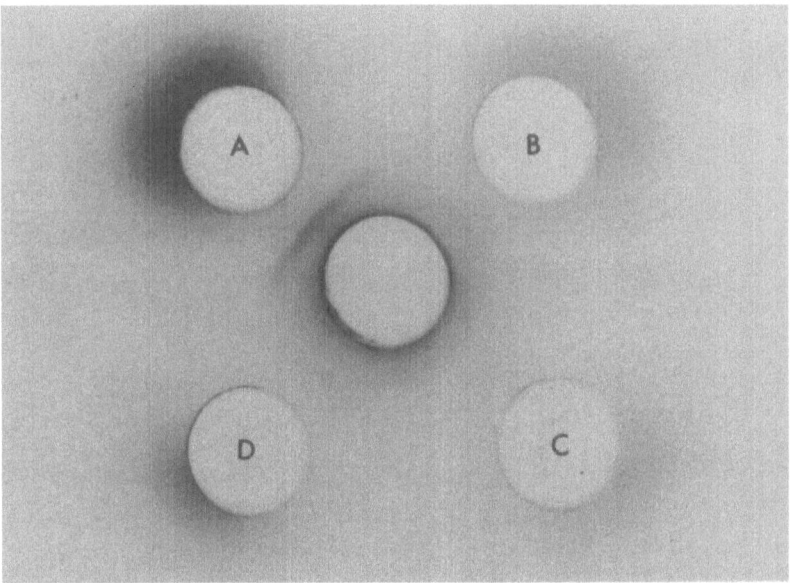

Fig. 3. Double-immunodiffusion of GAD antibodies against mouse and rat brain extracts. 20 μl of both antigen and antibody were applied. A and D rat brain extract (192 and 96 μg respectively); B and C mouse brain extract (114 and 57 μg protein respectively); 840 ug antibody was used.

immunoreactivity could be demonstrated after the exposure of the tissue sections to antibodies from either non-immunized or Sepharose-diamino-butyric-pyridoxal phosphate treated rabbits. Due to possible low GAD–apoenzyme concentration in the perykarya of the GABAergic neurons and/or low affinity, no cell bodies could be demonstrated. Ultrastructural studies (Johansson, O., unpublished) showed that the GAD immunoreactivity found with the present antibodies is confined to presynaptic nerve terminals. Since the distribtuion of the inhibitory GABA neurons in the cerebellum is very well known (11), we studied the distribution of the GABA neuron system in this region to obtain a further test for specificity. Numerous GAD immunoreactive terminals are present both in the cerebellar cortex and nuclei (Fig. 4A, B). In the cerebellar cortex fine, moderately GAD immunoreactive dots are found in all layers (Fig. 4A). In the molecular layer the GABAergic terminals are scattered all over following the distribution of the stellate cell terminals. In the Purkinje cell layer the GAD immunoreactive terminals, like the basket cell endings, surround the Purkinje cell bodies. In the granular layer the GABAergic terminals are found in the spaces between the granular cell bodies. Such GAD immunoreactivity may correspond to Golgi II type nerve terminals within the cerebellar glomeruli. In all the cerebellar nuclei large

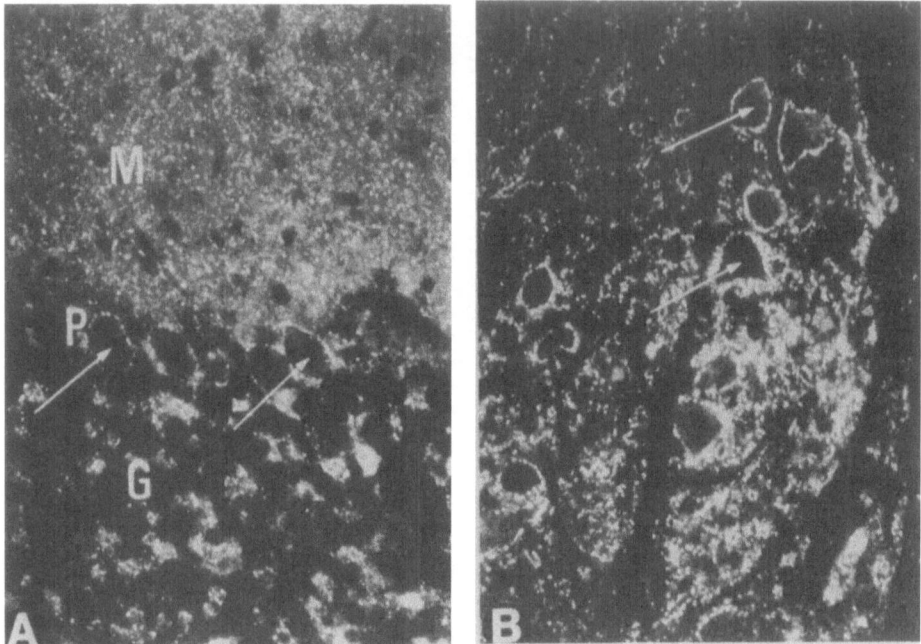

Fig. 4. GAD immunoreactivity in the cerebellar cortex and in the cerebellar nuclei. A. GAD immunoreactive dots are found in all layers. M=Molecular layer, P=Purkinje cells layer, G=granular layer. Arrows point Purkinje cell bodies surrounded by GAD immunoreactive terminals (120 X). B. Strongly immunoreactive terminals surround the nerve cell bodies (arrows) of a cerebellar nucleus (120 X).

strongly immunoreactive dots and varicose fibres surround the somata and processes of the nerve cells (Fig. 4B). The location of these GABAergic terminals correspond to the expected distribution of Purkinje cell endings within the cerebellar nuclei. Since the stellate, Purkinje, basket and Golgi cells are inhibitory and may release GABA at their terminals (14,33,41,58) the immunoreactivity found with these antibodies is compatible with a specific labeling of the GABAergic boutons in the cerebellum. The above results are identical to those reported by Saito et al. (54) in their studies on the distribution of the GABA system in the cerebellum. The distribution of the GABAergic synapses within the DA cell body rich area of the mesencephalon are shown in Fig. 5. Large numbers of GAD immunoreactive terminals were demonstrated in the substantia nigra, especially within the zona reticulata. The GABAergic terminals, as previously described (18,20) and in agreement with Ribak et al. (52) were fairly large, strongly immunoreactive and often forming rings surrounding non-immunoreactive processes, probably representing dendrites of DA and non-monoamine nerve cells (Fig. 5A). A medium

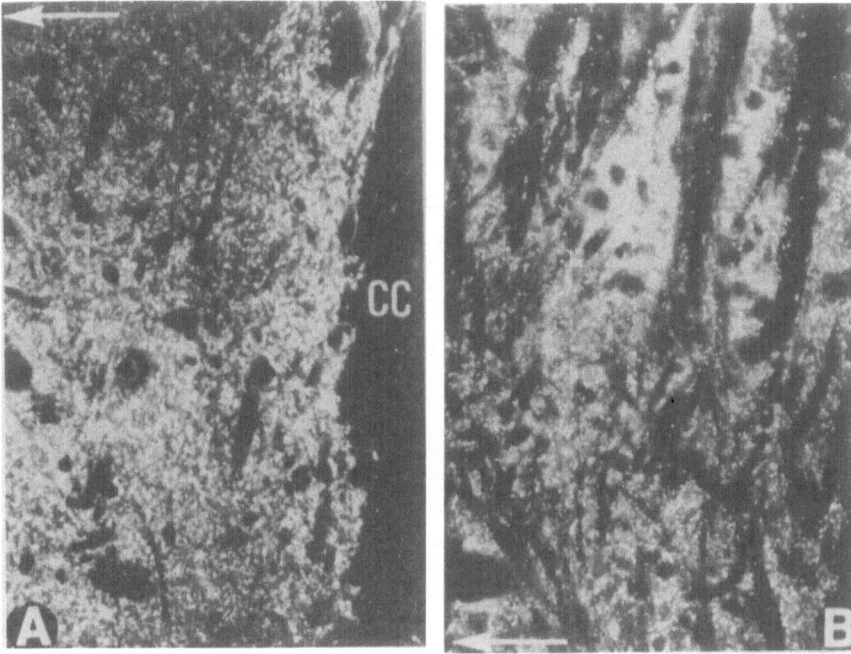

Fig. 5. GAD immunoreactivity within the DA-cell body rich areas of the mesencephalon. A. High density of strongly immunoreactive terminals is observed innervating the ventral part of the zona reticulata of the substantia nigra. CC=crus cerebri. 140 X. B. In the midline area overlying the anterior part of the nucleus inter-peduncularis a medium density of moderately GAD immunoreactive terminals is found. The arrow points toward the dorsal direction (350 X).

density of fine GABAergic terminals (18) was found within the paranigral area which surrounds the nucleus interpenducularis and contains the DA cell group A10 (Fig. 5B).

Within the terminal fields of the nigroneostriatal and meso-limbic DA systems diffuse networks of fine GABAergic terminals are found in the neuropil of nuc. caudatus and nuc. accumbens (16) and interspersed between the nerve cell bodies in the different limbic cortices. In contrast, within the globus pallidus and the substantia innominata (ventral pallidum) a high density of fairly large and strongly immunoreactive fibres surrounding processes and cell bodies were demonstrated. From the substantia innominata these large and strongly immunoreactive fibres seem to extend (Fig. 6 A,C) into the adjacent area of the medial forebrain bundle and the inner pole of Calleja's islands in the tuberculum olfactorium. A different

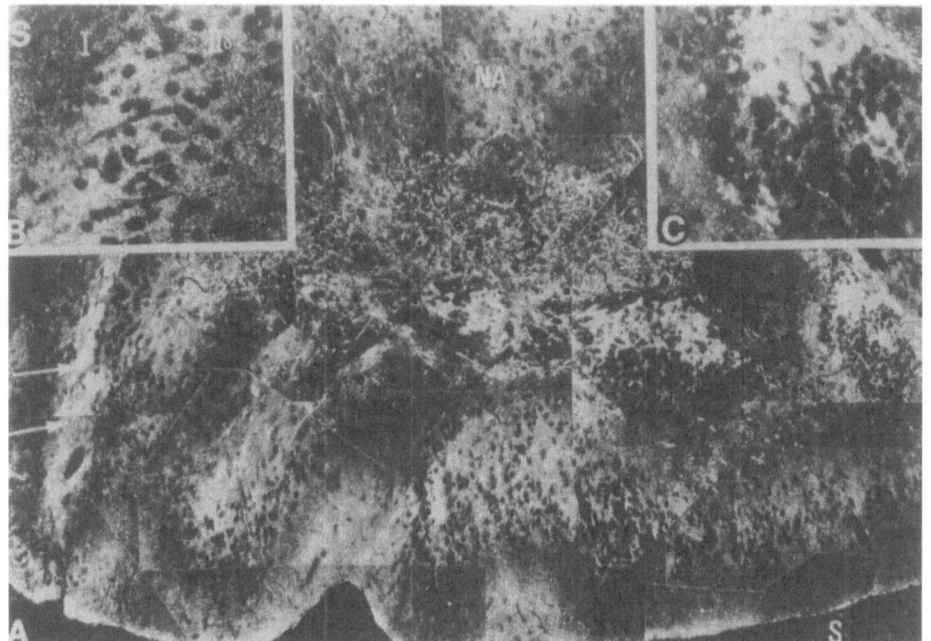

Fig. 6. GAD immunoreactivity in the tuberculum olfactorium regions.
A. High density of strongly immunoreactive fibres is observed in the
inner pole of Calleja's islands and within the area in between nucleus
accumbens (NA) and tuberculum olfactorium (corresponding in part to
the medial forebrain bundle area). Arrows point to strongly immunoreactive fibres entering the inner layer of the tuberculum
olfactorium. S= ventral surface of the brain (80 X). B. Higher
magnification to see the weakly to moderately GAD immunoreactive
terminals present in layers I and II of the tuberculum olfactorium.
S=ventral surface of the brain (210 X). C. Higher magnification of a
Calleja's island. A high density of strongly immunoreactive fibres
is seen in the inner pole of the island (210 X).

GABAergic system seems to innervate the outer layers of the tuberculum
olfactorium (Fig. 6B). Their GAD immunoreactive terminals are fine,
weakly to moderately immunoreactive and are located between the nerve
cell bodies.

        The previous and present results provide the anatomical basis
for the existence of a strong GABAergic control of the nigro-
neostriatal and mesocortical DA systems. Two types of GABAergic
terminals were demonstrated. One is fairly large, strongly immunoreactive and is distributed mainly in the deep cerebellar nuclei,
nuc. vestibularis lateralis, the substantia nigra, the globus
pallidus, the substantia innominata together with adjacent parts of

the medial forebrain bundle, and the dorsal parts of the tuberculum
olfactorium, especially in the deeper pole of Calleja's islands
(Fig. 6C). The second type is fine, weakly to moderately immunoreactive and found in most areas of the brain, e.g. the nuc. caudatus,
nuc. accumbens, limbic cortices and the outer layers of the tuberculum
olfactorium. The fact that the GABAergic neurons, which give rise to
the large and strongly immunoreactive terminals within the nuc.
vestibularis lateralis, the deep cerebellar nuclei, the substantia
nigra and the globus pallidus, belong to the Golgi type I neurons,
suggest that the GABAergic terminals demonstrated in the substantia
innominata and the dorsal tuberculum olfactorium area may originate
from neurons which have their cell bodies in some other regions of
the brain. The decrease in GAD activity in the substantia innominata
which follows lesions of nuc. accumbens (66) gives support to this
suggestion. Some experiments are currently being carried out in our
laboratories to study this possibility. Thus, Golgi type I and Golgi
type II GABAergic neurons seem to be regulating the activity of the
mesolimbic and nigro-neostriatal DA system. Their regulatory effects
could be direct or indirect. The direct influences may act at the DA
dendritic cell body or terminal level, while the indirect ones either
within the DA terminals or even beyond that level. Golgi I type
neurons originating in the nuc. caudatus-putamen (13,30,40,69) act
directly on the nigral DA cell bodies, and on putative GABA nerve
cells in the zona reticulata, while mainly Golgi type II local
interneurons (15) seem to influence the DA neurons which are the
main origin of the mesolimbic DA system within the A10 area.
Reciprocal interactions may exist in the substantia nigra through the
dendritic release of DA (9) which could, in turn, modulate the
activity of the GABAergic terminals. Golgi type II neurons, which may
give rise to the fine GABAergic terminals within the DA terminal
fields, and/or Golgi type I neurons which give rise to the large
and strongly immunoreactive terminals found forming baskets around
cells and main processes in the globus pallidus or the substantia
innominata, the adjacent medial forebrain bundle area and the inner
pole of Calleja's islands, may indirectly enhance or block the
effects of postsynaptic DA-receptor activation. Since the major
outflow of the basal ganglia must pass via the globus pallidus and
thalamus to reach the motor cortex, GABAergic mechanisms acting at
the globus pallidus and thalamus may modify the action of the DA
systems on motor activity (57). Similarly, since the outer layer
of the tuberculum olfactorium project into the deep structures, the
effects of the postsynaptic DA receptor activation in the olfactory
cortex can also be modified by GABAergic influences at the level of
the Calleja's islands. Although GABAergic influences acting at the
DA terminal level could act directly, via axo-axonic synapses, this
type of interaction may be minor. Thus, preliminary experiments in
our laboratory (43) indicate that GABA binding sites in the nuc.
caudatus and tuberculum olfactorium do not decrease ten days after
a 6 hydroxydopamine-induced lesion of the nigro-neostriatal and mesolimbic DA systems. Since at this time there is a large degeneration

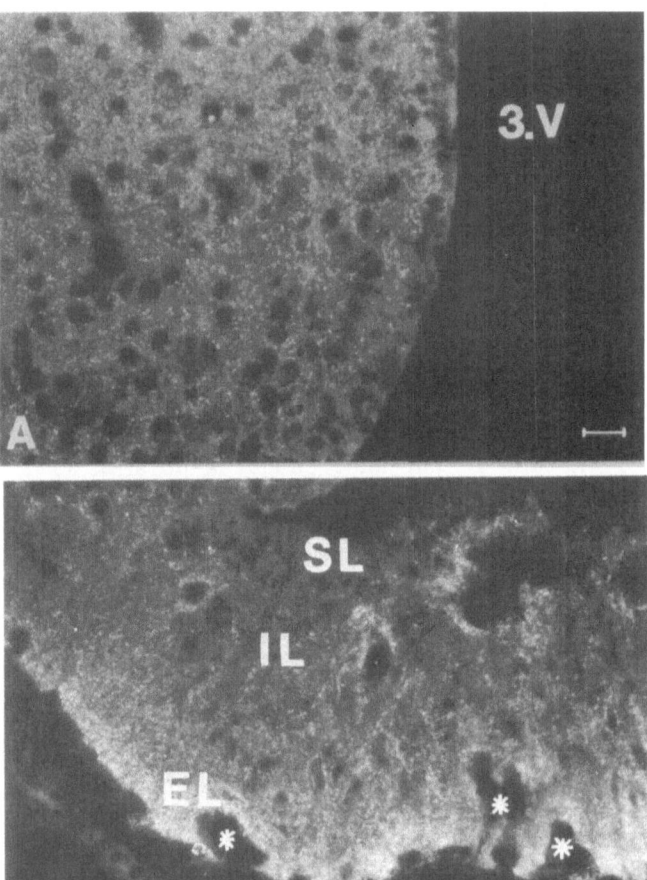

Fig. 7. Immunofluorescence micrograph of the basal hypothalamus
showing the arcuate nucleus (A) and the medial eminence (B) after
incubation with GAD antiserum. A dense plexus of fine GAD immunoreactive fibers is seen in the arcuate nucleus and periventricular
zone. In the median eminence a high density of fine GAD immunoreactive fibers is seen in the external layers (EL) e.g. around portal
vessels (*). The internal layer (IL) contains mostly a less dense
plexus and the subependymal layer (SL) contains only few GAD immunoreactive fibers. 3.V = third ventricle. Bar indicates 50 μm.

of the DA axons, the experiment indicates that the number of axo-
axonic GABA synapses with DA boutons is small in these areas. Furthermore, in the tuberculum olfactorium an increase in the GABA binding
sites seems to occur instead, suggesting that when the inhibitory
dopaminergic input is removed another inhibitory input can take over,
in this case a GABAergic input. Further studies are needed to show
the possible existence of GABA-DA-axo-axonic interactions in the
forebrain.

Hypothalamus. A large plexus of fine, moderately immunoreactive GABA-
ergic terminals is present in practically all nuclei of hypothalamus.
GABAergic mechanisms seem to influence the activity of the tubero-
infundibular DA system at the terminal and cell body levels (20,24)
since, in agreement with biochemical work (65), numerous GABAergic
terminals were found innervating the nuc. arcuatus (Fig. 7A) and the
medial and lateral parts of the median eminence (Fig. 7B). A low
density of GAD immunoreactive dots were found in the internal layers
with the exception of the regions which contain the capillary loops,
where large numbers of GABA terminals are seen surrounding the
capillaries. The GABAergic terminals demonstrated seem to belong to
an intrahypothalamic system, since the deafferentation of the
hypothalamus does not affect GAD activity in the median eminence
(8,62). These anatomical results underline the role of the GABA
system in the neuroendocrine functions of the hypothalamus. Thus,
GABAergic mechanisms could act on the tuberoinfundibular DA system
and in this way affect the secretion of several hormonal releasing
or inhibitory factors. The GABAergic terminals found in the median
eminence may also affect directly the peptidergic neuron terminals
located in the external layer, or release GABA into the primary
capillary plexus, and also in that way affect the secretion of
anterior pituitary hormones. The finding that GAD-immunoreactive
terminals surround the primary capillary plexus in the median eminence
is the first anatomical evidence, so far reported, that GABA might
be released into the bloodstream and, like a "hormone", affect the
release of some other hypophysiary hormones such as prolactin (49).

PHARMACOLOGICAL STUDIES ON GABA-DA PATHWAY INTERACTIONS

Effects of Drugs Influencing DA Receptor Activity on GABA Turnover

The DA receptor agonist apomorphine (4) and the DA receptor
antagonists pimozide, chlorpromazine and sulpiride (1,22,67) have
been used. GABA turnover was evaluated by following the accumulation
of GABA after GABA-T inhibition. This GABA accumulation seems to be
nerve impulse-dependent since it is considerably reduced by
interruption of the nerve impulse flow. Thus, the GABA accumulation
may reflect GABAergic neuronal activity (46). The DA receptor
stimulation by apomorphine (44) results in an increase on GABA turnover in the ventral midbrain tegmentum (Fig. 8), the nuc. caudatus

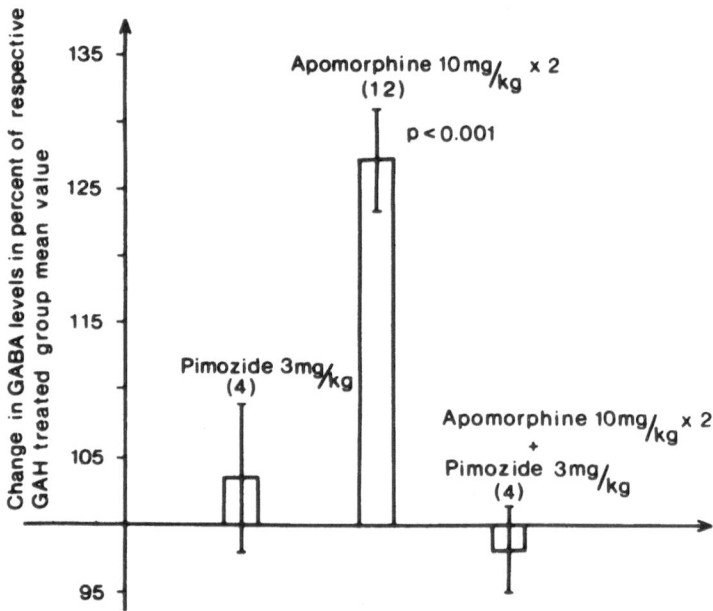

Fig. 8. Effect of apomorphine and pimozide on the GABA accumulation
in the DA-cell body rich area of the mesencephalon after GABA-T
inhibition. ɣ-glutamyl hydrazide (GAH), a GABA-T inhibitor (37), was
given i.p. at a dose of 160 mg/kg; apomorphine was given. i.p. 5 min
before and 1 h after GAH, and pimozide was given i.p. 2.5 h before
GAH. Rats were killed 2 h after GAH. Values are means $\pm$ SEM; number
of rats in parentheses. All comparisons have been made with the
group treated with GAH alone. Student's t-test was used for
statistical evaluation. GABA levels (μmoles/g wet weight) were for
the GAH-treated-control group 5.66 $\pm$ 0.28 in the apomorphine
experiments and 6.12 $\pm$ 0.20 in the experiments with pimozide and
apomorphine and apomorphine + pimozide (44).

and the subcortical limbic regions, namely nuc. accumbens and
tuberculum olfactorium (45). The effects of apomorphine are
counteracted by the previous administration of DA receptor antagonists
such as pimozide (Fig. 8). In view of the fact that apomorphine alone
does not modify neither GABA levels nor GABA-transaminase activity
(44) the above results suggest the existence of important GABA-DA
interactions within both extrapyramidal and limbic systems and within
the substantia nigra, where DA receptor activates and may enhance
GABA release (51). Furthermore, the fact that oxotremorine prevents
the effects of apomorphine on GABA turnover (42) may suggest the
participation of cholinergic neurons in these interactions (see also
7,27,32,35,50). Thus, GABA turnover might be increased by an

Table 1. Effects of sulpiride on GABA levels in several regions of the brain*

Treatment	Nuc. caudatus	Subcortical limbic regions	Prefrontal cortex	Midbrain ventral tegmentum
Saline	100 ± 8.9 (7)	100 ± 10.4 (7)	100 ± 9.4 (5)	100 ± 10.3 (7)
Sulpiride (15 mg/kg; 4h)	87.9 ± 0.16 (7)	69.1 ± 10.6 (7)	78.6 ± 4.2 (5)	62 ± 5.2 (6)
		0.05<p<0.1	0.05<p<0.1	p<0.01

*Values are in percent of the control saline-treated mean group ± SEM. Absolute GABA values (μmoles/g with tissue) for the control group were: nuc. caudatus, 183 ± 0.16; subcortical limbic regions (mainly nuc. accumbens + tuberculum olfactorium + the tractus diagonalis regions), 4.5 ± 0.47; prefrontal cortex, 1.17 ± 0.11; midbrain ventral tegmentum, 4.95 ± 0.5. Number of animals in parentheses. Student's t-test. Data taken from ref. 16.

inhibitory action of apomorphine on excitatory cholinergic interneurons in the forebrain, which in turn may lead to a reduced activity
in unknown inhibitory neurons controlling the striato-nigral GABA
pathway and the GABA interneurons in the A10 region, the nuc. caudatus
and the subcortical limbic regions studied. As indicated above, the
GABA turnover in the midbrain is also controlled directly or
indirectly by dendroaxonic DA synapses.

In contrast to apomorphine, the administration of DA receptor
blockers have little effects on GABA turnover in the above regions
(44). As seen in Fig. 8, pimozide does not affect GABA turnover.
Furthermore, the administration of chlorpromazine and sulpiride
produces only a trend for a selective decrease in GABA turnover
within the substantia nigra (16). These results are in line with the
absence of effects on GABA turnover observed by other workers (35,
36,50) with pimozide an haloperidol in the striatum and the substantia
nigra. Sulpiride, however (16), in a non-cataleptogenic dose (15 µg/
kg) produces a substantial depletion of the GABA stores within the
substantia nigra and trends for depletion of GABA stores in the
prefrontal cortex and in the limbic subcortical regions (see Table 1).

The reason why DA receptor blockade does not result in the
expected decrease in GABA turnover is unknown. More research on the
basic neuronal relationships operating within the extrapyramidal and
limbic systems is needed before these paradoxical results can be
satisfactorily explained. One can only speculate, on the basis of the
suppressive effects of oxotremorine on the apomorphine-induced GABA
turnover increase and the lack of effects of oxotremorine alone on
GABA turnover (42,50), that GABA turnover in regions rich in DA cell
bodies and DA terminals is increased only when cholinergic neuronal
activity is decreased. It may further be speculated that the reason
for the failure of DA receptor blockers to change GABA turnover could
be that under these circumstances other feedback loops are involved
in the regulation of GABA turnover, such as excitatory pathways
containing glutamate and/or substance P. The selective depletion of
nigral GABA turnover found after the administration of sulpiride and
chloropromazine seems to underline the importance of dendritic DA in
the facilitatory control of GABA synthesis and release in nigral GABA
terminals (9,51). Therefore, blockade of DA receptors in the
substantia nigra can lead to the reduction of GABA stores and turnover observed in the present experiments.

Effect of Drugs Modulating GABA Neurotransmission on DA Turnover

Presumable GABA agonists and antagonists, such as muscimol,
γ-hydroxy, γ-ethyl, γ-phenyl-butyramide (HEPB), γ-acetylenic GABA,
bicuculline and picrotoxin, were used. Muscimol is a well known GABA
receptor agonist (28,31). HEPB, previously called 5-ethyl, 5-phenyl-
2-pyrrolidinone (see 29) is a drug which seems to have some GABA-
ergic activity (47,60). γ-Acetylenic GABA is a potent GABA

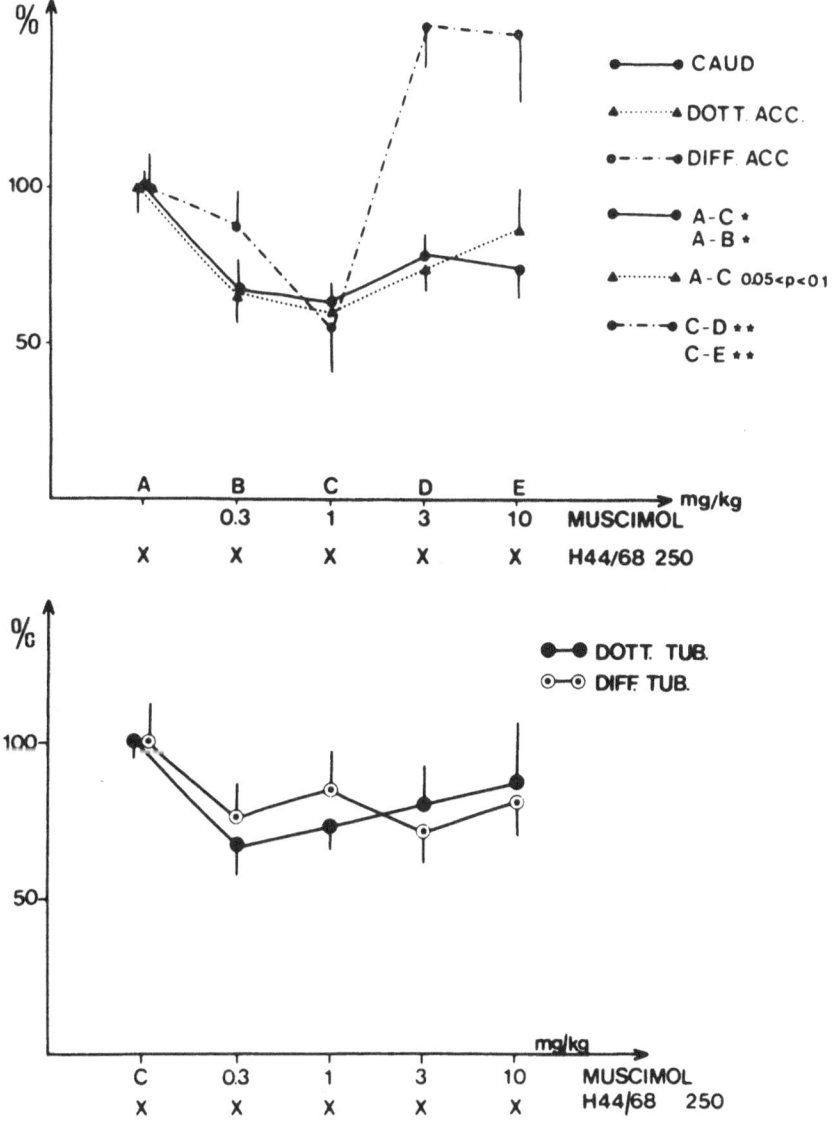

Fig. 9. Effect of muscimol on H44/68-induced fluorescence
disappearance in various regions of the forebrain. H44/68 was given
i.p. in a dose of 250 mg/kg 2 h before killing. Muscimol was
administered i.p. simultaneously. Doses of muscimol are shown on a
log scale. On the y axis the means ± SEM of the DA fluorescence in
each area is shown; the values are given in percent of the respective
H44/68 mean value. The Wilcoxon one-way classification comparing all
possible pairs of treatments was used. N=5. *p<0.05, **p<0.01. CAUD=
anterior and medial part of nuc. caudatus; DOTT.ACC.=dotted type of
DA terminal in posterior nuc. accumbens; DIFF ACC.=diffuse type of DA

aminotransferase inhibitor (56) and bicuculline and picrotoxin are
thoroughly studied GABA antagonists (28,59). DA turnover, a well known
index for dopaminergic activity, was evaluated by studying the DA
fluorescence disappearance after tyrosine hydroxylase inhibition
using α-methyltyrosine methylester (H44/68) (2). The DA stores were
measured by quantitative microfluorimetry (12) in paraffine sections
prepared according to the routine Falck-Hillarp technique. Most of
the results presented have been published elsewhere (16,19,23).

The effects of several doses of muscimol on DA turnover in the
forebrain are seen in Fig. 9. Muscimol at low doses (up to 1 mg/kg)
significantly increased DA turnover within the nucleus caudatus and
produced a trend for an increase in the posterior part [dotted type
of DA terminal; DA-CCK 8 (cholecystokinin 8) like immunoreactive]
of nuc. accumbens. In the diffuse type of DA terminals (not CCK 8
immunoreactive) of nuc. accumbens a biphasic dose effect curve was
observed, with a decreased DA turnover at higher doses. The increase
in DA turnover seen in the nuc. caudatus and posterior nuc. accumbens
was no longer observed when higher doses of muscimol were used. The
DA turnover within the tuberculum olfactorium was not affected. Within
the median eminence (16), muscimol at low doses increased the DA
turnover within the lateral palizade zone (LPZ), whereas no effects
could be observed at higher doses. The noradrenaline turnover within
the subependymal layer (SEL) was increased by muscimol in a dose of
3 mg/kg, while the increase of catecholamine turnover in the medial
palizade zone did not become significant. After pretreatment with
the DA receptor blocking agent pimozide, muscimol (0.75-3 mg/kg)
could not any longer increase DA turnover in the nuc. caudatus and nuc.
accumbens. Instead even a significant reduction of DA turnover was
observed in the tuberculum olfactorium [dotted type of terminals;
CCK 8 immunoreactive (see 16)].

In contrast to muscimol, γ-acetylenic GABA (Fig. 10) reduced DA
turnover in the nuc. caudatus and left the nuc. accumbens (diffuse
type of DA terminals; not CCK 8 immunoreactive) unaffected. After
pretreatment with pimozide, γ-acetylenic GABA preferentially reduced
DA turnover in the nuc. accumbens (dotted type of DA terminals; CCK
8 immunoreactive) and in the tuberculum olfactorium (diffuse type of
DA terminals; not CCK 8 immunoreactive) (16). In the median eminence
γ-acetylenic GABA reduced DA turnover in the LPZ, but like muscimol
increased noradrenaline turnover in the SEL. HEPB, like lioresal
(200 mg/kg) (19,23), affects preferentially the mesolimbic DA systems.
Accordingly, as seen in Fig. 11, a decrease in DA turnover was
produced in the tuberculum olfactorium and the nuc. accumbens while

---

terminal in the anterior nuc. accumbens; DOTT. TUB.=dotted type of DA
terminal in posterior tuberculum olfactorium; DIFF. TUB.=diffuse type
of DA terminal in anterior tuberculum olfactorium. (16).

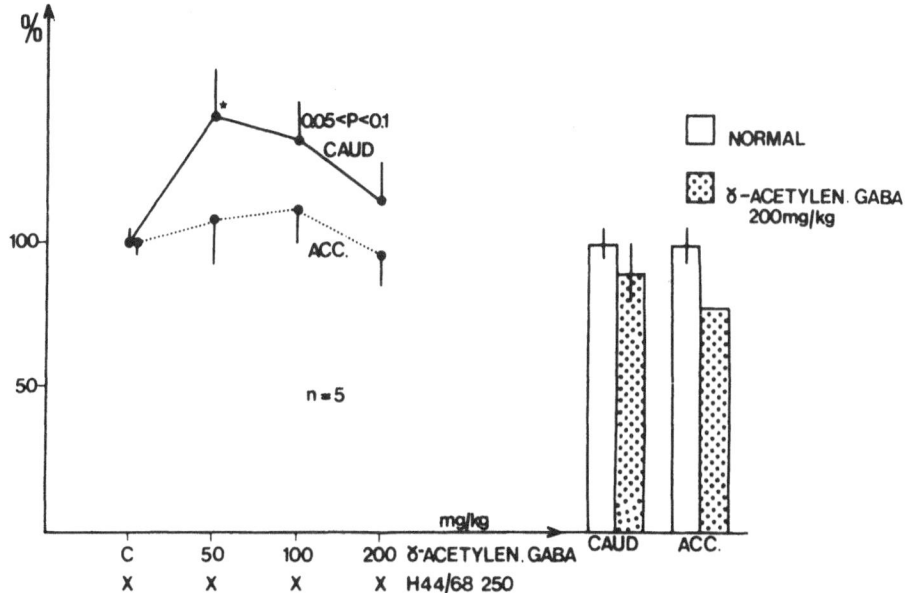

Fig. 10. Effect of γ-acetylenic GABA on the H44/68-induced
fluorescence disappearance in the n. caudatus and in the n. accumbens
diffuse type of DA terminal (anterior part). γ-Acetylenic GABA was
given i.p. at the same time as the H44/68. For further details see
the legend to Fig. 9. Significances refer to the H44/68 control
group. n=5 *=p<0.05. (16).

no effects were observed in the nuc. caudatus. After pretreatment
with pimozide, the administration of HEPB, like γ-acetylenic GABA and
muscimol, specifically reduced DA turnover within limbic regions
(nuc. accumbens and tuberculum olfactorium) (Fig. 12). These results
show that either excitatory or inhibitory effects on DA neurons can
be observed after the administration of GABAergic drugs. The
predominant effect is dependent upon the drug used, the dose and
the functional conditions of the DA receptors. Thus, in the forebrain
the excitatory effects are seen after the administration of muscimol
at low doses, while the inhibitory ones were observed following
treatment with γ-acetylenic GABA, HEPB, or high doses of muscimol.
The DA receptor blockade by pimozide facilitated the development of
the latter effects, since the excitatory effects of muscimol on DA
neurons were cancelled and even inhibitory effects showed up in one
limbic region. Similarly, GABA receptor blocking agents produced
either excitatory or inhibitory effects on DA turnover, depending
upon the route of administration (23). Thus, picrotoxin and
bicuculline when given systemically decreased DA turnover in the
limbic regions, leaving the nuc. caudatus unaffected. Statistically
significant changes were found in the tuberculum olfactorium with

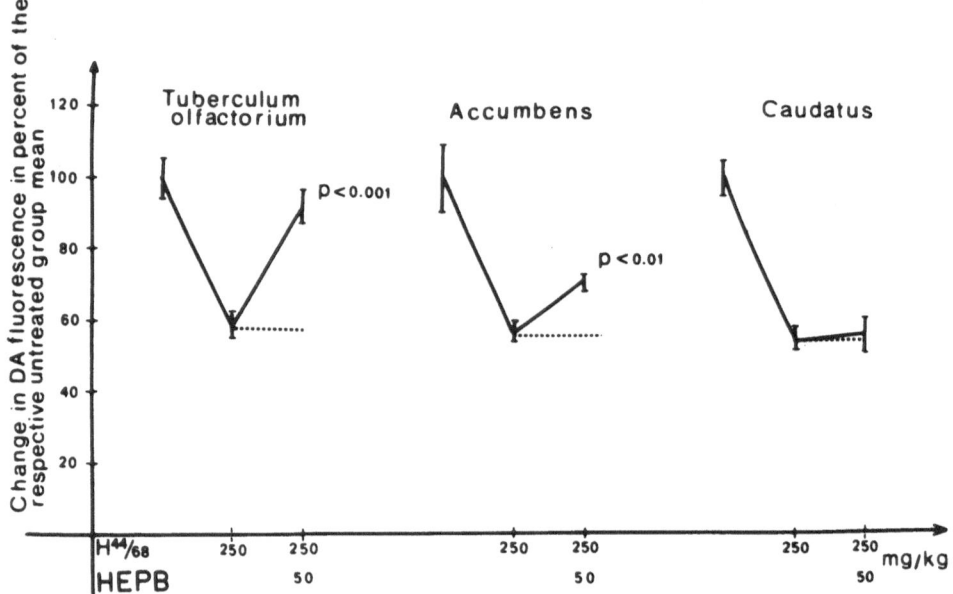

Fig. 11. Effect of HEPB on the H44/68 induced DA fluorescence
disappearance in various regions of the forebrain. HEPB was given
i.p. 15 min before H-44/68. Values are means $\pm$ SEM. Student's t-test.
N=5 (23).

bicuculline (Fig. 13), and in the nuc. accumbens with picrotoxin
(Fig. 14). However, when picrotoxin was infused slowly into the
ventral tegmental area (VTA) (23) a significant increase was observed
in the tuberculum olfactorium and a trend in the same direction in
the nuc. accumbens. As with systemic administration, no effects were
observed in the nuc. caudatus following injection of picrotoxin in
VTA.

        The effects of GABA drugs on DA turnover can be explained on the
basis of the existence of two different GABA receptors having
different overall actions on the DA systems (16). One type of receptor
may have a strong inhibitory action and could partly be located in
the ventral midbrain tegmentum (A9 and A10 areas). The other may have
an "excitatory" effect and could be located beyond the DA receptors.
Thus, muscimol, at low doses, may activate only the "excitatory"
type, leading, probably through the action of several synapses, to
an increase in DA turnover. Higher doses of muscimol may, however,
activate also the inhibitory type of GABA receptors counteracting
in this way its excitatory effects and even eliciting in some areas
a decrease in DA turnover, as observed in the nuc. accumbens (Fig. 9).
On the other hand, after GABA aminotransferase inhibition, GABA may
leak out from the GABA terminals (see Bradford, this Symposium) to

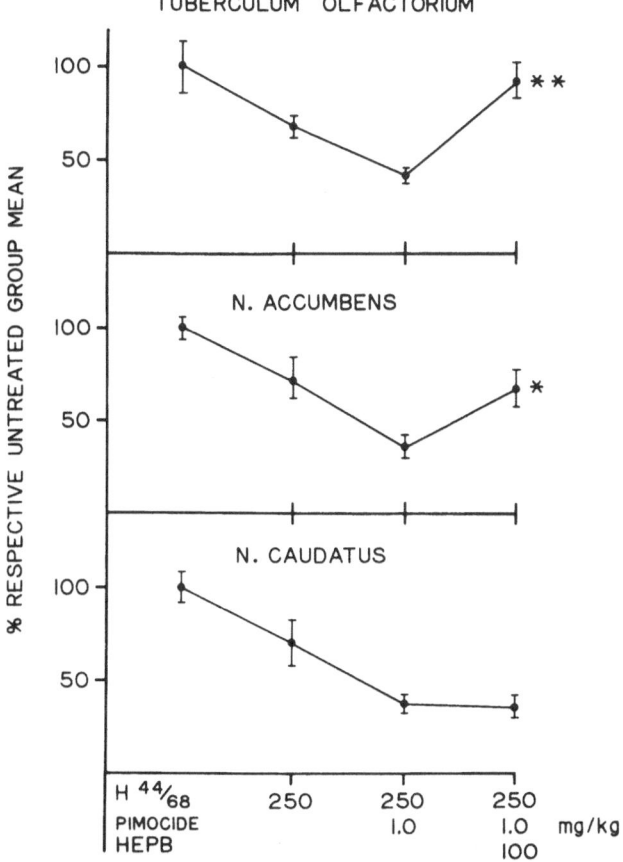

Fig. 12. Effect of HEPB on the pimozide-induced increase in the H44/68 induced DA fluorescence disappearance in the forebrain. Pimozide was given i.p. 2 hr. before the H44/68. HEPB was administered i.p. 15 min before H44/68. Values are means $\pm$ SEM. Student's t-test. N=4-5. Statistical comparisons were made against the H44/68 + pimozide group. *=p<0.05, **=p<0.01.

reach GABA receptors in all areas, leading to a decrease of GABA turnover (Fig. 10). DA receptor blockade might preferentially potentiate the effects of the inhibitory GABA receptors in the limbic system. HEPB (Fig. 11 and 12) and Lioresal (19,23) like γ-acetylenic GABA may preferentially act on the GABA receptors having inhibitory influences on the limbic DA pathways. The results obtained with the GABA receptor antagonists can also be explained by the above interpretation. Thus, systemically injected bicuculline or picrotoxin (Figs. 13 and 14) may preferentially block the excitatory limbic GABA receptors leading to a decrease in DA turnover within nuc. accumbens or tuberculum olfactorium, while local injection of picrotoxin into

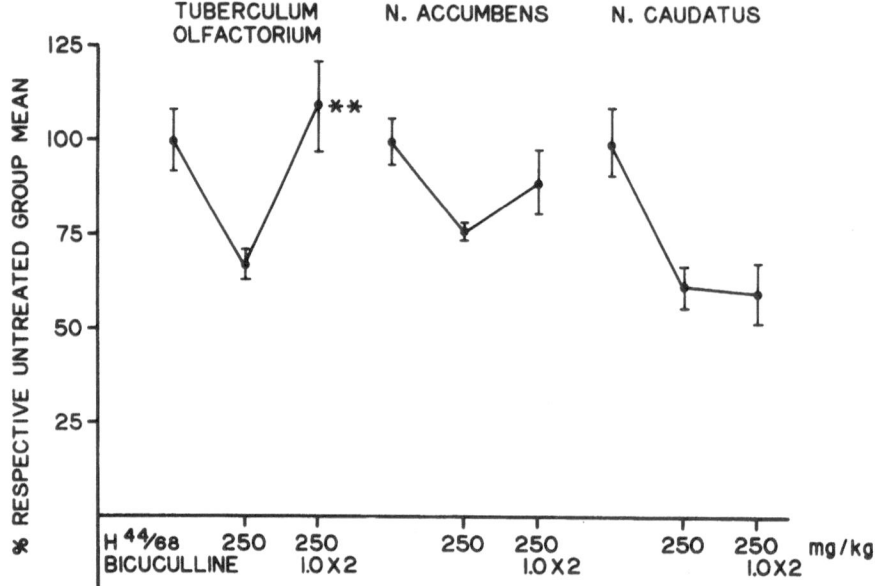

Fig. 13. Effect of bicuculline on the H44/68-induced fluorescence disappearance in the forebrain. Bicuculline was given s.c. at the same time and 1 h after H44/68 (250 mg/kg, 2 hr). Values are means ± SEM. Student's t-test. N=4-5. **=p 0.01.

the ventral midbrain tegmental area may block only the inhibitory receptors located within this area, eliciting therefore an increase of DA turnover in subcortical limbic regions (23). The fact that muscimol (low doses) and γ-acetylenic GABA are able to prevent the apomorphine induced locomotion and stereotypes (see 16 and 57), suggests the existence of GABAergic influences acting beyond the postsynaptic DA receptors, since the behavioral expression of DA receptor activity is blocked. Furthermore, the rich GABAergic innervation of the globus pallidus and the deeper structures of the tuberculum olfactorium region (see above) give anatomical support to the existence of GABA receptors located beyond the DA receptors which may modulate the biological actions of postsynaptic DA receptor activities in the neostriatum, nuc. accumbens and the tuberculum olfactorium proper.

The above results also give evidence for the existence of important GABA-catecholamine interactions within the hypothalamus. The fact that, within the LPZ of the median eminence, muscimol increases DA turnover, whereas -acetylenic GABA elicits an opposite effect (16), may indicate the existence also of excitatory and inhibitory GABA receptors in the control of the activity of the tubero--infundibular DA neurons.

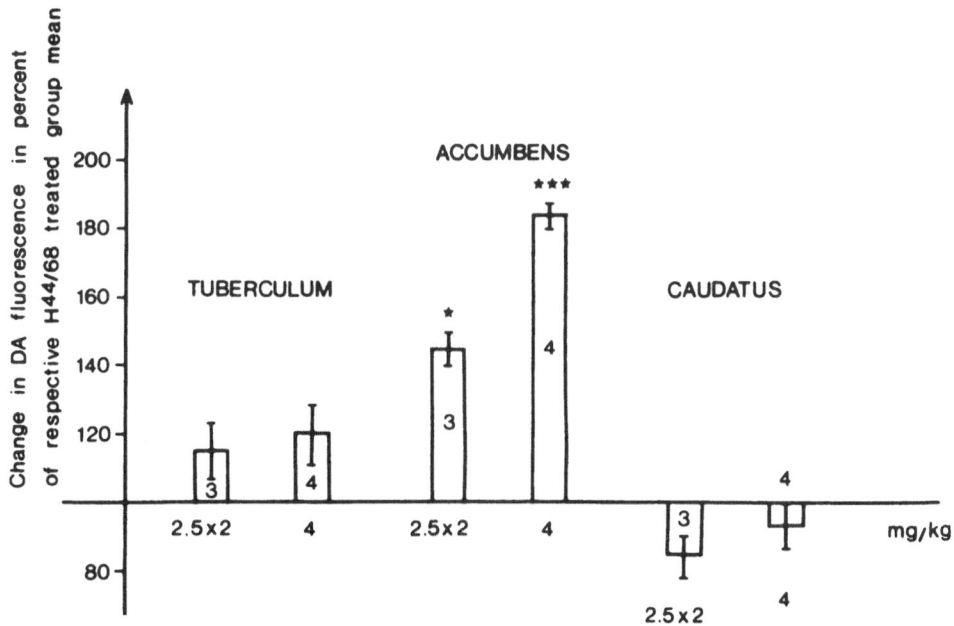

Fig. 14. Effect of picrotoxin on the H44/68-induced fluorescence disappearance in the forebrain. Picrotoxin was given i.p. as a single dose of 4 mg/kg or in two doses of 2.5 mg. The first dose was administered simultaneously with H44/68 (250 mg/kg; 2 hr) and the second dose one h afterwards. Values are means ± SEM. Number of rats is shown on the bars. Student's t-test. *p<0.05;***p<0.001 (23).

GABAergic Interactions with Other Neurotransmitter Systems

Our studies give experimental evidence for the existence of GABA-noradrenaline and GABA-opioid peptide interactions in the brain (43). The fact that both muscimol and γ-acetylenic-GABA increase the noradrenaline turnover within the SEL of the median eminence (16) may indicate the existence of a local GABAergic control of this noradrenaline terminal system.

Studies on the effect of peptides on GABA turnover (Fig. 15) showed that the intraventricular injection of β-endorphin, but not of substance P, increased GABA turnover within the nuc. caudatus and within the substantia nigra. These results may indicate the existence of interactions between GABA and enkephalin immunoreactive neurons in these areas.

CONCLUSIONS

The above anatomical, biochemical and pharmacological results

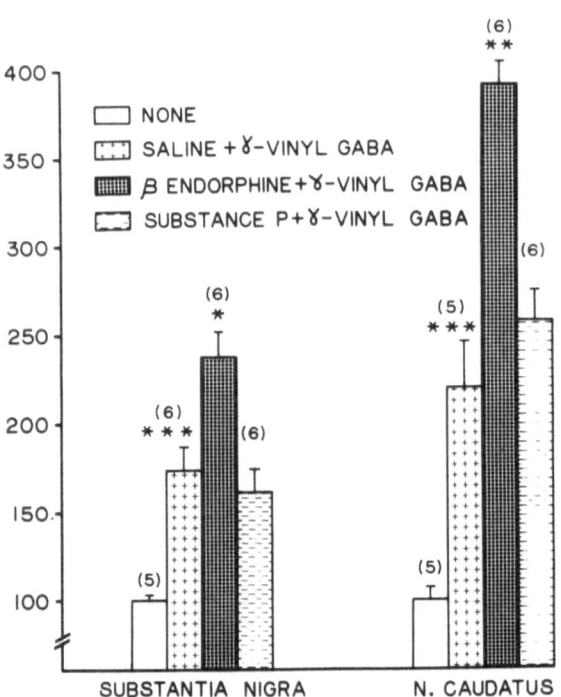

Fig. 15. Effect of β-endorphine and substance P on the γ-vinyl GABA-
induced GABA accumulation. β-endorphine (5 μg) and substance P (10 μg)
were given intraventricularly (30 μl/3 min) in awake rats, through a
cannula chronically implanted into the left lateral ventricle. γ-Vinyl
GABA was given (750 mg/kg) s.c. simultaneously with the intraventricular injection of either the saline or each of the peptides.
Rats were killed 3 hr after the administration of the drugs. On the
x-axis the treatments and areas are shown. On the y-axis the values
are given in percent ± SEM of untreated group means. Absolute values
for GABA in untreated rats were (μmoles/g wt tissue) 5.45 ± 0.02 and
1.12 ± 0.08 in the substantia nigra and n. caudatus, respectively.
Number of animals in parentheses. The statistical analysis was done
in two steps: 1) untreated group was tested against saline + γ-vinyl-
GABA treated group according to Mann-Whitney U-test,***p<0.001.
2) γ-vinyl-GABA-treated groups were compared to each other according
Wilcoxon one-way classification, comparing all possible pairs of
treatments: β-endorphin vs saline in substantia nigra, *p<0.05; in
nuc. caudatus, **p<0.01.

provide experimental evidence for the existence of important interactions between GABA neurons and monoaminergic and, possibly, enkephalin neurons in the brain. Cholinergic neurons also seem to play an important role in the GABA-DA interactions. GABAergic regulatory actions on the activity of the DA pathways may be exerted directly or indirectly at several levels, namely at the level of the DA dendrites, cell bodies and terminals and in more remote regions, such as the globus pallidus or the substantia innominata to which the DA innervated structures project. It is postulated that the effects of the GABAergic actions on the DA systems could be either inhibitory e.g. by an action directly on the DA cell bodies and dendrites, or excitatory if the GABAergic influences are indirect. Finally, the fact that after the administration of neuroleptics GABAergic drugs preferentially reduce DA turnover in the mesolimbic DA system suggest their therapeutic use, in combination with low doses of neuroleptics, in schizophrenia, in which a specific decrease of the mesolimbic DA systems is desired (19,23).

Acknowledgements. This work was supported in part by a grant from Consejo Nacional de Ciencia y Tecnología, México, D.F. (Project PCCBNAL 790253) by grants (04X-715, 04X-1010) from the Swedish Medical Research Council, by a grant from Magnus Bergvalls Stiftelse and by a grant (MH 25504-06) from the National Institutes of Health, USA. The excellent technical assistance of Mr. J. Méndez Franco is gratefully acknowledged. γ-acetylenic GABA and γ-vinyl GABA were generously provided by the Centre de Recherche Merrel International Strasbourg, France.

REFERENCES

1. Andén, N.-F., Butcher, S.G., Corrodi, H., Fuxe, K. and Ungerstedt U., Receptor activity and turnover rate of dopamine and noradrenaline after neuroleptics, J. Pharmacol. 11 (1970) 303-314.
2. Andén, N.-E., Corrodi, H. and Fuxe, K. Turnover studies using synthesis inhibition, In: G. Hooper, (Ed.), Metabolism of Amines in The Brain, McMillan, London, 1969, pp. 38-47.
3. Andén, N.-E., Dahlström, A., Fuxe, K., Larsson, K., Olson, L. and Ungerstedt, U., Ascending monoamine neurons to the telencephalon and diencephalon, Acta physiol. scand., 67 (1966) 313-326.
4. Andén, N.-E., Rubenson, A., Fuxe, K., Hökfelt, T., Evidence for dopamine receptor stimulation by apomorphine, J. Pharm. Pharmacol. 19 (1967) 627-629.
5. Andén, N.-E. and Wachtel, H. Some effects of GABA and GABA-like drugs on central catecholamine mechanisms. In S. Grattini, J.F. Pujol and R. Samanin (Eds.), Interactions Between Putative Neurotransmitters in the Brain, Raven Press, New York, 1978, pp. 161-174.
6. Barber, R. and Saito, K. Light Microscopic Visualization of GAD and GABA-T in Immunocytochemical Preparations of Rodent CNS. In

E. Roberts, T.N. Chase and D.B. Tower (Eds.), <u>GABA in Nervous System Function</u>, Raven Press, New York. 1976, pp. 113–132.

7. Bartholini, G. Interaction of dopamine, acetilcholine, and ɣ-aminobutyric acid in limbic and striatal structures: Relation to therapy of schizophrenia. In De Ajuriaguerra, J. and Tissot, R. (Eds.) <u>Rhinencephale, Neurotransmetterus et Sychoses.</u> Masson, Paris, 1977 pp. 241–251.

8. Brownstein, M., Palkovits, M., Tappaz, M., Saavedra, J. and Kizer, S., Effect of surgical isolation of the hypothalamus on its neurotransmitter content, <u>Brain Res.</u> 117 (1976) 287–295.

9. Cuello, A.C. and Iversen, L.L. Interactions of dopamine with other neurotransmitters in the rat substantia nigra: A possible functional role of dendritic dopamine. In Grattini, S., Pujol. P.J. and Samanin, R. (Eds.) <u>Interactions Between Putative Neurotransmitters in the Brain</u>, Raven Press, New York, 1978, pp. 127–149.

10. Dahlstrom, A. and Fuxe, K., Evidence for the existence of monoamine containing neurons in the central nervous system. I, Demonstration of monoamines in the cell bodies of brain stem neurons. <u>Acta Physiol. Scand.</u>, 62, Suppl. 232 (1964) 1–55.

11. Eccles, J.C., Ito, M. and Szentagothai, J., <u>The Cerebellum as a Neural Machine.</u> Springer-Verlag, New York, 1967.

12. Einarsson, P., Hallman, P. and Jonsson, G., Quantitative microfluorimetry of formaldehyde-induced fluorescence of dopamine in the caudate nucleus. <u>Med. Biol.</u> 53 (1975) 15–24.

13. Fonnum, F., Grofová, I., Rinvik, E., Storm-Mathisen, J. and Walberg. F., Origin and distribution of glutamate decarboxylase in substantia nigra of the cat, <u>Brain Res.</u>, 71 (1974) 77–92.

14. Fonnum, F., Storm-Mathisen, J. and Walberg, F., Glutamate decarboxylase in inhibitory neurons. A study of the enzyme in Purkinje cell axons and boutons in the cat <u>Brain Res.</u>, 20 (1970) 259–275.

15. Fonnum, F., Walaas, I. and Iversen, E. Localization of gabaergic, cholinergic and aminergic structures in the mesolimbic system, <u>J. Neurochem.</u>, 29 (1977) 221–230.

16. Fuxe, K., Andersson K., Ogren S-O, Pérez de la Mora M., Schwarcz R., Hökfelt T., Eneroth P., Gustafsson J-A & Skett P., GABA neurons and their interaction with monoamine neurons. An anatomical, pharmacological and functional analysis. In P. Krogsgaard-Larsen, J. Scheel-Kruger and H. Kofod (Eds.) <u>GABA Neurotransmitters. Pharmacological, Biochemical and Pharmacological Aspects</u>, Munksgaard, Copenhagen, 1979, pp. 74–94.

17. Fuxe, K., Andersson, K., Schwarcz, R., Agnati, L.F., Pérez de la Mora, M. Hökfelt, T., Goldstein M., Ferland, L., Possani, L. and Tapia, R., Studies on different types of dopamine nerve terminals in the forebrain and their possible interactions with hormones and with neurons containing GABA, glutamate and opioid peptides, In L.J. Poirier, T.L. Sourkes and P.J. Bedard (Eds.), <u>Advances in Neurology, Vol. 24</u>, Raven Press, New York, 1979, pp. 199–216.

18. Fuxe, K., Hökfelt, T., Agnati, L.F., Johansson, O., Goldstein, M., Pérez de la Mora, M., Possani, L., Tapia, R., Terán, L. and Palacios, R. Mapping out central catecholamine neurons: Immuno-histochemical studies on catecholamine-synthesizing enzymes. In M.A. Lipton, A. Di Mascio and K.F. Killam (Eds.), Psycho-pharmacology. A Generation of Progress, Raven Press, New York, 1978, pp. 67-94.

19. Fuxe, K., Hökfelt, T., Agnati, L., Ljungdahl, A., Johansson, O. and Pérez de la Mora, M., Evidence for an inhibitory gabaergic control of the mesolimbic dopamine neurons: possibility of improving treatment of schizophrenia by combined treatment with neuroleptics and gabergic drugs, Med. Biol., 53 (1975) 177-183.

20. Fuxe, K., Hökfelt, T., Johansson, O., Ganten, D., Goldstein, M., Pérez de la Mora, M., Possani, L., Tapia, R., Terán, L., Palacios, R., Said, S. & Mutt, V. Monoamine neuron system in the hypo-thalamus and their relation to the GABA and peptide-containing neurons. In Mornex, R. and Barry, J. Neuromediateurs et Poly-peptides Hypothalamiques a Action Relachante ou Inhibitrice, Institut National de la Sante et de la Recherche Medicale, Paris, 1977, pp. 17-40.

21. Fuxe, K., Hökfelt, T. and Ungerstedt, U., Morphological and functional aspects of central monoamine neurons. In International Review of Neurobiology, Vol. 13, Academic Press, New York, 1970, pp. 93-126.

22. Fuxe, K., Ogren, S.-O., Fredholm, B., Agnati, L., Hokfelt, T. and Pérez de la Mora, M., Possibilities of a differential blockade of central monoamine receptors. In De Ajuriaguerra, J. and Tissot, R. (Eds.) Rhinencéphale Neurotransmetteurs et Sychoses, Masson, Paris, 1977, pp. 253-289.

23. Fuxe, K., Pérez de la Mora, M., Hökfelt, T., Agnati, L., Ljungdahl, A., and Johansson, O. GABA-DA interactions and their possible relation to schizophrenia. In C. Shagaas, S. Gershon and A.J. Friedhof (Eds.) Psychopathology and Brain Dysfunction, Raven Press, New York, 1977, pp. 97-111.

24. Hökfelt, T., Elde, R., Fuxe, K., Johansson, O., Ljungdahl, A. Goldstein, M., Luft, R., Efendic, S., Nilsson, G., Terenius, L., Ganten, D., Jeffcoate, S.L., Rehfeld, J., Said, S., Pérez de la Mora, M., Possani, L., Tapia, R., Terán, L. and Palacios, R. Aminergic and peptidergic pathways in the nervous system with special reference to the hypothalamus. In S. Reichlin, R.J. Baldessarini and J.B. Martin (Eds.), The Hypothalamus, Raven Press, New York, 1978, pp. 69-135.

25. Hökfelt, T. and Ljungdahl, A. Uptake mechanisms as a basis for the histochemical identific ation and tracing of transmitter specific neuron populations, In W.M. Cotman and M. Cuénod (Eds.), The Use of Axonal Transport for Studies of Neuronal Connectivity, Elsevier Scientific Publishing Co., Amsterdam, 1975, pp. 251-305.

26. Iversen, L.L. and Schon, F. The use of autoradiographic techniques for the identification and mapping of transmitter specific neurons in CNS. In A. Mandel and D. Segal (Eds.) New Concepts

of Transmitter Regulation, Plenum Press, New York, 1973, pp.
153-193.

27. Javoy, F., Agid, Y. and Glowinski, J., Oxotremorine and atropine-
    induced changes of dopamine metabolism in the rat striatum. J.
    Pharm. Pharmacol. 27 (1975) 677-681.

28. Johnston, G.A.R. Physiologic pharmacology of GABA and its
    antagonists in the vertebrate nervous system. In Roberts, E.
    Chase, T.N. and Tower, D.B. (Eds.) GABA in Nervous System Function,
    Raven Press, New York, 1976, pp. 395-411.

29. Joseph-Nathan, P., Massieu, G., Carvajal, P. and Tapia, R.,
    γ-Hydroxy-γ-Phenyl, Caproamide, an anticonvulsant molecule, Rev.
    Latinoamer. Quim. 9 (1978) 90-92.

30. Kim, J.S., Bak, I.J., Hassler, R. and Okada, Y., Role of
    γ-aminobutyric acid (GABA) in the extrapyramidal motor system.
    2. Some evidence for the existence of a type of GABA-rich
    strionigral neurons, Exp. Brain Res., 14 (1971) 95-104.

31. Krogsgaard-Larsen, P., Johnston, G.A.R., Curtis, D.R., Game,
    C.J.A., and McCulloch, R.M., Structure and biological activity
    of a series of conformotionally restricted analogues of GABA, J.
    Neurochem. 25 (1975) 203-209.

32. Ladinsky, H., Consolo, S., Bianchi, S. and Jori, A., Increase
    in striatal acetylcholine by picrotoxin in the rat: evidence
    of a gabaergic-dopaminergic-cholinergic link, Brain Research,
    108 (1976) 351-361.

33. Lasher, R.S., The Uptake of (^{3}H) GABA and differentiation of
    stellate neurons in cultures of dissociated postnatal rat
    cerebellum, Brain Res. 69 (1974) 235-254.

34. Lipton, M.A., DiMascio, A. and Killam, K.F. (Eds.), Psycho-
    pharmacology: A Generation of Progress, Raven Press, New York,
    1978.

35. Mao, C.C., Cheney, D.L., Marco, E., Revuelta, A. and Costa, E.
    Turnover times of gamma-aminobutyric acid and acetylcholine in
    nucleus caudatus, nucleus accumbens, globus pallidus and
    substantia nigra: Effects of repeated administration of
    haloperidol, Brain Res. 132 (1977) 375-379.

36. Mao, C.C., Marco, E., Revuelta, A. and Costa, E., Antysychotics
    and GABA turnover in mammalian brain nuclei. In Grattini, S.,
    Pujol, J.F. and Samanin, R. (Eds.), Interactions Between Putative
    Neurotransmitters in the Brain, Raven Press, New York, 1978, pp.
    151-160.

37. Massieu, G.H., Tapia, R., Pasantes, H.O. and Ortega, B.G.
    Convulsant effect of L-glutamic acid-γ-hydrazide by simultaneous
    treatment with pyridoxal phosphate, Biochem. Pharmacol. 13 (1964)
    118-120.

38. Matsuda, T., Wu, J.-Y. and Roberts, E., Immunochemical studies on
    glutamic acid decarboxylase (EC 4.1.1.15) from mouse brain, J.
    Neurochem. 21 (1977) 159-166.

39. Matsuda, T., Wu, J.-Y. and Roberts, E., Electrophoresis of
    glutamic acid decarboxylase (EC 4.1.1.15) from mouse brain in
    sodium dodecyl sulphate polyacrylamide gels, J. Neurochem., 21

(1973) 167-172.

40. McGeer, P.L., Fibiger, H.C., Maler, M., Hattori, T. and McGeer E.G., Evidence for descending pallido-nigral GABA-containing neurons. In F. MacDowell and A. Barbeau (Eds.), Advances in Neurology, Vol. 3, Raven Press, New York, 1974, pp. 153-163.

41. Obata, K. Ito, M., Ochi, R. and Sato, N. Pharmacological properties of the postsynaptic inhibition by Purkinje cell axons and the action of γ-aminobutyric acid on Deiters neurons, Exp. Brain Res. 4 (1967) 43-57.

42. Pérez de la Mora, M. and Fuxe, K. Brain GABA, dopamine and acetylcholine interactions. I. Studies with oxotremorine, Brain Res., 135 (1977) 107-122.

43. Pérez de la Mora, M., Fuxe, K., Andersson, K., Hökfelt, T., Ljungdahl, A. Possani, L. and Tapia, R. Studies on GABA-monoamine and GABA-endorphin interactions. In E. Usdin, I.J. Kopin and J. Barchas (Eds.), Catecholamines: Basic and Clinical Frontiers, Pergamon Press, New York, 1979, Vol. II, pp. 1032-1034.

44. Pérez de la Mora, M., Fuxe, K., Hökfelt, T. and Ljungdahl, A. Effect of apomorphine on the GABA turnover in the DA cell group rich area of the mesencephalon. Evidence for the involvement of an inhibitory gabergic feedback control of the ascending DA neurons, Neurosci. Lett., 1 (1975) 109-114.

45. Pérez de la Mora, M., Fuxe, K., Hökfelt, T. and Ljungdahl, A. Further evidence that apomorphine increases GABA turnover in the cell body rich and DA nerve terminal rich areas of the brain, Neurosci. Lett. 2 (1976) 239-241.

46. Pérez de la Mora, M., Fuxe, K. Hökfelt, T. and Ljungdahl, A. Evidence for a nerve impulse-dependent GABA accumulation in the substantia nigra after treatment with γ-glutamylhydrazide, Neurosci. Lett. 5 (1977) 75-82.

47. Pérez de la Mora, M., and Tapia, R. Anticonvulsant effect of 5-ethyl, 5-phenyl, 2-pirrolidinone and its possible relationship to γ-aminobutyric acid-dependent inhibitory mechanisms, Biochem. Pharmacol. 22 (1973) 2635-2639.

48. Possani, L.D., Bayón, A. and Tapia, R., Synthesis of affinity chromatographic resins for the purification of brain glutamate decarboxylase, Neurochem. Res. 2 (1977) 51-57.

49. Racagni, G., Apud, J.A., Locatelli, V. Cocchi, D., Nistico, G. di Giorgio, R.M. and Muller, E.E. GABA of CNS origin in the rat anterior pituitary inhibits prolactin secretion. Nature (Lond.) 281 (1979) 575-578.

50. Racagni, G. Bruno, F., Cattabeni, F., Maggi, A. Di Giulio, A.M. and Groppeti, A. Interactions among dopamine, acetylcholine, and GABA in the nigrostriatal system, In: Grattini, S. Pujol, J.F. and Samanin, R. (Eds.) Interactions Between Putative Neuro-transmitters in the Brain, Raven Press, New York, 1978, pp. 61-72.

51. Reubi, J.C., Iversen, L.L. and Jessel, T.M. Dopamine selectively increases ^{3}H-GABA release from slices of rat substantia nigra in vitro, Nature 268 (1977) 652-653.

52. Ribak, C.E., Vaughn, J.E., Saito, K., Barber, R. and Roberts, E.,
    Immunocytochemical localization of glutamate decarboxylase in
    rat substantia nigra, Brain Res. 116 (1976) 287-298.
53. Roberts, E., Chase, T.N. and Tower, D.B. (Eds.), GABA in Nervous
    System Function, Raven Press, New York, 1976.
54. Saito, K., Barber, R., Wu, J-Y., Matsuda, T., Roberts, E. and
    Vaughn, J.E., Immunohistochemical localization of glutamate
    decarboxylase in rat cerebellum, Proc. Nat. Acad. Sci. USA.,
    71 (1974) 269-273.
55. Saito, K. Wu, J.-Y., Matsuda, T. and Roberts, E. Immunochemical
    comparisons of vertebrate glutamic acid decarboxylase, Brain Res.
    65 (1974) 277-285.
56. Schechter, P., Tranier Y., Jung, M.J. and Bohlen, P. Audiogenic
    seizure protection by elevated brain GABA concentration in mice:
    Effects of γ-acetylenic GABA and γ-vinyl GABA, two irreversible
    GABA-T inhibitors. Europ. J. Pharmacol. 45 (1977) 319-328.
57. Scheel-Kruger, J., Arnt, J., Braestrup, C., Christensen, A.V.
    and Mageluno, E., Development of new animal models for GABAergic
    actions using muscimol as a tool. In P. Krogsgaard-Larsen, J.
    Scheel-Kruger and H. Kofod (Eds.), GABA-Neurotransmitters,
    Pharmacochemical, Biochemical and Pharmacological Aspects,
    Munksgaard, Copenhagen, 1979, pp. 447-464.
58. Sotelo, C. Privat, A. and Drian, M.-J. Localization of (^{3}H) GABA
    in tissue culture of rat cerebellum using electron microscopy
    radioautography, Brain Res., 61 (1972) 379-384.
59. Tapia, R., Biochemical Pharmacology of GABA in CNS In: Iversen
    L.L., Iversen, S.D. and Snyder, S.H. (Eds.) Handbook of
    Psychopharmacology Vol. 4. Plenum Publishing Co., New York, 1975,
    pp. 1-58
60. Tapia, R., Drucker-Colín, R.R., Meza-Ruiz, G., Durán, L. and Levi
    G. Neurophysiological and neurochemical studies on the action of
    the anticonvulsant γ-hydroxy, γ-ethyl, γ-phenyl-butyramide,
    Epilepsia 20 (1979) 135-145.
61. Tappaz, M.L., Brownstein, M.J. and Palkovits, M., Distribution of
    glutamate decarboxylase in discrete brain nuclei, Brain Res.
    108 (1976) 371-380.
62. Tappaz, M.L. and Brownstein, M.J., Origin of glutamate
    decarboxylase (GAD) containing cells in discrete hypothalamic
    nuclei, Brain Res. 132 (1977) 95-106.
63. Ungerstedt, U. Stereotaxic mapping of the monoamine pathway in
    the rat brain. Acta physiol. scand., 82 (1971) 1-48.
64. Usdin, E., Kopin, I.J. and Barchas J. (Eds.), Catecholamines:
    Basic and Clinical Frontiers. Pergamon Press, New York, 1979.
65. Walaas, I. and Fonnum, F. The effect of parenteral glutamate
    treatment on the localization of neurotransmitters in the
    mediobasal hypothalamus, Brain Res., 153 (1978) 549-562.
66. Waalas, I. and Fonnum, F., The distribution and origin of
    glutamate decarboxylase and choline acetyltransferase in ventral
    pallidum and other basal forebrain regions, Brain Res. 117 (1979)

325-336.

67. Wiesel, F.A. and Seduall, G. The effect of antipsychotic drugs on homovanillic acid levels in striatum and olfactory tubercle of the rat, Europ. J. Pharmacol. 30 (1975) 364-367.

68. Wood, J.G. McLaughlin, B.J. and Vaughn, J.E. Immunocytochemical localization of GAD in electron microscopic preparations of rodent CNS. In E. Roberts, T.N. Chase and D.B. Tower (Eds.), GABA in Nervous System Function, Raven Press, New York, 1976, pp. 133-148.

69. Yoshida, M. and Precht, W., Monosynaptic inhibition of neurons of the substantia nigra by caudate-nigral fibers, Brain Res. 32 (1971) 225-228.

70. Wochwa, S., Immunoelectrophoresis (including zone electrophoresis), In N.R. Rose and H. Friedman (Eds.), Manual of Clinical Immunology, American Society for Microbiology, Washington, D.C. 1976, pp. 17-35.

71. Wu, J.-Y., Matsuda, T. and Roberts, E., Purification and characterization of glutamate decarboxylase from mouse brain, J. Biol. Chem. 248 (1973) 3029-3034.

GABA RELEASE IN VIVO AND IN VITRO: RESPONSES TO PHYSIOLOGICAL AND

CHEMICAL STIMULI

Harry F. Bradford

Department of Biochemistry
Imperial College
London S.W.7 2AZ, England

INTRODUCTION

Studying the release of neurotransmitters from in vitro
preparation presents special problems of possible artefact. The
preparation of tissue slices, isolated ganglia, synaptosomes and
their like always involves damaging tissue elements, whose partial
or substantial recovery, usually by membrane-resealing, during
subsequent incubation, is their limiting feature. It is unlikely
that this recovery is ever complete and therefore the performance
of such preparations, however sophisticated must be regarded with
suspicion. In this respect amino acid neurotransmitters are a
special case due to their ubiquitous presence in neural cells and
their high concentrations (x 10^3) compared to other neurotransmitters.
This renders more likely the possibility of artefact due to leakage
through damaged membranes or diffusional loss down high concentration
gradients enhanced by high fluid: tissue ratios. Where release is
evoked by depolarizing agents such as veratrine or tityustoxin which
stimulate active Na^+ channels, and whose actions are entirely
suppressed by tetrodotoxin, a much greater degree of confidence is
engendered in the relevance of these transmitter release signals to
the synaptic events which occur in the intact nervous system.

However, in experimental biology judgements must be based on the
actual performance of the preparations rather than on prospective
criticism alone. Applying this criterion it is clear that a wealth
of important information has been obtained with such preparations on
the identification, biosynthesis, storage, transport, and release
requirements of transmitters in various PNS and CNS regions. The
in vitro approach, with all its limitantions, is therefore vindicated.

Monitoring release in vivo with the aid of superfusion techniques
can produce persuasive complementary evidence for a transmitter role
for physiologically active amino acids or other putative neurotransmitters in various brain regions. This is particularly so where
an adequate recovery period intervenes between the surgery and the
investigation of transmitter release, and where neurally-mediated
stimuli can be used to activate the brain region being superfused.
Such a system has been developed by our group at Imperial College.
This allows superfusion of specific surface regions of cerebral
cortex (4 mm diameter) of awake, unrestrained and behaviourally
normal rats using a specially constructed 'swivel' cannula.
Electrodes implanted around the brachial plexus of either side allows
activation of sensorimotor cortex by physiological routes and
chemical depolarizing agents can be introduced into the superfusion
stream. The actions of drugs, applied by either systemic or cannula
routes, on neurally or chemically-evoked release of transmitters
can also be studied with this system, and examples of this approach
are described in this paper.

EXPERIMENTAL PROCEDURES

Superfusion Experiments

The superfusion experiments were carried out on adult female
hooded Rowett rats (220-250 g body weight). Anaesthesia was induced
with Avertin (250 mg/kg) and atropine (1.7 mg/kg) given intraperitoneally and repeated to maintain light anaesthesia as tested by
a positive corneal reflex. The rats were kept in a 0-4°C cold room
until the body temperature decreased to 22°C. This greatly reduced
the risk of brain oedema during surgery. After implantation of the
superfusion cannula (Ref. 12 for full details) the animals were
either left at least 1 h for recovery before all acute experiments,
or left for 24 h before chronic experiments were begun.

In some experiments normal saline was employed as the superfusion
fluid, and in others artificial cerebrospinal fluid (CSF) solution
was employed, containing the following components (37) (mM): NaCl,
123.5; KCl, 2.4; KH_2PO_4, 0.5; Na_2SO_4, 0.16; $MgCl_2$, 0.9; $CaCl_2$, 1.47;
$NaHCO_3$, 22.0; glucose, 5.0: fructose, 0.22; urea, 4.2; adenosine,
0.1; guanosine, 0.1; creatine, 0.1; m-inositol, 0.15; sodium citrate,
0.32; sodium lactate, 1.34; sodium pyruvate, 0.08, sodium ascorbate,
0.06. Neutral pH was maintained by gassing the solution with a 5%
CO_2-95% O_2 gas mixture, and the colloidal osmotic pressure was
maintained by adding 20 mg% inulin instead of protein. When acetylcholine was measured the collecting tubes contained eserine sulphate
to give a final concentration of 1 mg/ml and enough HCl to provide
a final pH of 3 to 4.

For these experiments a special superfusion method was used
which is described in detail by Dodd et al. (20). This method used
a specially constructed "swivel" cannula which allows study of the
release of cerebral amino acid transmitters in sub-acute and chronic
experiments from the cortex of anaesthetised or unrestrained,
behaviourally normal rats. The cannula (4 mm diameter) is made from
non-toxic materials (Teflon) and has a very small dead space (15 µl).
It withstands normal intracranial pressure without leaking, and
embodies facilities which prevent the tubing leads from twisting
whilst allowing the animals to move freely.

The small size of the cannulae allows superfusion of small
discrete areas of the brain surface and therefore permits study of
the effects of specific forms of sensory stimulation on the release
of amino acids from these small localised regions of cerebral cortex.
The sensorimotor strip and the visual cortex were chosen for study.
The cortical area superfused was the S_1 forelimb sensorimotor area
which was located by drilling a 4 mm diameter hole in the skull with
a centre 3 mm from the central (saggital) suture, and centred on the
bregma suture. The accurate location of the cannula over the forelimb
sensorimotor area was checked by stimulating the cortex either
chemically (50 mM KCl), or electrically with pulses applied to an
incorporated stimulating wire. This evoked clear contralateral forelimb movements. Also cobalt powder discs implanted on this area
caused clonic limb-jerks of principally the forelimb.

Brachial Plexus Stimulation. After exposing the brachial plexus
of both forelimbs and separating the nerve trunks from adjacent blood
vessels a small electrode of the kind developed by Dobkin was placed
around all the nerves of the plexus on each side (4). Stimulation
applied to the forelimb contralateral to the cannulated hemisphere
activated the cortex being superfused through the cannula. Ipsilateral
stimulation should affect mainly the cortex in the uncannulated
hemisphere, though a later interhemispherically conducted response
could also have occurred.

By comparing the effects of these two forms of stimulation it
was possible to test whether any changes in transmitter release was
the direct result of stimulation of the cortical area, or due to
other bodily stresses caused by the stimulation. Electrical
stimulation of the brachial plexus was by square wave pulses of the
following specifications: amplitude 2-7 V; current 1-3 mA; pulse
duration 1 msec; frequency 2-3 Hz or 50 Hz. Five to seven minute
periods os stimulation were applied and always evoked clearly
visible muscular jerking in the ipsilateral forelimb. Clonic
seizures were never seen to occur.

    In vivo application of GABA analogues, tityustoxin, tetrodotoxin,
morphine and naloxone. At a specific time γ-vinyl-GABA (4-amino-hex-
5-enoic acid, RMI 71754) 1000-1500 mg/kg or γ-acetylenic-GABA (4-
amino-hex-5-ynoic acid, RMI 71645) 100 mg/kg were administered
(i.p.) as aqueous solutions (0.6 ml/100 g body weight). The drugs,
which were synthesized as described by Metcalfe and Casara (45),
were supplied by Centre de Recherche Merrell International,
Strasbourg (France).

    Tityustoxin and tetrodotoxin were introduced directly onto
superfused sensorimotorcortex dissolved in superfusate in the
concentration range 100 nM to 1 µM as indicated. Morphine (10 mg/kg
and Naloxone (2 mg/kg) were administered by intraperitoneal injection.

Experiments With Synaptosomes

    Adult female Sprague-Dawley rats were used (200-250 g) for all
experiments. Synaptosomes were prepared from cerebral cortex by the
method of Gray and Whittaker (28), as modified by Bradford et al. (7).

    Experiments with tityustoxin. Tityustoxin is a basic polypeptide
(mol. wt. 6995 daltons; pI, 8.25), which was purified from the venom
of a Brazilian yellow scorpion (Tityus serrulatus) by the method of
Coutinho-Netto and Diniz (15). The procedure involved extraction and
chromatographic separation on Sephadex G-50 and CM-Cellulose-52. The
toxin was homogenous as judged by electrophoresis in polyacrylamide
gels with or without added sodium dodecyl sulphate. Only one N-
terminal (L-lysine) was detected in the preparation. It showed an
LD_{50} of 18 µg/kg in mice when administered intraperitoneally. The
preparations were found to be relatively free of contamination by
amino acids (see Results).

    Synaptosome fractions were gently resuspended at 3-4 mg protein/
ml in a Krebs-phosphate medium of composition (mM): NaCl, 124; KCl,
5; KH_2PO_4, 1.2; $MgSO_4$, 1.3; $CaCl_2$, 0.75; Na_2HPO_4, 20; pH 7.4,
containing 10 mM glucose and gassed with O_2. For Ca^{2+}-free media,
Ca^{2+} was omitted and EGTA (0.5 mM) was added. For respiratory
measurement incubation was at 37°C in Warburg respirometer vessels
for 45 minutes. This period included 35 minutes preincubation
followed by 10 minutes exposure to the toxin or other agents which
were added from the side arm. For studies of amino acid release,
incubation was at 37°C in Krebs-phosphate medium in 5.0 ml conical
flasks which were gassed with O_2. Preincubation was for 35 min before
addition of tityustoxin and a further 10 minute exposure period.
When tetrodotoxin (1 µM) was used, addition was 10 sec before
adding tityustoxin. After separating the tissue from the medium by
brief centrifugation in a bench ultracentrifuge (18,000 g, 1 min)
each fraction was extracted with ice-cold tricholoroacetic acid
(5% w/v for supernatants; 10% w/v for tissue), and the extracts were

purified on Zeocarb 225, as previously described (8,48). Purified
extracts were analysed using the autoanalyser described below.

In the GABA uptake experiments synaptosome suspensions (2-3 mg
protein/ml) were incubated at 37°C in 5.0 ml conical flasks in Krebs-
phosphate medium. The flasks were gassed with O_2. Tityustoxin was
added and incubation was continued for 10 minutes before addition of
0.5 µCi of [U-^{14}C] -GABA to give a final concentration of 2.2 x 10^{-3}
µM and specific radioactivity of 224 µCi/µmol. Incubation was
continued for 5 minutes before centrifugation in a bench ultracentrifuge to deposit tissue. The deposited tissue was extracted with
1.0 ml 10% trichloroacetic acid and radioactivity in the extract
was measured by liquid scintillation counting using a toluene-PBD-
methoxyethanol scintillation mixture. Where tetrodotoxin was used,
it was added to a final concentration of 1 µM, 10 seconds before
addition of tityustoxin.

Experiments with alkyl GABA-analogues. For the studies of
[U^{14}C] GABA release, synaptosomes were incubated (3 mg protein/ml
phosphate medium) for 20 minutes in the presence of [U^{14}C] -GABA
(final sp. act. 0.25 µCi/ml) and sedimented in a bench ultracentrifuge prior to resuspension in phosphate medium containing either
Ca^{++} or the calcium chelating agent, 5- [(3, 4-dimethoxy phenethyl)
methyl amino] -2-(3, 4-dimethoxyphenyl)-2-iso propyl valeronitrile
(Verapamil) (100 µM). These suspensions were incubated for 30 minutes
before tetrodotoxin (TTX; 1.0 µM), γ-acetylenic-GABA (0.5-5 mM) or
γ-vinyl-GABA (0.25-5 mM) were added as appropriate. Incubation
continued for a further 10 minutes prior to sedimentation of the
synaptosomes in a bench ultracentrifuge. The supernatants were added
to vials. A toluene-based liquid scintillant was added and the vials
were taken for radiolabel counting.

For the release of endogenous GABA, rats were injected intraperitoneally with γ-acetylenic-GABA (100 mg/kg) or γ-vinyl-GABA
(1500 mg/kg) in saline or with a similar volume of saline in control
animals at 6 hours or 14 hours respectively before preparation of
cerebrocortical synaptosomes. These synaptosomes were incubated in
Krebs-phosphate medium for 30 minutes prior to addition of tetrodotoxin (1 µM) or veratrine (75 µM) as appropriate. Incubation
continued for a further 10 minutes. Synaptosomes were sedimented in
a bench ultracentrifuge and the supernatants and pellets were
extracted with 10% TCA. The TCA extracts were purified on a semi-
automatic ion exchange apparatus to collect amino acids. The eluates
containing amino acids were taken to dryness on a Buchler Evapomix.
Samples were then dissolved in 0.025 M HCl and analysed on the
automatic amino acid analyser (8,48).

In GABA uptake studies synaptosome suspension volumes of 1 ml
containing approximately 3 mg/ml protein were pre-incubated with or

without the drugs as appropriate at 37°C for 5 minutes. Then $(U^{14}C)$-
GABA (final sp. act. 0.25 µCi/ml) was added to a final concentration
of 1 µM and this incubation proceeded for a further 1, 2.5, or 5
minutes. The suspensions were then sedimented in a bench ultracentrifuge and the clear supernatants were discarded. The synaptosomal
pellets were extracted with 1 ml 10% (w/v) TCA and resedimented. The
supernatants were dissolved in 20 ml of liquid scintillant (toluene,
1 litre; methoxyethanol, 800 ml; 2 (4-tert-butylphenyl)-5-(4-
biphenylyl)-1, 3, 4-oxadiazole, 6 g) and counted using a Packard-
Tricarb scintillation counter.

   Other procedures. The amino acids were measured by autoanalysis
using the analyser described by Bradford and Thomas, (8,48) but with
a fluometric attachment. This technique allows rapid (35 minutes)
measurement of 1-5 pmol of amino acids in 200 µl sample volumes.
Calculation of the results was aided by a small on-line computer
(Digico 16S) by relating the values to standard norleucine signals
added to the samples. The high sensitivity of the technique allowed
direct measurement of amino acids in the superfusion fluid after
acidification, without concentration or purification. The latter
was collected in 6-8 minute fractions (about 1 ml) in an automatic
fraction collector from the output tube of the superfusion line.
γ-Vinyl-GABA was completely separated from GABA in the analysis
system, whilst γ-acetylenic-GABA was not. No GABA contamination of
the γ-vinyl-GABA used could be detected when 5 mM samples were
analysed. The concentration of the amino acids present in the superfusate was expressed as pmol/min/sq.cm of the cortical surface
exposed. The statistical significance of the differences between
stimulated and control in the same animal was evaluated using a
"paired t-test".

   Potassium determination was performed on 10% (w/v) trichloracetic
acid extract of tissue samples, employing an EEL (Evans Electro-
selenium Ltd.) flame photometer. Mixtures of K^+ in 30 mM NaCl were
employed as standard. Protein was determined by the method of Lowry
et al. (42) using bovine serum albumin as standard. Acetylcholine
was bioassayed using the guinea-pig ileum preparation (55), and
samples were prepared as previously described.

   Tetrodotoxin was purchased from the Sigma Chemical Co. (London).
The calcium blocking agent Verapamil was a gift from Abbot
Laboratories, Queenborough, Kent. At 100 µM it had no measurable
effect on the performance of synaptosomes in the control condition,
but did prevent the depolarisation-induced release of amino acid
neurotransmitters from synaptosomes caused by Veratrine (75 µM) or
by K^+ (56 mM).

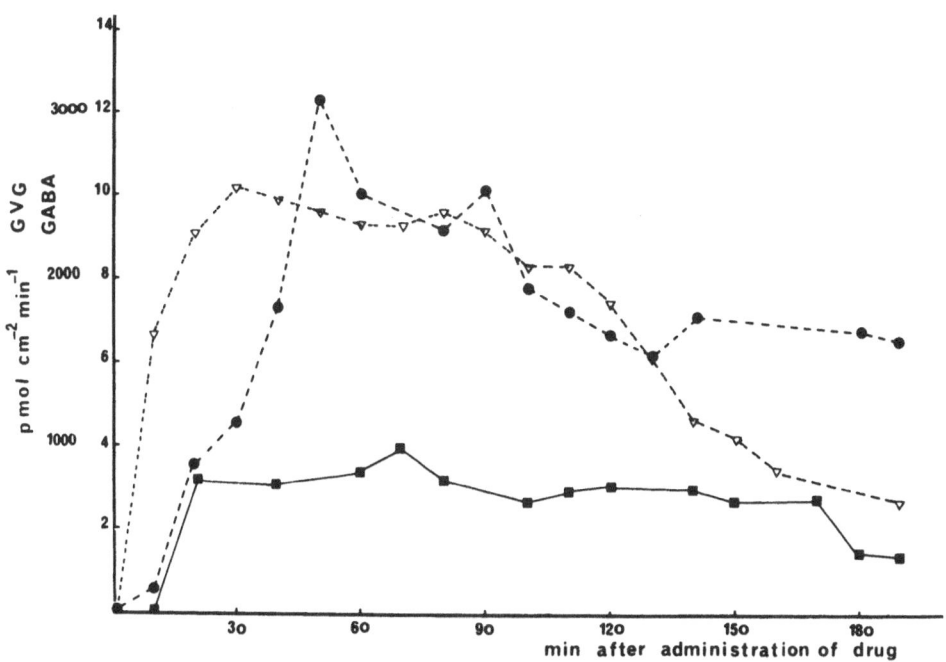

Fig. 1. Release of GABA from sensorimotor cortex following administration of γ-vinyl-GABA. Released GABA (●) was determined in 8 minute fractions of superfusate collected immediately prior to, and up to 3 h after a single intraperitoneal injection of γ-vinyl-GABA (1500 mg/kg). The release of γ-vinyl-GABA (GVG, ∇) was also followed. Alternatively, γ-vinyl-GABA (100 μM in artificial CSF) was delivered continuously onto the cortex through the superfusion line and GABA was similarly measured (■). GABA and γ-vinyl-GABA release rates following i.p. injection of the drug are mean values from 8 experiments. GABA release rates following γ-vinyl-GABA injection into the cannulae are mean values from two experiments. Standard error values were less than 17%. (From Ref. 2).

EXPERIMENTAL OBSERVATIONS

Alkyl GABA Analogues

    The effect of γ-vinyl-GABA and γ-acetylenic-GABA release by cerebral cortex in vivo. As reported previously (1,18) no GABA could be detected in superfusate of sensorimotor cortex. However, following a single intraperitoneal dose of γ-vinyl-GABA (1500 mg/kg), GABA began to appear in the superfusate of cerebral sensori-motor cortex and gradually increased in concentration reaching a maximum rate of release (9 pmol/min/sq.cm) within 50-90 minutes. γ-Vinyl-

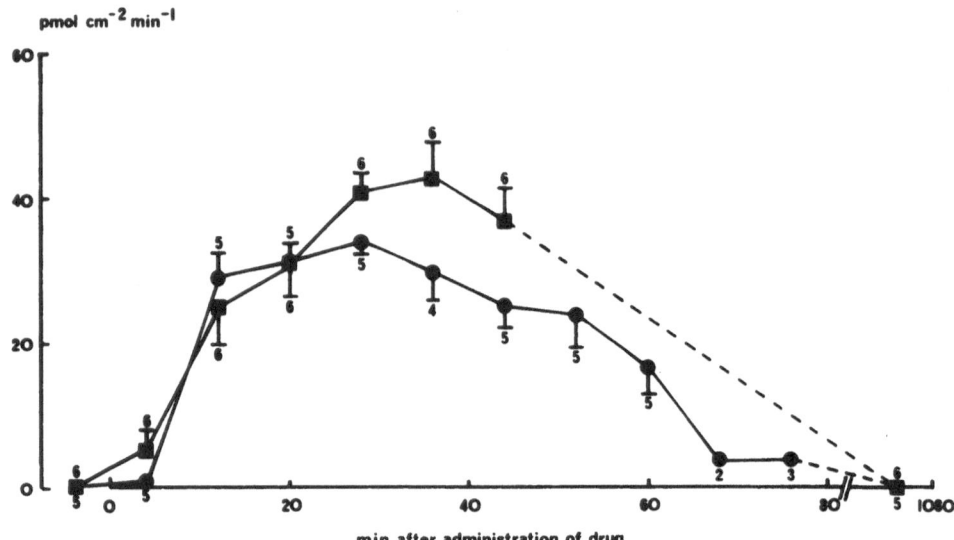

Fig. 2. Release of GABA from two regions of cerebral cortex
following administration of γ-acetylenic-GABA. Release of GABA was
determined in 8 minute fractions of superfusate collected
immediately prior to, and for up to 18 h after, injection of
γ-acetylenic-GABA (100 mg/kg i.p.). Release was measured from
sensorimotor cortex (■) and visual cortex (●). Results are mean
± S.E.M. for the number of animals indicated. (From Ref. 2).

GABA itself was released at a rate which became steady by 20 minutes
(2200 pmol/min/sq.cm cortex) (Fig. 1). Two hours after a single i.p.
injection, γ-vinyl-GABA release began to decline and was at low
levels after 3 h. However, GABA release was still high at this time
(6 pmol/min/sq.cm) (Fig. 1).

        Following a single intraperitoneal dose of γ-acetylenic-GABA
(100 mg/kg), the mixed peaks of GABA and γ-acetylenic-GABA (we were
not able to separate them) increased, gradually reaching a maximum
within 25-45 minutes (40-45 pmol/min/sq.cm cortex). The changes
were reversible and were close to zero after 2 h (Fig. 2). Neither
drug influenced the "resting" release from the cortex of the other
amino acids measured (Fig. 3).

        The possible breakdown of γ-vinyl-GABA and γ-acetylenic-GABA to
GABA was examined. For this purpose synaptosomes (3 mg protein/ml)
were incubated at 37°C for 20 minutes. Tissue was then sedimented
and lysed in cold water (1.5 ml water/3 mg protein). Membranes and
lysate were collected and stored. The drugs were then incubated for
5 minutes at 37°C with each of these fractions including the

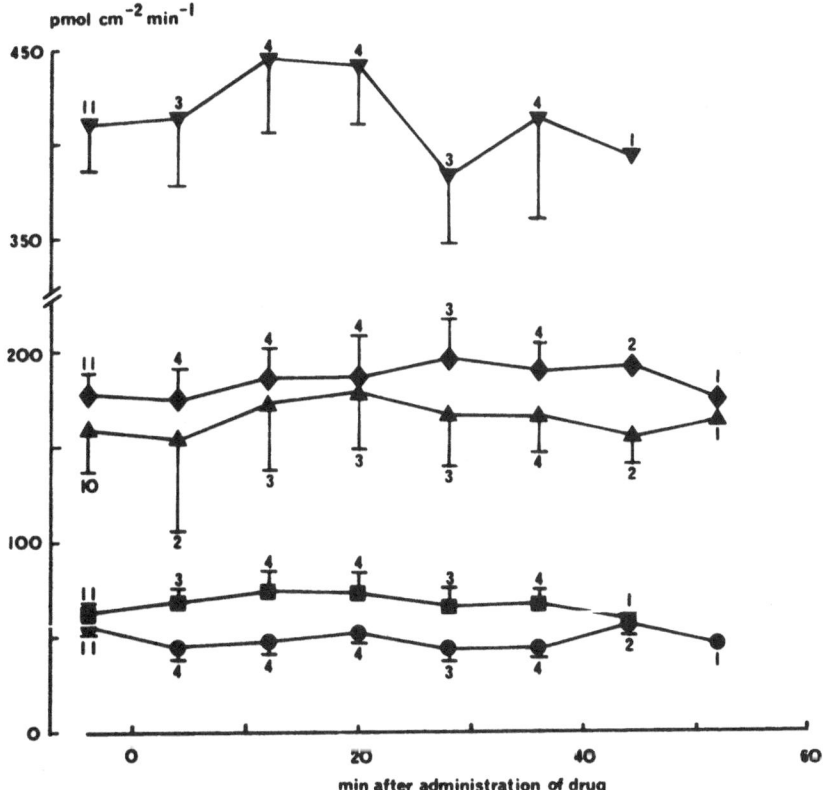

Fig. 3. Release of amino acids from sensorimotor cortex following administration of γ-acetylenic-GABA. Release of amino acids from sensorimotor cortex was determined in 8 min (1 ml) fractions of superfusate collected immediately prior to, and up to 56 min after injection of γ-acetylenic-GABA. Glutamate, ● ; valine, ■ ; alanine, ▲ ; glycine, ◆; glutamine and serine ▼. (From Ref. 2).

original incubation medium. No detectable GABA was generated in this period. Also, γ-vinyl-GABA was introduced directly onto the cortex through the cannula line at 100 µM continuously for 170 minutes. No larger GABA peak was observed (4 pmol/min/sq.cm) than followed i.p. injection of γ-vinyl-GABA (i.e. 9 pmol/min/sq.cm^2; Fig. 1), when lower concentrations prevailed in tissue and superfusate.

The possibility that the GABA present was a contaminant of the drug samples was examined by analysing 5 mM samples of γ-vinyl-GABA under conditions where 5-50 pmol of GABA could be detected. None was found. This implies that any contamination was less than one in 10^5 parts of the drug.

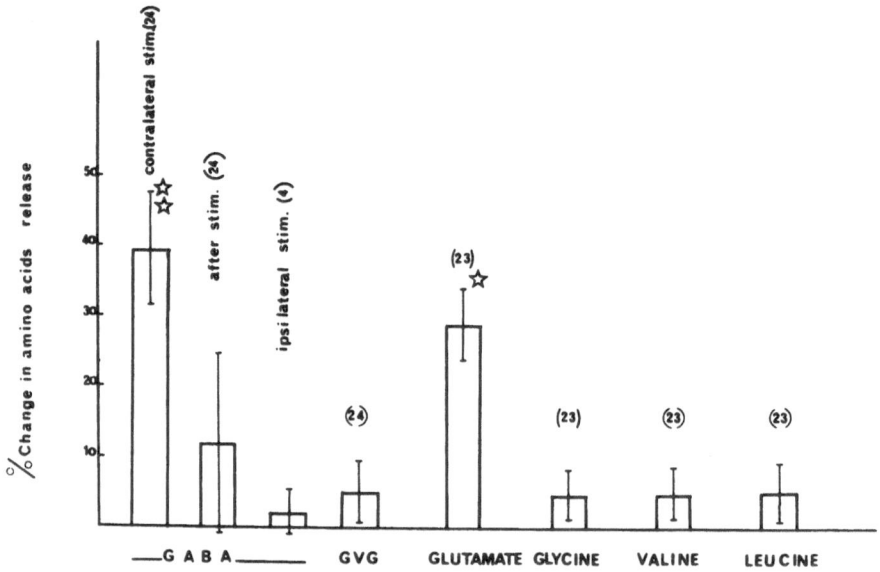

Fig. 4. Release of GABA from sensorimotor cortex during brachial
plexus stimulation. Fractions of superfusate (8 min, 1 ml) were
collected from animals after a single intraperitoneal injection
of γ-vinyl-GABA (1500 mg/kg). After at least 60 minutes, brachial
plexus stimulation was applied (see legend to Figure 5) to either
side as indicated. Histograms represent percentage changes relative
to unstimulated levels. Values are mean ± S.E.M. for the number of
stimulations indicated in brackets. *P<0.05, **P<0.001. (From Ref.2).

Brachial plexus stimulation and amino acid release from
sensorimotor cortex. GABA was not normally detected in superfusates
and stimulation of the brachial plexus of either forelimb was also
without effect on its release from sensorimotor cortex. However,
when stimulation was applied to the plexus of the contralateral
limb of anaesthetized animals 30-120 minutes after i.p. administration of γ-vinyl-GABA (when spontaneous GABA release was well
established) a substantial increase (40%, P<0.001) in GABA release
was observed (Fig. 4). This effect was reversible, and was not
produced by ipsilateral stimulation. Using the same stimulus
parameters, no significant changes were obtained in the rate of
release of γ-vinyl-GABA.

At 25-45 minutes after i.p. injection of γ-acetylenic-GABA,
stimulation of the contralateral brachial plexus gave an increase
(37%, P<0.001) in the rate of release of the mixed peaks of GABA
and γ-acetylenic-GABA (Fig. 5). No significant increase in the rate
of GABA efflux from sensorimotor cortex followed the ipsilateral

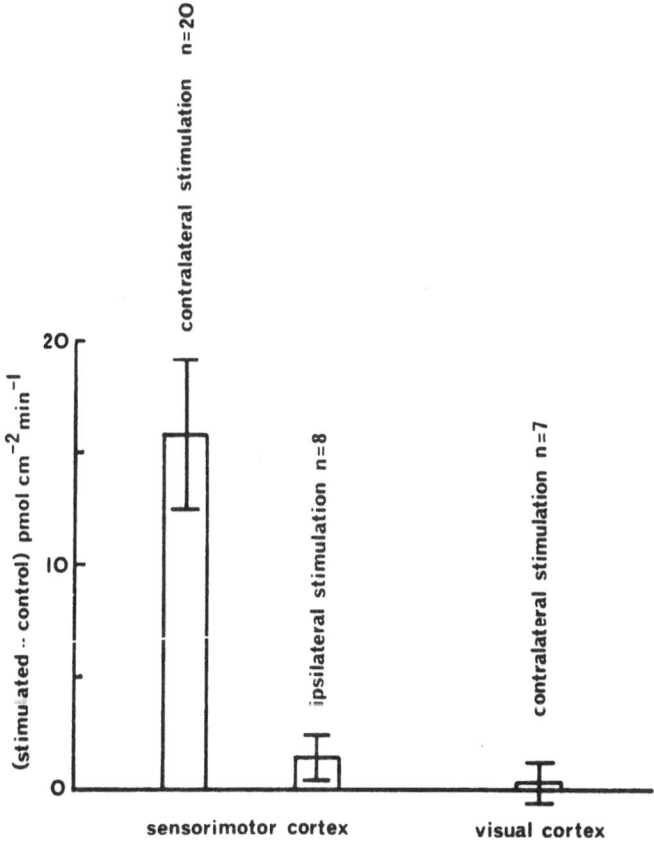

Fig. 5. Effect of brachial plexus stimulation on the release of GABA
from two regions of cerebral cortex following administration of
γ-acetylenic-GABA. In the case of sensorimotor cortex, 5 minute
periods of brachial plexus stimulation (5-12 V, 3-7 mA, 1 m sec
pulse duration, 2-3 or 50 Hz) were given between 25 and 45 min after
injection of γ-acetylenic-GABA (100 mg/kg i.p.). In the case of
visual cortex, stimulations were given 16-40 min after injection of
drug. Results shown are the means of the differences between release
of GABA during stimulation and release during an equivalent control
period preceding stimulation expressed as mean $\pm$ S.E.M. Only
contralateral stimulation of sensorimotor cortex produced a
significant increase in GABA release (37%, P<0.001 by paired t-test).
The average level of GABA (plus γ-acetylenic-GABA) in the pre-
stimulus period was 42.3 $\pm$ 4.2 (20) pmol/min/sq.cm (contralateral)
and 42.8 $\pm$ 1.2 (8) (ipsilateral) for sensorimotor cortex; and 30.1 $\pm$
2.1 (7) for visual cortex. (From Ref. 2).

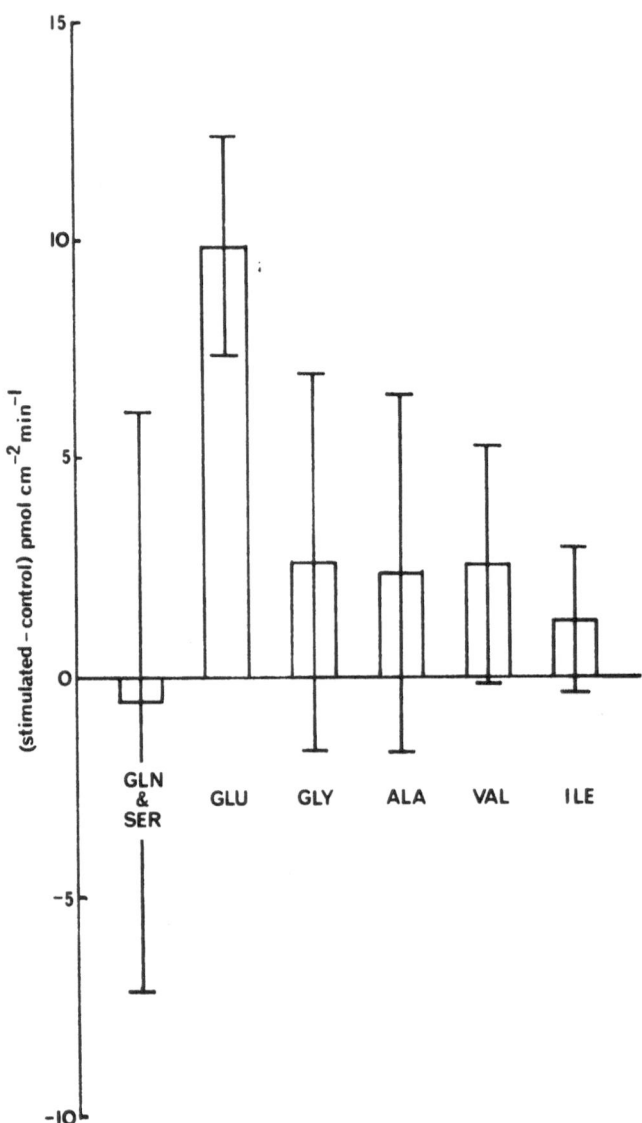

Fig. 6. Effect of brachial plexus stimulation on release of amino acids from sensorimotor cortex following administration of γ-acetylenic-GABA. Conditions as in Fig. 4; results are presented similarly (n = 14 control and stimulated pairs for each amino acid). Only glutamate release was significantly increased by contralateral afferent stimulation (19%, P<0.001). The average level of glutamate was 50.8 ± 3.9 (14) pmol/min/sq.cm and of glutamine + serine was 332 ± 22 (14) during the pre-stimulus period. (From Ref. 2).

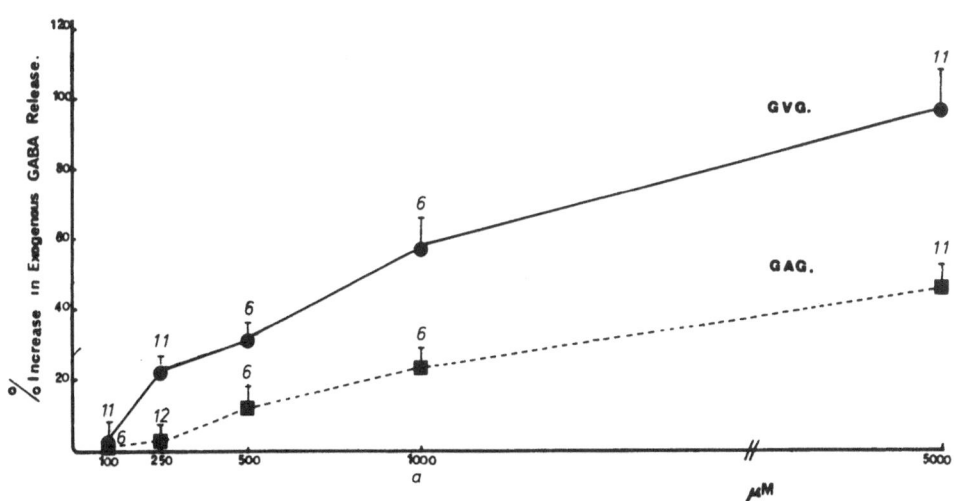

Fig. 7. Effect of γ-vinyl-GABA (GVG) and γ-acetylenic-GABA (GAG)
on release of preloaded [U14C] -GABA. Experimental details are as
for Fig. 2. The values are mean ± SEM for the number of experiments
indicated above the data points. (From Ref. 5).

stimulation. Also, no increased GABA release from superfused visual
cortex followed stimulation of the brachial plexus of either limb,
even though the resting rates of GABA release were elevated by
γ-acetylenic-GABA.

With both GABA-T inhibitors, γ-vinyl-GABA and γ-acetylenic-GABA,
the rate of release of glutamic acid following contralateral
stimulation of the brachial plexus was increased significantly by
30% (P<0.05) and 19% (P<0.01) respectively (Figs. 4 and 6). All
the changes were reversible at the cessation of the stimulation.
Under the same conditions of stimulation no clear changes were
obtained with any of the other amino acids measured, i.e. aspartate,
glutamine, serine, glycine, alanine, valine, leucine, tyrosine and
phenylalanine, all measured at the 50 pmol level.

Release of preloaded [U14C] -GABA from synaptosomes. γ-Acetylenic
GABA γ-vinyl GABA at 1 mM concentrations caused a significant
increase of preloaded [U14C] -GABA release from incubated synaptosomes
corresponding to 23% (p<0.025) and 57% (p<0.005) respectively
(Fig. 7). This effect increased with drug concentration to 46% with
γ-acetylenic GABA (5 mM) and to 96% with γ-vinyl GABA (5 mM). The
minimum effective dose with γ-acetylenic GABA was 0.5 mM (12%) and
with γ-vinyl GABA, 0.25 M (22%).

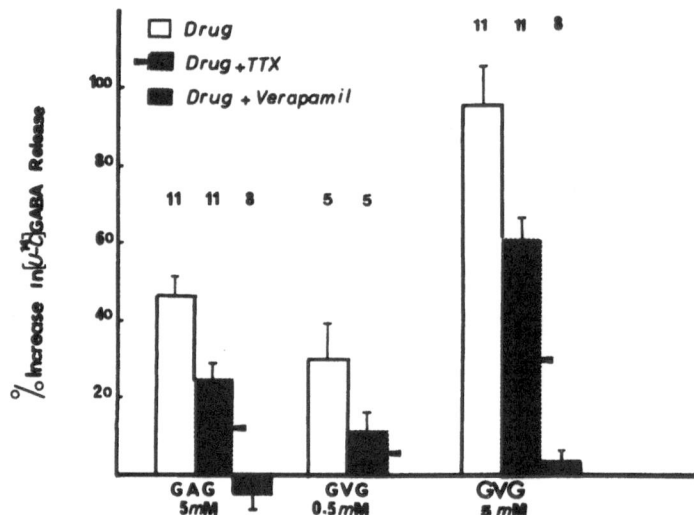

Fig. 8. Effect of tetrodotoxin and Verapamil on the release of preloaded [U14C] -GABA from synaptosomes incubated with GABA-analogues.
Synaptosomes were incubated (3 mg protein/ml phosphate medium) for
20 min in the presence of [U14C] -GABA (final specific activity 0.25
µCi/ml) and sedimented in a bench centrifuge prior to resuspension
in phosphate medium containing either Ca++ or 100 µM Verapamil.
These suspensions were incubated for 30 min., before addition of TTX
(1 µM), γ-vinyl GABA (0.25-5 mM), or γ-acetylenic GABA (0.5-5 mM)
as appropriate. Incubation continued for a further 10 min. prior to
sedimentation of the synaptosomes in a bench ultracentrifuge. The
supernatants were decanted into radioactive counting vials and the
pellets were extracted with 10% TCA. The extracted supernatants were
also decanted into vials. A toluene based liquid scintillant was
added, and the vials were taken for radiolabel counting. Data
represent mean ± SEM for the number of experiments indicated above.
(From Ref. 5).

        Tetrodotoxin (1 µM) added to the incubated synaptosomes one
minute before γ-acetylenic GABA (5 mM) and γ-vinyl GABA (5 mM and
0.5 mM) significantly inhibited the release of [U14C] -GABA by 46%
35%, and 60% respectively (Fig. 8). The release of [U14C-] GABA from
synaptosomes incubated in the presence of either drug at 5 mM concentration was inhibited in the absence of Ca++ when Verapamil was
added to 100 µM.

        Release of endogenous GABA from synaptosomes. When γ-vinyl GABA
at a final concentration of 5 mM was added at the start of a 5 min
incubation period to suspensions of cerebral cortical synaptosomes,
GABA was greatly increased in concentration in the medium, from

78 ± 41 nmoles/100 mg protein to 593 ± 60 nmoles/100 mg protein.
However, no change occurred in the release of other amino acids.
Similar results were obtained with γ-acetylenic GABA, although it
was not possible to separate γ-acetylenic GABA from GABA itself in
the analytical system employed.

Following intraperitoneal administration of γ-acetylenic GABA
(100 mg/kg) or γ-vinyl GABA (1000 mg/kg) for periods of 6 and 14
hours respectively, animals were sacrificed and cerebral cortical
synaptosomes were prepared and incubated in a Krebs-phosphate medium
for 30 min. before addition of veratrine and/or tetrodotoxin. The
release of endogenous GABA from synaptosomes of animals pretreated
with one of the drugs was much higher than from controls treated
with saline. No change in release was observed in any of the other
amino acids measured.

In the presence of depolarising concentrations of veratrine
(Fig. 9), a large increase in the release of glutamate, aspartate
and GABA occurred. Other amino acids remained unaffected. However,
this well established veratrine-stimulated release of putative neurotransmitter amino acids was larger (150-200 %) when the animals
were pretreated with γ-acetylenic GABA or γ-vinyl GABA. The
stimulated release of endogenous glutamate, aspartate and GABA in
all cases was totally inhibited when the synaptosomes were incubated
with tetrodotoxin (1 μM; Fig. 9).

Measurement of the total amino acid content of cerebrocortical
synaptosomes from treated and untreated rats showed that γ-acetylenic
GABA and γ-vinyl GABA administered in vivo, caused an increase of
3 to 4-fold in the content of GABA in the synaptosome preparation.
No detectable changes were found with other amino acids. Veratrine

---

Fig. 9. Influence of the GABA-analogues on endogenous amino acid
release. Fourteen hours prior to synaptosome preparation, rats were
injected with γ-vinyl GABA (1000 mg/kg) or γ-acetylenic GABA (100 mg/
kg, 6 hrs) or with equivalent volumes of saline. Cerebral cortical
synaptosomes were subsequently prepared and incubated (3 mg protein/
ml medium) in Krebs-phosphate medium for 30 min. prior to addition of
TTX (1 μM) or Veratrine (75 μM) as appropriate. Incubation continued
for a further 10 min. Synaptosomes were sedimented on a bench
ultracentrifuge, and the supernatants and pellets were extracted with
10% TCA. The TCA extracts were purified on a semi automatic ion
exchange apparatus to collect amino acids. The eluates containing
amino acids were taken to dryness on a Buchler Evapomix. Samples
were then dissolved in 0.025M HCl and analysed on an automatic amino
acid analyser (19). Values are mean ± SEM for four experiments. (From
Ref. 5).

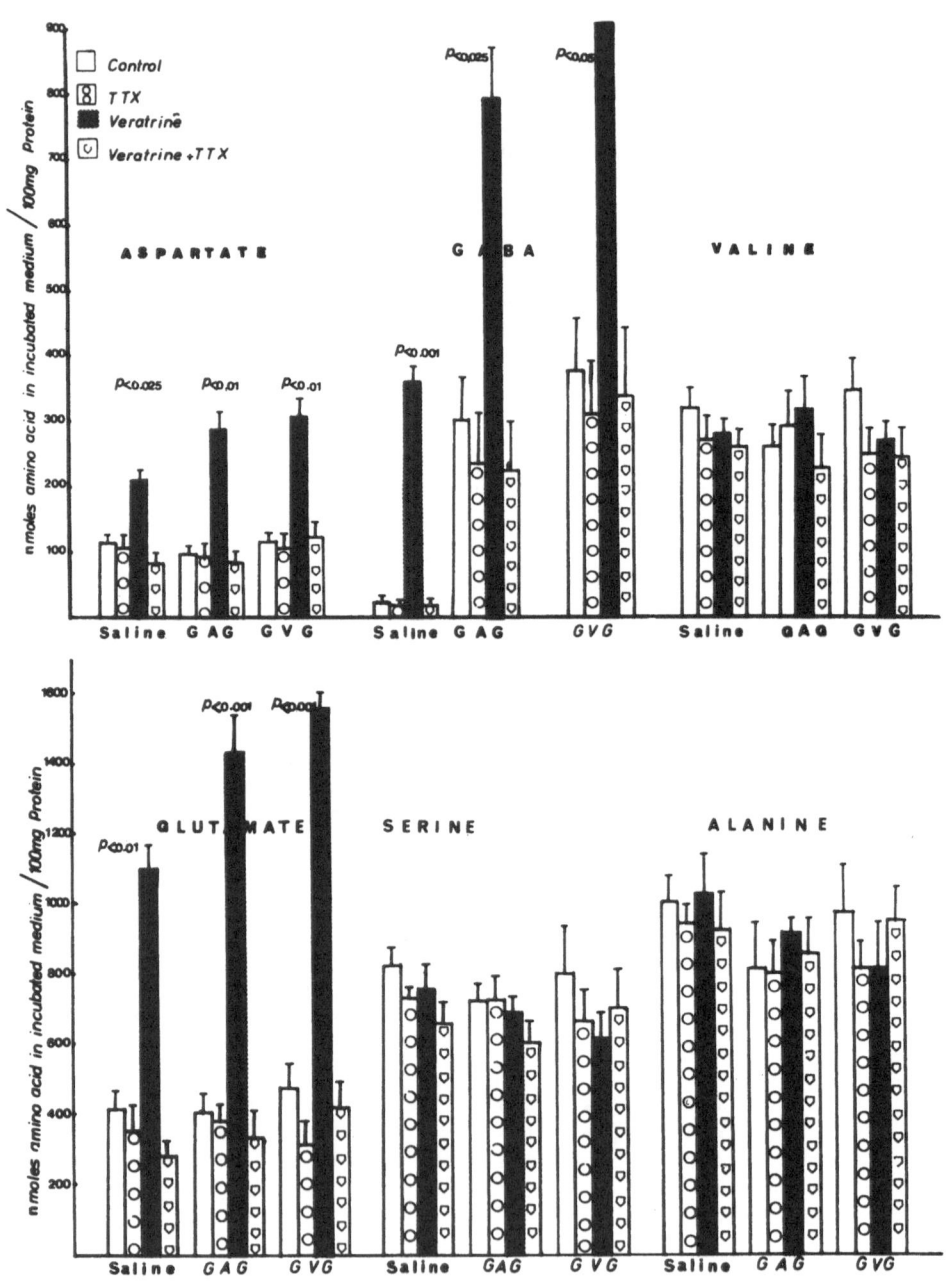

Fig. 9. The legend is in the preceding page

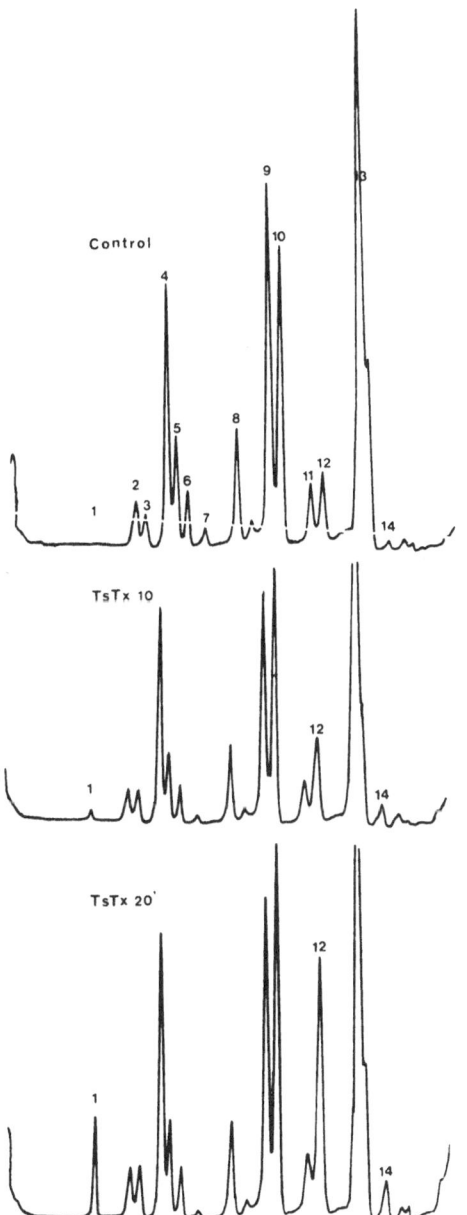

Fig. 10. Patterns of amino acids released from superfused rat sensorimotor cortex. Representative chromatograms of control and stimulated samples collected 10 min and 20 min after beginning super-fusion with saline containing tityustoxin (1 μM) are shown. Amino acids: 14, asparate; 13, threonine, glutamine and serine; 12, glutamate; 11, citrulline; 10, glycine; 9, alanine; 8, valine; 7, methionine; 6, isoleucine; 5, leucine; 4, norleucine (50 pmol/100 μl of sample); 3, tyrosine; 2, phenyl-alanine; 1, GABA. (From Ref. 14).

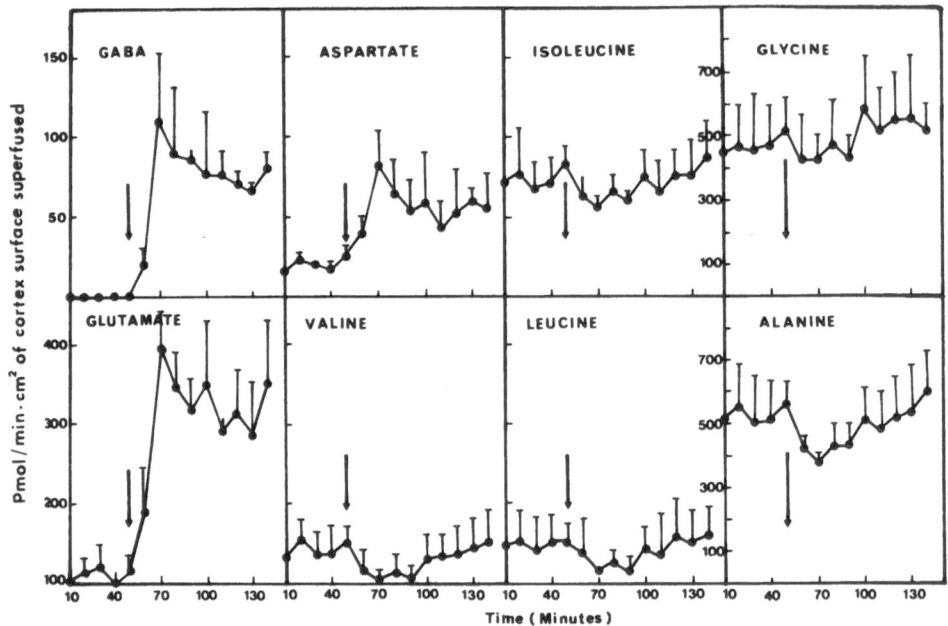

Fig. 11. Release of amino acids from superfused rat sensorimotor
cortex during superfusion with saline containing tityustoxin. Five
1 ml control samples (10 min each) were collected before tityustoxin
(1 µM, final concentration) was added to the superfusion fluid
(indicated by the arrow). The values represent the mean ± S.D. for
7 animals superfused, one observation being made with each animal.
(From Ref. 14).

stimulation lowered the amounts of both GABA and aspartate in the
tissue in parallel to increasing their release to the medium.

   [U14C] -GABA uptake. In addition to their effect on its release,
γ-acetylenic-GABA and γ-vinyl-GABA at 1.0 mM concentrations caused
a significant (p<0.001) inhibition of [U14C]-GABA uptake corresponding
to 21% and 36% respectively. This inhibition was increased with
rising drug concentration to a plateau of 70% with γ-vinyl GABA in
the 3-5 mM range, and up to 50% with γ-acetylenic GABA. No significant
effect on uptake was observed with concentrations below 0.5 mM of
either drug.

   Similar effects on uptake were obtained after 1.0 min. and 2.5
min. of incubation with [U14C] -GABA using both drugs and both
concentrations. This apparent inhibitory action of the drugs was
reduced by only 30 to 50% by the addition of tetrodotoxin or
Verapamil (to block Ca^{2+} entry). In view of this, a true transport

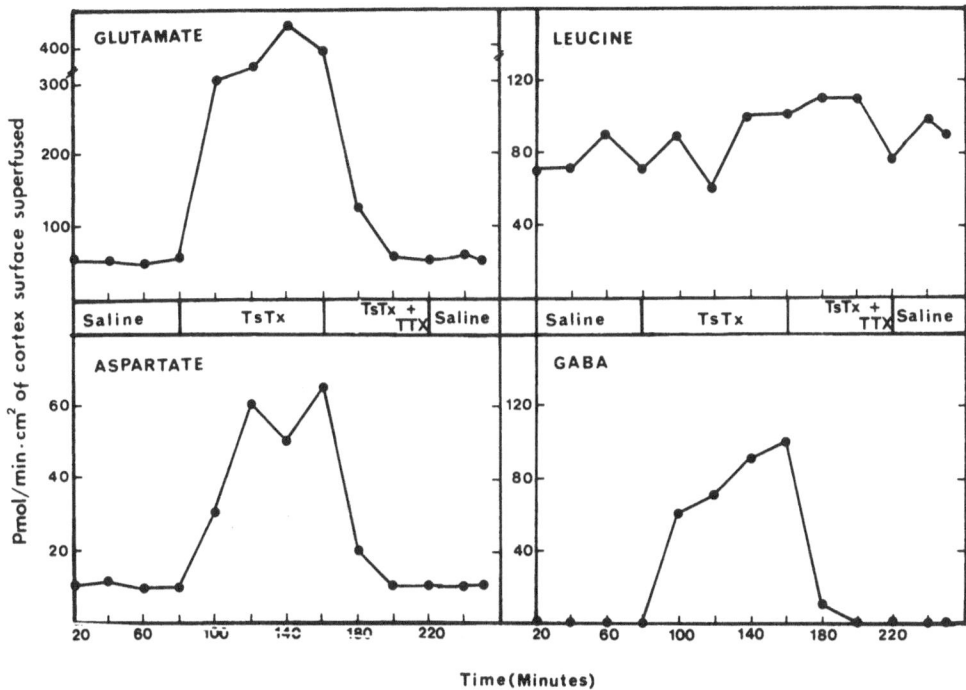

Fig. 12. Effect of tetrodotoxin on the amino acid-releasing action of
tityustoxin. In each animal eight 1 ml control samples were collected
(10 min. each) from sensorimotor cortex before tityustoxin (1 µM
final concentration) was added to the superfusion fluid for 80 min.
Then, both tetrodotoxin (0.1 µM) and tityustoxin (1 µM) were
introduced for 60 min. This superfusate mixture was then replaced by
saline. The values given represent the mean of 3 values from three
animals superfused. Only alternate samples were analysed for amino
acid content. (From Ref. 14).

block in addition to a GABA-releasing action during the same 1 to 5
minute period appears to be responsible for the effects observed.

Tityustoxin

        Behavioural effects of in vivo administration of toxin. About 5
min after introducing tityustoxin into the superfusion stream of
awake animals (1 nmol/ml superfusate) the rats showed characteristic
movements which included frequent manipulation or scratching of the
cannula, principally with the left forelimb. The cannula was
positioned over the right sensorimotor cortical region. Other
responses included intensive grooming movements and chattering of
the jaw. Myoclonic jerks of the left forelimb began at about 15 min
after administration of the toxin, and their frequency increased

Table 1. Effect of tityustoxin and veratrine on the respiratory rate
and potassium content of cortical synaptosomes

	Potassium (μ equiv./100 mg Synaptosomal Protein)	Respiratory Response (% Control)
Control	33 ± 3 (6)	100 (18)
Veratrine	23 ± 4 (6)	176 ± 34 (6)
Tityustoxin	22 ± 3 (6)	377 ± 96 (5)

Synaptosomes were incubated at 37°C for 35 min in Krebs-phosphate
medium. Tityustoxin (50 nM) or veratrine (75 μM) were then added and
incubation continued for a further 10 min. Synaptosomes were
deposited by centrifugation and the K^+ content of the synaptosomes
was measured. Respiration was measured for 20 min. after equilibration. Agents were then added from the side arm and respiration
continued for 10-15 min. The data show the respiratory rate during
the second period as a percentage of that during the first period.
Values are means ± S.D. of samples in parentheses.

with time. This focal pattern of epileptic limb-jerking continued
throughout the period of infusion of the toxin (60-90 min.), but
stopped entirely 20-30 min after introducing toxin-free saline.

In vivo release of amino acids. During the period of toxin
administration, the superfusate showed a gradual but large increase
in its content of only three of the thirteen endogenous amino acids
measured. These were the putative amino acid neurotransmitters
glutamate, aspartate and GABA (Figs. 10 and 11). Since the toxin
preparations were themselves free from contamination by these
compounds, this increase was clearly contributed by the sensorimotor
cortex itself. Thus, saline containing 1 μM toxin contained very
small quantities of phenyl-alanine (2 pM), other amino acids being
absent. The release of glutamate and GABA from sensorimotor cortex
were by far the largest changes observed, the release of aspartate
being smaller but still significant (p<0.001) (Fig. 10). Other
changes which were significant were decreases in valine (p<0.001),
leucine (p<0.001), alanine (p<0.001) and isoleucine (p<0.01) release
during the first 40 min.

Action of tetrodotoxin. After applying tityustoxin and
observing the pattern of response described above, a mixture of

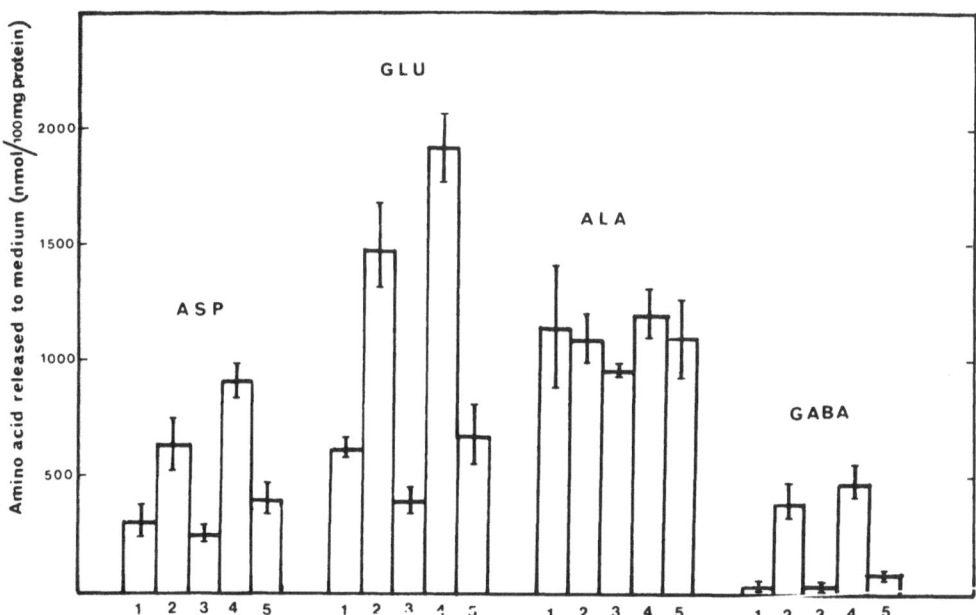

Fig. 13. Effect of Ca^{2+} and tetrodotoxin on the amino acid-releasing action of tityustoxin on rat cortical synaptosomes. Synaptosomes were incubated at 37°C for 35 min. in 3 ml of Krebs-phosphate medium with Ca^{2+} (or without Ca^{2+} + 0.5 mM EGTA); tityustoxin (50 nM) or tityustoxin (50 nM) + tetrodotoxin (1 μM) were then added and incubation continued for a further 10 min. Synaptosomes were then centrifuged and amino acids were determined in the supernatant. Values represent the mean $\pm$ S.D. for 2 or 3 determinations in different experiments, as follows: 1 = control - Ca^{2+}; 2 = TsTx (50 nM) - Ca^{2+}; 3 = control + Ca^{++}; 4 = TsTx (50 nM) + Ca^{2+}; 5 = TsTx (50 nM) + TTx (1 μM) + Ca^{2+}. Numbers of samples were as follows: condition 1 = 3 values; 2 = 3 values; 3 = 3 values; 4 = 9 values; 5 = 3 values. (From Ref. 14).

tityustoxin (1 μM) and tetrodotoxin (0.1 μM) was introduced. The myoclonic limb jerks either decreased dramatically in frequency (to 10% or 1%) or ceased entirely. The animals became quiet and motion-less during the 40 min period of infusion. Subsequent 20 or 40 min periods of superfusion with saline alone resulted in the return of the contralateral limb jerks to their original frequency for 10-15 min before they ceased again entirely. Tetrodoxin applied alone (0.1 μM) in saline also caused the animals to become quiescent and reduce their locomotor activity. This observation was repeated 2 or 3 times in each of the 3 animals.

When the pattern of release of amino acids was examined, it was

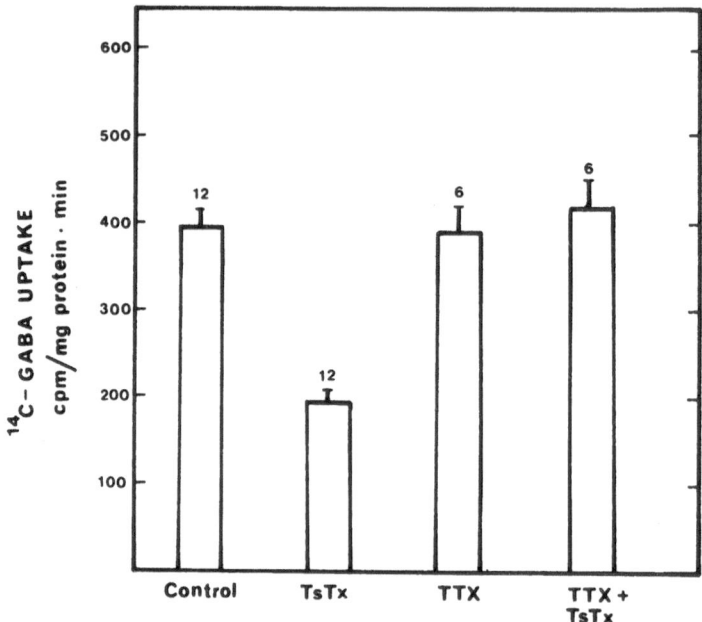

Fig. 14. Effect of tityustoxin on the uptake of $[U^{14}C]$ -GABA by rat cortical synaptosomes. Synaptosomes were incubated at 37°C for 20 min in Krebs-phosphate medium containing 10 mM glucose and where indicated tityustoxin (100 nM) alone, tetrodotoxin (1 μM) alone, or a mixture of the two. At the end of this period $[U^{14}C]$ -GABA was added and incubation continued for a further 5 min before centrifugation to deposit synaptosomes which were then analysed for their ^{14}C content. Values are mean $\pm$ S.D. for 12 or 6 experiments as indicated. Differences between control, tetrodotoxin added, and tetrodotoxin plus tityustoxin added, are insignificant. Difference between control and tityustoxin added is significant with p<0.001 (From Ref. 14).

found that during the period when both tityustoxin and tetrodotoxin were being applied together, no enhanced release of physiologically active amino acids occurred (Fig. 12). Tetrodotoxin applied alone did not influence the pattern of amino acid release.

Action of tityustoxin on incubated synaptosomes. Synaptosomes were preincubated for 35 min. before addition of the toxin (50 nM) for periods of 10 min. This caused a 33% loss of K^+ content and a 277% stimulation of respiratory rates (Table 1). Veratrine (75 μM) caused a similar effect on K^+ and a somewhat smaller (76%) effect on respiration over the same time period (Table 1).

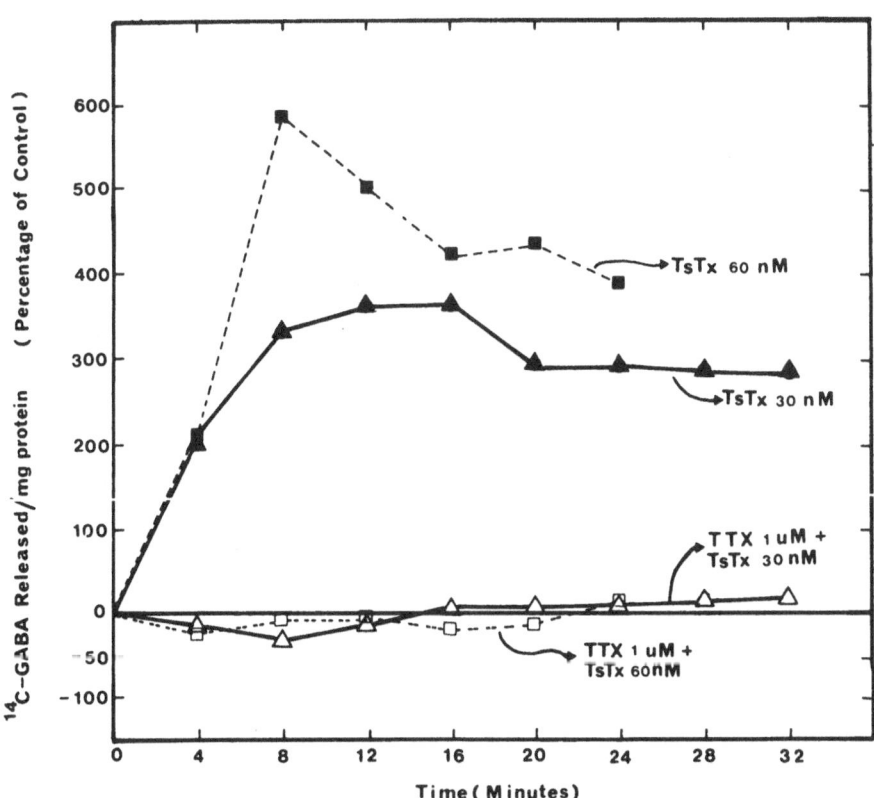

Fig. 15. Effect of tityustoxin (TsTx) concentration and time of action
on the release of $[U^{14}C]$ -GABA from rat cortical synaptosomes.
Synaptosomes were incubated in Krebs-phosphate medium at 37°C for
15 min. with 30 µCi of $[U^{14}C]$ -GABA and were then centrifuged, washed,
and resuspended in fresh medium. Following a further 15 min. incubation period, TsTx (30 and 60 nM) or tetrodotoxin (TTX; 1 µM) + TsTx
(30 nM or 60 nM) were added, and at 4 min intervals, a sample was
removed and centrifuged to deposit synaptosomes. Radioactivity was
determined in the supernatant. TTX was added in all experiments 10
sec before TsTx. Results represent the change in medium content of
$[U-^{14}C]$ -GABA expressed as a percentage of control levels. Values
are mean for two experiments involving 4 samples for each condition.
(From Ref. 14).

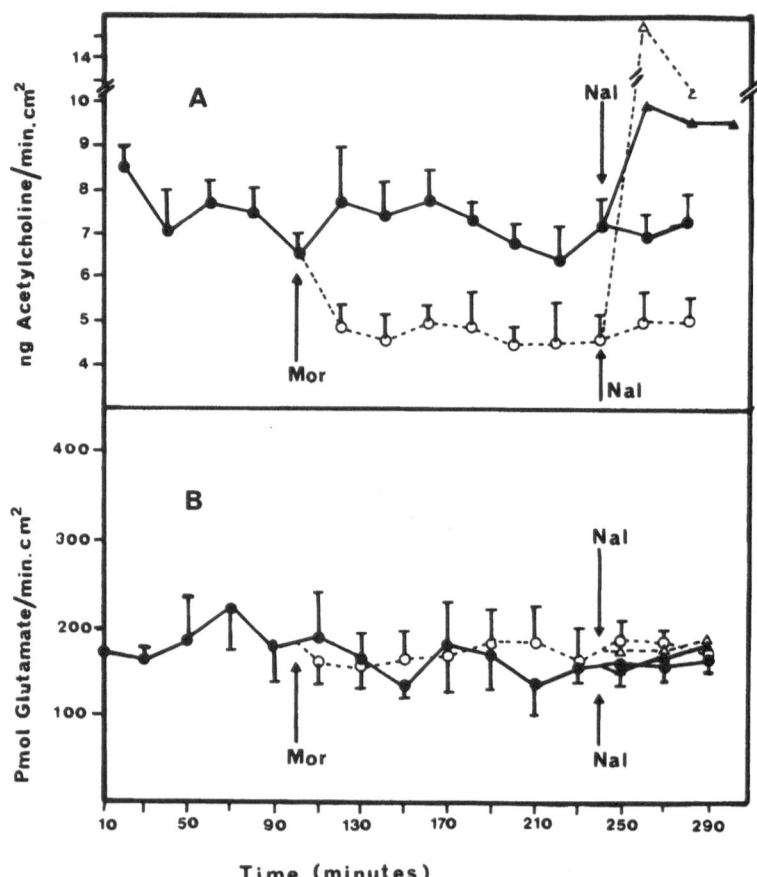

Fig. 16. Effects of morphine and naloxone on spontaneous release of
acetylcholine (A) and glutamate (B) from superfused rat sensorimotor
cortex. Morphine (Mor; 10 mg/kg) and naloxone (Nal; 2 mg/kg) were
administered intraperitoneally at the points indicated. ● = control
(n = 7); 0 = morphine treated (n = 7); ▲ = naloxone treated control
controls; Δ = naloxone given after morphine (n = 2). Data represent
the amount of neurotransmitter released to superfusion fluid over a
10 min period ± S.E.M. (flow rate = 6 ml/h). (From Ref. 13).

     Measurement of amino acids in synaptosomes and incubation media
after exposure to the toxin showed a similar pattern of release to
that observed in vivo. Thus, of the thirteen amino acids measured,
only glutamate, GABA and aspartate showed an increase in the extent
of their release, glutamate and GABA showing the largest
proportional changes (Fig. 13). The lack of effect on alanine is
given as representative of the other amino acids measured.

When tissue levels of thirteen amino acids were measured after incubation, no significant fall in the levels of glutamate, GABA and aspartate acids could be detected. This suggests that considerable resynthesis of amino acids occurs in the tissue during the 30 min. incubation period to compensate for their release. When Ca^{2+}-free, EGTA (0.5 mM)-containing media were used, the action of tityustoxin was considerably reduced. Thus, release of glutamate (42%; p<0.01), GABA (8%; p<0.1) and aspartate (48%; p<0.01) were all diminished, and there was a tendency (p<0.05) for control levels in the absence of Ca^{2+} to be raised (Fig. 13). Tetrodotoxin (1 µM) prevented the releasing action of tityustoxin on all three transmitter amino acids, but was without action in the absence of the toxin (Fig. 13).

$[U^{14}C]$-GABA uptake. In the presence of tityustoxin synaptosomes in suspension showed a reduced capacity to accumulate $[U^{14}C]$-GABA over a 5 min period. This effect was found to be maximal at about 30 nM tityustoxin in dose-response studies, and this concentration produced 42 ± 9 (3)% inhibition of GABA (at 3 µM) uptake over 5 min and 49 ± 5 (3)% inhibition of uptake at 10 min; no greater inhibition was observed during the subsequent 25 min. A reduction of GABA uptake into the tissue measured in this way over short incubation periods would normally be attributed to an inhibition of uptake by the toxin. However, tetrodotoxin prevented this apparent uptake inhibition by the toxin, suggesting that the effect was due to depolarisation-induced release preventing the accumulation of $[U^{14}C]$-GABA from occurring (Fig. 14). This was demonstrated to be the case by examining the effect of tityustoxin over an equivalent 5 min period on the release of $[U^{14}C]$-GABA previously accumulated in the absence of toxin (Fig. 15). These experiments showed that 75% of the $[U^{14}C]$-GABA release occurred during the first 5 min of exposure to the toxin. A comparison of the time course of the inhibition of $[U^{14}C]$-GABA uptake and its release from preloaded synaptosomes (Fig. 15) shows that the two events are very similar, strongly supporting the conclusion that the $[U^{14}C]$-GABA-releasing property explains the apparent inhibition of uptake by tityustoxin. The same conclusion was reached by Dolly et al. (21) using a similar experimental approach. These authors also showed that 70% of the radioactivity released by tityustoxin was recoverable as $[^{14}C]$-GABA, which means that in the present experiment GABA was the major substance being counted in the incubation medium.

Actions of Morphine and Naloxone in Vivo

Morphine given i.p. immediately depressed the spontaneous release of ACh, an effect which lasted for at least 3 hours (Fig. 16A), but it did not affect the spontaneous release of glutamate (Fig. 17A) or any of the other amino acids measured (i.e. aspartate, GABA, glutamine, serine, threonine, glycine, alanine, valine methionine, leucine, tyrosine and phenylalanine). In fact, one

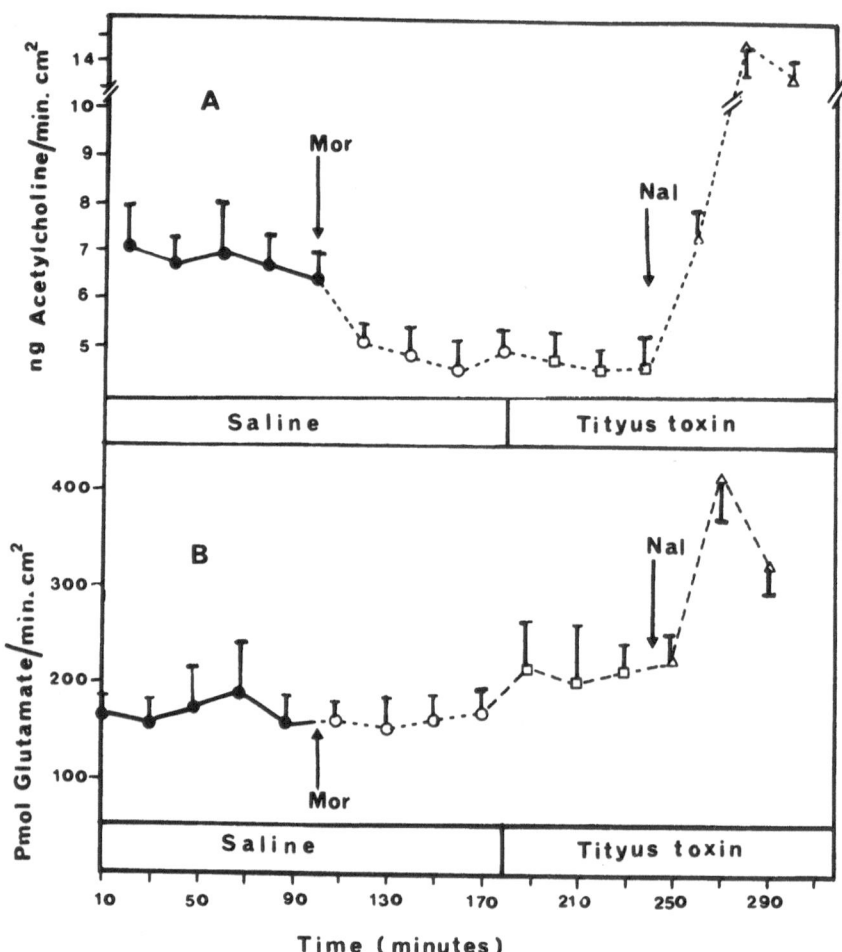

Fig. 17. Effect of morphine and naloxone on tityustoxin-evoked release
of acetylcholine (A) and glutamate (B) from superfused rat sensorimotor cortex. Superfusion was started with saline + Ca^{2+} and the
same plus tityustoxin (1 µM) was introduced into the superfusion
stream where indicated. Morphine (Mor; 10 mg/kg) and naloxone (Nal;
2 mg/kg) were administered intraperitoneally when indicated by the
arrows. ● = control (n = 7); 0 = morphine treated (n = 7); □ =
tityustoxin added (n = 3); Δ = naloxone (n = 2). Data represent the
amounts of each transmitter collected during 10 min of superfusion
± S.E.M. (flow rate = 6 ml/h). (From Ref. 13).

out of seven animals showed a significantly increased release of glutamate, aspartate and GABA, without change in other amino acids.

In morphine-treated rats, naloxone given i.p. produced a large rebound effect (34), with spontaneous ACh exceeding its original control levels (Fig. 16A). Again, no effect on amino acid release was detected (Fig. 17A). However, in both of the two experiments performed, naloxone given alone caused some increase (25%) of spontaneous ACh release (Fig. 16A).

As described in the previous section, tityustoxin evokes release of transmitter amino acids from sensorimotor cortex when introduced into the superfusion stream. It also releases acetylcholine (ACh) (13). This is accompanied by stereotyped grooming movements of the forelimbs, jaw-chatter and myoclonic jerks of the contralateral forelimb which began 10 min after introduction of the toxin and continued during its application. All of these effects were abolished by tetrodotoxin.

In the present experiments, morphine given i.p. (10 mg/kg) entirely prevented the release to the superfusion stream of ACh and transmitter amino acids induced by tityustoxin (Figs. 16B, 17B). These actions of morphine were fully reversed by naloxone given i.p. (2 mg/kg) at various times during tityustoxin infusion (e.g. 140 - 150 min after morphine; Figs. 16A, 16B, 17B).

The behavioural and physiological responses showed a correlating pattern. Thus, in the presence of morphine, tityustoxin was unable to produce the effects on limb movements and behaviour described above. However, following naloxone administration, all of these characteristic effects reappeared, demonstrating that they had been evoked by the agents (presumably the neurotransmitters) released by the toxin.

When morphine (131 µM) was added to incubated synaptosomes it appeared to have no influence on the extent of amino acid neurotransmitter release induced by tityustoxin (50 nM). This was in contrast to its actions in vivo when given i.p.

DISCUSSION

Effects of Alkyl Analogues on Spontaneous GABA Release

Unlike glutamate and aspartate, GABA is normally present in superfusion fluid only at the limits of detection (1,11,18,31,32) and is mostly reported as absent.

The flow rates used here (about 10 ml/h) did not allow detection of GABA until the animals were treated with γ-vinyl-GABA or

γ-acetylenic-GABA, the irreversible inhibitors of GABA-T, that
elevate GABA levels in the brain up to 5-fold (35,36). The low
toxicity of these drugs make them ideal research tools for the
purpose of raising tissue GABA levels and studying changes in the
inhibitory transmitters in different physiological states.
Unfortunately, the specificity (35-38, 41,51) of these agents is
not as great as was first thought and this must limit the extent
to which their effects can be ascribed to an action on GABA-
transaminase, or even on the GABA system. However, after intraperitoneal injection of a single dose of γ-vinyl-GABA (1500 mg/kg),
GABA, but not other amino acid measured, was increased to a steady
rate of release. The pattern of GABA release was different to that of
γ-vinyl-GABA itself. Thus, GABA reached its maximum rate of release
after 50 minutes, whilst γ-vinyl-GABA achieved this after only 20
minutes. Also, the rate of decline of concentration of γ-vinyl-GABA
was considerably faster than that of GABA itself.

The GABA peak which was detected in the superfusate is unlikely
to be present due to either contamination of the drug samples or
breakdown of γ-vinyl-GABA, since incubation at 37°C for 5 min with
supernatant or with deposited tissue from lysed cortical synaptosomes,
caused no breakdown to GABA even though 5-50 pmol of GABA would have
been readily detectable. Also, no GABA contamination of the γ-vinyl-
GABA samples used could be detected at a limit of 10^{-5}% which is
well below the levels of GABA found to be reached in vivo. Since
γ-vinyl-GABA (100 μM) when introduced continuously into the superfusion fluid, produced the appearance of GABA in the outgoing
stream at a level equivalent to, or lower than, that seen to follow
intraperitoneal administration, in vivo breakdown also seems not
to be occurring.

It was not possible to separate the peaks of GABA and
γ-acetylenic-GABA in our system. However, the mixed peak of GABA
and γ-acetylenic-GABA followed the pattern for GABA alone in the
experiments with γ-vinyl-GABA, and we conclude that this drug
therefore has similar actions.

Brachial Plexus Stimulation

We have previously reported (1,2) that brachial plexus
stimulation caused consistent increases in the rate of release of
glutamate and glutamine when applied contralaterally but not
ipsilaterally, but GABA release was never detected under these
conditions.

However, after elevating GABA levels in the brain by
administering γ-vinyl-GABA, contralateral afferent stimulation of
this kind increased the rate of GABA release by 40% (i.e. from
8.31 ± 0.87 pmol/min/sq.cm to 11.92 ± 1.22), a significant change

that was reversible 10 min after cessation of the stimuli. Similar
changes were observed with the mixed peak of GABA and γ-acetylenic-
GABA after treatment with γ-acetylenic-GABA, suggesting a similar
pattern of action, i.e. increased release of GABA.

Contralateral brachial plexus stimulation also caused a
consistent and significant increase in the release of glutamate
from sensorimotor cortex under the conditions of drug treatment
used here and in their absence (1,2). Similar patterns of glutamate
and GABA release were detected in visual cortex in response to photic
stimulation (1). The present results and those of others who have
followed in vivo release (discussed in Ref. 1) complement the large
volume of vivo evidence that glutamate and GABA are neurotransmitters
in the cerebral cortex.

Implications for the Anticonvulsant Action of the Drugs

The two drugs used in this study raise brain GABA to high
levels (5-fold) and are also potent anticonvulsants. How raised
brain GABA levels lead to an anticonvulsant action is not clear since
it is unlikely to increase the rate or quantum content of
synaptically released GABA. Our results suggest that the agents have
in addition a GABA-releasing action (from neurones or glial cells in
vivo) and the extracellular GABA which results could be responsible
for the observed anticonvulsant action, through its interaction with
GABA receptors located in inhibitory synapses and elsewhere. Against
this conclusion, however, is the fact that the extracellular GABA
was only detectable in our system for 3-4 h whilst the anticonvulsant
action in mice (and also the effect on raised brain GABA levels)
lasts for some 48 h or more. It remains possible that the extracellular GABA levels remain high in deeper regions of the cortex but
are not detectable in superfusates which wash the cortical surface
and rely on diffusion from these deeper layers for detection to
occur.

Action of GABA-Analogues in Vitro

Some parallelism exists between the actions of these agents in
vivo and their actions on incubated synaptosomes. Thus, both
preloaded ^{14}C-GABA and endogenous GABA are released from synaptosomes
by γ-vinyl GABA above 0.25 mM and γ-acetylenic-GABA above 0.5 mM.
This implies that these concentrations of the drug are reached in
the brain following in vivo administration, though the longer period
of exposure in vivo may render low concentrations more effective.
Certainly, the GABA release observed was not due to breakdown of
the analogues as judged by the inability of synaptosomal lysates or
membranes to generate free GABA from the drugs.

The apparent specificity of the releasing action of the agents
towards GABA in vitro is surprising in view of its tetrodotoxin and

calcium sensitivity. This suggests that part of the action is due
to Na^+-channel activation, and this would be expected to lead
to Ca^{2+} dependent release of all neurotransmitters. A possible
explanation comes from the apparent GABA-uptake blocking action of
the drugs. If the active Na^+-channel activation capacity of the
drugs is weak then their depolarizing action would be minimal and
would lead to only moderate degrees of transmitter release. Such
relatively small signals of aspartate and glutamate could be
rapidly re-absorbed into synaptosomes by their high affinity
transport systems, but GABA would remain and accumulate in the
incubation medium because of the inhibitory action of the agents on
its high-affinity transport. Such "partial" Na^+-channel activation
has been described for other agents (e.g. the dinoflagellate toxin,
gymnodinium breve) (58). The amplified signals of GABA from
synaptosomes isolated from animals pretreated with the drugs and
treated with veratrine appears to correlate with the 3 to 4-fold
larger GABA content of these synaptosomes. However, the veratrine-
evoked release of aspartate and glutamate (but no other amino acids)
was also enhanced by this pretreatment, and to the same extent as
was GABA. This could be due to a similar "enhancing" effect of the
residual drugs on the Na^+-channel activation property of veratrine,
though more experimental evidence is required to establish the true
reason for this enhancement phenomenon.

In summary, both in vivo superfusion and studies with incubated
synaptosomes indicate that GABA-release to the extracellular space
induced by these alkyl GABA analogues is due to their capacity to:
a) raise brain GABA content (including nerve-terminal content),
which leads to an increased "spontaneous" and stimulus-evoked rate of
efflux; b) an inhibitory action on GABA uptake, and c) possibly
a mild Na^+-channel activitating property leading to a tetrodotoxin-
and-Ca^{2+}-sensitive efflux of GABA. These effects, leading to extracellular accumulation of GABA could also be responsible for the
anticonvulsant action of these compounds. Of course all of the in
vitro results were obtained with drug concentrations at 0.5 or 0.25
mM and above, lower levels being ineffective. The concentrations of
these drugs reached in the brain following intraperitoneal injection
is not known, but their hydrophobic nature and the similarities
between the in vivo and in vitro effects make it likely that the
brain levels reached are in the order of those found to be necessary
to affect incubated synaptosomes.

Actions of Tityustoxin

Application of tityustoxin directly onto the sensorimotor
cortex produced contralateral myoclonic forelimb jerks, which is the
characteristic response to similar application of established
depolarising agents, e.g. veratrine and electrical pulses, (19) and
therefore supports a depolarising action for tityustoxin. Application

of cobalt metal produces equivalent limb-jerks but over a much longer (e.g. 5-10 days) period. The reversibility of the action of tityus-toxin following saline washes suggests that the binding of the agent to its sites of action is not irreversible, and the relatively long (15 minutes) period before onset of limb jerks means either that these sites are not readily accessible from the cortex surface, or that barriers such as pial membranes, or cell membranes must be crossed before interaction can occur.

In vivo release of amino acids. The specificity of action of the toxin in releasing only three of thirteen endogenous amino acids is clearly seen in the results presented in Fig. 10. Since these three compounds are all putative transmitters in the cortex, this suggests that transmitter-releasing systems have been activated by a depolarising action of the toxin. This is supported by the effect of tetrodotoxin in preventing the toxin-induced release and the in vivo limb-jerking response from occurring, and indicates that Na^+-channel activation is the mode of action. The transmitters released by this action would be expected to produce additional actions.

In vitro release of amino acids. Dolly et al (21) have previously reported a tetrodotoxin-sensitive release of radiolabelled glutamate and GABA from incubated cortical synaptosomes induced by low concentrations of tityustoxin. Our results show that endogenous glutamate, together with endogenous aspartate and GABA are also released by the toxin whilst thirteen other amino acids present in a freely-soluble state in synaptosomes are not released. However, a 33% reduction in soluble K^+-content did occur, (Table 1) showing that the ion-fluxes associated with depolarisation had been set into operation. This was confirmed by the two-fold stimulation in respiratory rates. The Ca^{2+}-dependence of this release was far from absolute and was similar in extent to that found previously (21) for release of preloaded exogenous $[U^{14}C]$-glutamate and $[U^{14}C]$-GABA.

This strong and rapid releasing action of the toxin clearly explains its apparent $(U^{14}C)$-GABA-uptake blocking action over a 5 min test period, as was seen by the parallel time course and potency of its action in releasing preloaded $(U^{14}C)$-GABA from synaptosomes. The tetrodotoxin-sensitivity of the apparent uptake-inhibition by the toxin also shows that its predominant action is to release $(U^{14}C)$-GABA. These results emphasise the need to check for other compounds (by preloading studies and tetrodotoxin and Ca^{2+}-sensitivity) that their apparent uptake-blocking action over the usual 1-5 min periods of incubation does not contain a releasing element which would add to their apparent potency as uptake blockers.

Source of released amino acids. Whether the neurotransmitters released are from neurones or glial cells cannot be clearly ascertained from the present data. The tetrodotoxin-sensitivity of

the effects does not answer the question since depolarizing levels
of K^+ released by active neurones could cause secondary release
from glial cells. Using radiolabelled precursors such as acetate and
glucose which are cell-specific could help to provide the answer.
The tetrodotoxin sensitivity does however rule out an extracerebral
origin of released amino acids.

Hydrolysis of the added toxin does not appear to be a significant
source of the amino acids recovered in the medium since calculation
shows a maximum of only 2 to 4% of glutamate and asparate respectively,
could be generated this way, and GABA is not a constituent amino
acid of the toxin.

In summary, our results show that both in vivo and in vitro
tityustoxin action is restricted to releasing only the transmitter
candidates among thirteen endogenous amino acids studied, and in vivo
myoclonic limb jerks induced by the toxin are associated with release
of these same transmitter amino acids. However, a direct depolarising
action of the toxin caused by activation of Na^+-channels could be
the mode of action producing the limb-jerks, and the release of
transmitter may be only a secondary action.

The Actions of Morphine and Naloxone on Transmitter Release

Morphine and other opiates have been shown to depress the
spontaneous release of acetylcholine in vitro (39,57,50,52,53) and
in vivo (33,34,43) and depress the release of dopamine (29) and the
turnover of biogenic amines (54).

It was recently reported that low concentrations of narcotic
analgaesics as well as the endogenous opiates, met-and leu-enkephalin,
hyperpolarise a proportion of the neurones in both the guinea-pig
myentenic plexus (49) and the frog sympathetic ganglion (59). Such
membrane hyperpolarisation may reflect the primary action of morphine
and indicate the basis for its inhibitory action on the firing of
neurones induced by a wide range of stimuli, including the application of aspartate and glutamate (9,17,22,23,60).

The anti-convulsant actions of opiates and an enkephalin
analogue (FK-33824) on reflex epilepsy in the baboon have recently
been reported (44). In these experiments the drugs were given intra-
cerebrally or intramuscularly, as their effects were reversed by
naloxone. These results would correlate well with our own reported
here, but contrast with the many previous reports of a convulsant
action of the opiates, usually given in high doses (e.g. up to 50
mg/kg) and causing both behavioural and EEG manifestations of
seizures (6,12,27,55,56).

We interpret the ineffectiveness of morphine or naloxone to
influence the level of spontaneous release of amino acid

neurotransmitters as an indication that these compound are continuously effluxing from cellular compartments other than nerve terminals, and by non-synaptic mechanisms. The greater part of these compounds would not, in this case, be influenced by agents modulating transmitter release. The much lower level of spontaneous release of ACh appears, on the other hand, to be mostly synaptic in origin (i.e. ACh is present at only a quarter per cent of the glutamate level in molecular terms).

In conclusion, our results support an inhibitory action for morphine in the cerebral cortex, and suggest that this is due to a supression of neurotransmitter release. This could be due to a strong hyperpolarizing effect, powerful enough to overcome, for instance, the potent depolarizing action of tityustoxin (3,14) (see above). Equally and perhaps more credibly, morphine could be working by blocking the calcium influx which would normally occur during the depolarization induced by tityustoxin. Such an action of morphine has been demonstrated (24,46) on the calcium current associated with action potentials. However, the naloxone sensitivity of the effect emphasises the involvement of receptors and limits the appeal of this explanation. The alternative view is that presynaptic opiate receptors are controlling transmitter release and a case for the existence of such receptors in spinal cord and pituitary gland has gained considerable momentum recently (30,40).

Since naloxone very effectively prevents this supression, the morphine must be acting via specific receptor-binding sites and these could be sited at any point on the neurones provided interaction leads to reduced firing or to reduced transmitter release (e.g. presynaptically). However as reduced neuronal firing appears to be a characteristic effect of endogenous opiate peptides (16,26,47,61), the former explanation is the more attractive, and in this case both morphine and naloxone could be producing their primary effects at cell bodies located at some distance from the cortical area being superfused, possibly even in different brain regions.

Since we have some evidence that naloxone given alone also enhanced spontaneous ACh release, it seems likely that the supression of transmitter release by morphine is revealing an influence normally exerted by endogenous opiate peptides in cerebral cortex. Though these agents are not found excessively concentrated in this CNS region, they are present, and morphine-type opiate receptors also abound in sensorimotor cortex (10,25).

It was noticeable that these in vivo effects of morphine and naloxone were not reproduced in vitro using synaptosomes under conventional conditions. This is in contrast to the effects of the GABA-analogues and tityustoxin itself, all of which showed comparable effects (at least at the qualitative level) both in vitro and in vivo.

This must sound a note of caution when drawing conclusions from
experiments employing shorter term in vitro preparations and shows
the value of having the dual approach as outlined in this paper.

Acknowledgements. The work described here was performed in collaboration with my colleagues (A. S. Abdul-Ghani, J. Coutinho-Netto, D.R.
Cox, P. R. Dodd and P. J. Norris) as part of several projects either
completed and published or still in progress at Imperial College,
London

REFERENCES

1. Abdul-Ghani, A.S., Bradford, H.F., Cox, D.W.G. and Dodd, P.R.,
    Peripheral sensory stimulation and the release of transmitter
    amino acids in vivo from specific regions of cerebral cortex,
    Brain Res. 171 (1979) 55-66
2. Abdul-Ghani, A.S., Coutinho-Netto, J. and Bradford, H.F., The
    action of γ-vinyl GABA and γ-acetylenic GABA on the resting
    and stimulated release of GABA in vivo, Brain Res. (in press
    1980).
3. Abdul-Ghani, A.S., Coutinho-Netto, J. and Bradford, H.F., In
    vivo release of acetylcholine evoked by brachial plexus stimula-
    tion and tityustoxin, Biochem. Pharmacol. 29 (1980) 2179-2182.
4. Abdul-Ghani, A.S., Marton, M. and Dobkin, J., Studies of the
    transport of glutamine in vivo between the brain and blood in
    resting state and during afferent electrical stimulation, J.
    Neurochem. 31 (1978) 541-546.
5. Abdul-Ghani, A.S., Norris, P.J., Smith, C.C.T. and Bradford,
    H.F., Effects of γ-acetylenic GABA and γ-vinyl GABA on
    synaptosomal release and uptake of GABA, Biochem. Pharmacol. (in
    press 1980).
6. Albruz, K. and Herz, A., Inhibition of behavioural and EEG
    activation induced by morphine acting on lower brain-stem
    structures, Electroenc. Clin. Neurophysiol. 33 (1972) 579-590.
7. Bradford, H.F., Bennett, G.W. and Thomas, A.J., Depolarizing
    stimuli and the release of physiologically active amino acids
    from suspension of mammalian synaptosomes, J. Neurochem. 21
    (1973) 495-505.
8. Bradford, H.F. and Thomas, A.J., Metabolism of glucose and
    glutamate by synaptosomes from mammalian cerebral cortex,
    J. Neurochem. 16 (1969) 1495-1504.
9. Calvillo, O., Henry, J.L. and Neuman, R.S., Effects of morphine
    and naloxone on dorsal horn neurones in the cat, Can. J,
    Physiol. Pharmac. 52 (1974) 1207-1211.
10. Chang, K-J., Cooper, B.R., Hazum, E. and Cuatrecasas, P.,
    Multiple opiate receptors: different regional distribution in the
    brain and differential binding of opiates and opioid peptides,
    Mol. Pharmacol. 16 (1979) 91-104.

11. Clark, R.M. and Collins, C.G.S., The release of endogenous amino acids from the rat visual cortex, J. Physiol. (Lond.), 262 (1976) 383-400.

12. Corrado, A.P. and Longo, V.G., An electrophysiological analysis of the convulsant action of morphine, codeine and thebaine, Arch. Int. Pharmacody. Ther. 132 (1961) 255.

13. Coutinho-Netto, J., Abdul-Ghani, A.S. and Bradford, H.F., Suppression of evoked and spontaneous release of neurotransmitters in vivo by morphine, Biochem. Pharmacol. (in press 1980).

14. Coutinho-Netto, J., Abdul-Ghani, A.S. Norris, P.J., Thomas, A. J. and Bradford, H.F., The effects of scorpion venom toxin on the release of amino acid neurotransmitters from cerebral cortex in vivo and in vitro, J. Neurochem. (in press 1980).

15. Countinho-Netto, J. and Diniz, C.R., 5th International Symposium on animal, plant and microbial toxins, (1976).

16. Denavit-Saubie, M., Champagnat, J. and Zieglgänsberger., Effects of opiates and methionine-enkephalin on pontine and bulbar respiratory neurones of the cat, Brain Res. 155 (1978) 55-67.

17. Dingledine, R. and Goldstein, A., Single neuron studies of opiate action in the guinea pig myenteric plexus, Life Sciences 17 (1975) 57-62.

18. Dodd, P.R. and Bradford, H.F., Release of amino acids from chronically superfused cerebral cortex, J. Neurochem. 23 (1974) 289-292.

19. Dodd, P.R. and Bradford, H.F., Release of amino acids from the maturing cobalt-induced epileptic focus, Brain Res. 111 (1976) 377-388.

20. Dodd, P.R., Pritchard, M.J. Adams, R.C.F., Bradford, H.F., Hicks, G. and Blanshard, K.C., A method for the continuous, long-term superfusion of the cerebral cortex of unanaesthetised, unrestrained rats, J. Scientific Inst. 7 (1974) 897-901.

21. Dolly, J.O., Chun, K., Tse, V., Spokes, J.W. and Diniz, C.R., β-bungarotoxin and tityustoxin on uptake and release of neuro-transmitters, Biochem. Soc. Trans. 6 (1978) 652-654.

22. Dostrovsky, J. and Pomeranz, B., Morphine blockade of amino acid putative transmitters on cat spinal cord sensory interneurones, Nature New Biol. 246 (1973) 222.

23. Duggan, A.W., Hall, J.G. and Headley, P.M. Suppression of transmission of nociceptive impulses by morphine: selective effects of morphine administered in the region of the substantia gelatinosa, Br. J. Pharmac. 61 (1977) 65-76.

24. Dunlap, K. and Fischbach, G.D. Neurotransmitters decrease the calcium component of sensory neurone action potentials, Nature (Lond.) 276 (1978) 837-839.

25. Fields, H.L., Emson, P.C., Leigh, B.I.C., Gilbert, R.F.T. and Iversen, L.L., Multiple opiate receptor sites on primary afferent fibres, Nature (Lond.) 284 (1980) 351-353.

26. Frenk, H., McCarty, B.C., Liebeskind, J.C., Different brain areas mediate the analgesic and epileptic properties of

enkephalin, Science 200 (1978) 335-337.

27. Frenk, H., Urca, G. and Liebeskind, J.C., Epileptic properties of leucine- and methionine-enkephalin: comparison with morphine and reversibility by naloxone, Brain Res. 147 (1978) 327-337.

28. Gray, E.G. and Whittaker, W.P., The isolation of nerve endings from brain, an electron microscopic study of cell fragments divided by homogenization and centrifugation, J. Anat. 96 (1962) 79-87.

29. Gudelsky, G.A. and Porter, J.C., Morphine and opioid peptide-induced inhibition of the release of dopamine from tubero-infundibular neurons, Life Sciences 25 (1979) 1697.

30. Iversen, L.L., Iversen, S.D. and Bloom, F.E., Opiate receptors influence vasopressin release from nerve terminals in rat neurohypophysis, Nature (Lond.) 284 (1980) 350-351.

31. Iversen, L.L., Mitchell, J.F. and Srinivasan, V., γ-amino-butyric acid, during inhibition in the cat visual cortex. J. Physiol. (Lond.) 212 (1971) 519-534.

32. Jasper, H.H. and Koyama, I., Rate of release of amino acids from the cerebral cortex in the cat as affected by brain stem and thalamic stimulation, Canad. J. Physiol. Pharmacol. 47 (1969) 889-905.

33. Jhamandas, K., Pinsky, C. and Phillis, J.W., Effects of morphine and its antagonists on release of cerebral cortical acetylcholine, Nature (Lond.) 288 (1970) 176.

34. Jhamandas, K. and Sutak, M., Modification of brain acetylcholine release by morphine and its antagonists in normal and morphine-dependent rats, Br. J. Pharmac. 50 (1974) 57-62.

35. Jung, M.J., Lippert, B., Metcalfe, B.W., Böhlen, P. and Schechter, P.J., γ-vinyl GABA (4-amino-hex-5-enoic acid), a new selective irreversible inhibitor of GABA-T: effects on brain GABA metabolism in mice J. Neurochem. 29 (1977) 797-802.

36. Jung, M.J., Lippert, B., Metcalfe, B.W., Schechter, P.J., Böhlen, P. and Sjoerdsma, A., The effect of γ-acetylenic GABA, catalytic inhibitor of GABA-T, on brain metabolism in vivo, J. Neurochem. 28 (1977) 717-723.

37. Jung, M.J. and Metcalfe, B.W., Catalytic inhibition of GABA-T of cerebral origin by 4-amino-hex-5-ynoic acid. A substrate analog, Biochem. Biophys. Res. Commun. 67 (1975) 301-306.

38. Jung, M.J. and Seiler, N., Enzyme-activated irreversible inhibitors of L-ornithine: 2-oxoacid aminotransferase, J. Biol. Chem. 253 (1978) 7431-7439.

39. Kaneto, H., in Narcotic Drugs: Biochemical Pharmacology (Ed. D. H. Clonet) (1971), p 300, Plenum Press, New York,

40. Lamotte, C., Pert, C.B. and Snyder, S.S., Opiate receptor binding in primate spinal cord: distribution and changes after dorsal root section, Brain Res. 112 (1976) 407-412.

41. Lippert, B., Metcalfe, B.W., Jung, M.J. and Casara, P., 4-amino-hex-5-enoic acid, a selective catalytic inhibitor of GABA-T in mammalian brain, Europ. J. Biochem. 74 (1977) 441-445.

42. Lowry, O.H., Rosebrough, N.J., Farr, A.L. and Randall, R.J., Protein measurement with the folin phenol reagent, J. Biol. Chem., 193 (1951) 265-275.

43. Matthews, J.D., Labrecque, G. and Domino, E.P., Effects of morphine, nalorphine and naloxone on neocortical release of acetylcholine in the rat, Psychopharmacologia 29 (1973) 113-120.

44. Meldrum, B.S., Menini, Ch., Stutzman, J.M. and Naquet, R., Effects of opiate-like peptides, morphine and naloxone in the photosensitive baboon, Papio papio, Brain Res. 170 (1979) 333-348.

45. Metcalfe, B.W. and Casara, P., Region specific 1.4 addition of a proparglylic anion. A general synthon for 2-substantial propargylamine as potential catalytic irreversible enzyme inhibitors, Tetrahedron Lett. 38 (1975) 3337-3340.

46. Mudge, A.W., Leeman, S.E. and Fischbach, G.D., Enkephalin inhibits release of substance P from sensory neurons in culture and decreases action potential duration, Proc. Natl. Acad. Sci. U.S.A. 76 (1979) 526-530.

47. Nicoll, R.A., Siggins, G.R., Ling, N., Bloom, F.E. and Guillemin, R., Neuronal actions of endorphins and enkephalins among brain regions: a comparative microiontophoretic study, Proc. Nat. Acad. Sci. U.S.A. 74 (1977) 2584-2588.

48. Norris, P.J., Smith, C.C.T., de Belleroche, J., Bradford, H.F., Mantle, P.G., Thomas, A.G. and Penney, R.H.C., Actions of tremorgenic fungal toxins on neurotransmitter release, J. Neurochem. 34 (1980) 33-42.

49. North, R.A. and Tonini, M., The mechanism of action of narcotic analgesics in the guinea-pig ileum, Br. J. Pharmac. 61 (1977) 541-549.

50. Paton, W.D.M., The action of morphine and related substances on contraction and on acetylcholine output of coaxially stimulated guinea-pig ileum, Br. J. Chemother. 12 (1957) 119-127.

51. Perry, T.L., Kish, S.J., Sjaastrad, O., Gjessing, L.R., Nesbakken, R., Schrader, H. and Loken, A.C., γ-vinyl GABA: effects of chronic administration on the metabolism of GABA and other amino compounds in rat brain, J. Neurochem. 32 (1979) 1641-1645.

52. Schaumann, W., Influence of atropine and morphine on the liberation of acetylcholine from the guinea pig's intestine, Nature (Lond.) 178 (1956) 1121.

53. Schaumann, W., Inhibition by morphine of the release of acetylcholine from the intestine of the guinea-pig, Br. J. Pharmac. Chemother. 12 (1957) 115.

54. Takemori, A.C., Neurochemical bases for narcotic tolerance and dependence, Biochem. Pharmacol. 24 (1975) 2121-2125.

55. Urca, G., Frenk, H., Liebeskind, J.C. and Tailor, A.N., Morphine and enkephaline: analgesic and epileptic properties, Science 197 (1977) 83-86.

56. Verdeaux, G. and Manty, R., Action sur l'électroencéphalogramme

de substances pharmacodynamiques d'interet clinique, <u>Rev. Neurol.</u>
91 (1954) 405–427.

57. Weinstock, M., in <u>Narcotic Drugs: Biochemical Pharmacology</u>
(Ed. D.H. Clonet) (1971), p 254, Plenum Press, New York.

58. Westerfield, M., Moore, J.W., Kim, Y. S. and Padilla, G.M., How
<u>Gymnodinium breve</u> red tide toxin(s) produces repetitive
firing in squid axons, <u>Am. J. Physiol.</u> 232 (1977) C23– C29.

59. Wourters, W. and Van den Bercken, J., Hyperpolarisation and
depression of slow synaptic inhibition by enkephalin in frog
sympathetic ganglion, <u>Nature, (Lond.)</u> 277 (1979) 53.

60. Zieglgänsberger, W. and Bayerl, H., The mechanism of inhibition
of neuronal activity by opiates in the spinal cord of cat,
<u>Brain Res.</u> 115 (1976) 111–128.

61. Zieglgänsberger, W., Fry, J.P., Herz, A., Moroder, L. and
Wünsch, E., Enkephalin-induced inhibition of cortical neurones
and the lack of this effect in morphine tolerant/dependent rats,
<u>Brain Res.</u> 115 (1976) 160.

TAURINE AS A NEUROMODULATOR: ITS ACTION ON CALCIUM FLUXES AND NEUROTRANSMITTER RELEASE

Herminia Pasantes-Morales and Julio Morán

Departamento de Neurociencias
Centro de Investigaciones en Fisiología Celular
Universidad Nacional Autónoma de México
México 20, D. F.

INTRODUCTION

Taurine is an ubiquitous constituent of animal tissues (25); its highest levels are found in excitable organs like brain, heart and secretory glands, where it may reach concentrations as high as 20-100 mM (8,25,27). In spite of its wide distribution and of recent interest in studies on taurine, the elucidation of its functional role has been particularly difficult. Taurine is not associated to proteins and except for the formation of taurocholic acid it does not participate in metabolic reactions. Therefore, it could be thought that the presence of taurine in a tissue or organ should be related only to the specific function subserved by the amino acid. In this communication we will consider some possible roles of taurine in nervous tissue.

TAURINE DISTRIBUTION IN NERVOUS TISSUE

The regional and subcellular distribution of amino acids is an important parameter in the study of their functional role in nervous tissue. Taurine is unevenly distributed in the central nervous system, with concentrations varying from 1.8 mM in the spinal cord to 6-8 mM in the striatum and cerebellum (34). Studies on the microdistribution of taurine in the spinal cord and thalamus have shown that it is evenly distributed in all the sections examined (30). In contrast, in cerebellum taurine appears to be mainly localized in the molecular layer (36).

Studies on the subcellular distribution of taurine might also be useful in the elucidation of its functional role in nerve cells.

Taurine shows the distribution expected for a diffusible compound after subcellular fractionation, about 70% being recovered in the soluble fraction (3). Taurine is also present in synaptosomes, where it is mainly localized in synaptic vesicles (31), in which it is the most abundant amino acid (15,28).

Taurine concentrations in brain are markedly constant; taurine levels remain unchanged in a variety of neurological alterations as well as under many different experimental conditions. A severe deficiency of vitamin B6 in the diet, substantially reduces taurine biosynthesis, yet brain taurine concentrations are maintained (20). Concomitantly, a marked decrease of taurine excreted in urine and feces is observed (45). Supplementation of the diet with taurine also has no effect on brain taurine concentrations; the excess taurine is simply excreted (32). There is one physiological condition in which taurine levels in brain markedly change and this is during development. The taurine concentration in fetal nervous tissues is more than 4 times higher than in the adult (1,2). With development taurine concentrations decrease in all areas of brain although the magnitude of decrease varies from area to area (9). No correlation has been established between the decrease of taurine levels and a particular function in which taurine might be involved during brain development. Among various possibilities, a role in microtubule establishment or in axonal transport deserves consideration.

TAURINE ACTIONS IN NERVOUS TISSUE

Pharmacological and physiological studies in vertebrate and invertebrate preparations have shown that taurine has actions similar to those of the inhibitory amino acids glycine and GABA. Taurine has a marked inhibitory effect on neuronal activity when it is applied via microiontophoresis (12-14,38). Taurine produces a hyperpolarization of the neuronal membrane by modifying its permeability to chloride or potassium ions. In most brain areas, however, taurine has only a rather weak depressant action. Moreover, the specificity of taurine effects on neuronal activity has not been proved until now, due to the lack of specific antagonists for taurine actions. Strychnine, which is considered an antagonist of glycine effects, also blocks taurine induced depression in the spinal cord, thalamus and cerebral cortex (10). Bicuculline, a GABA-antagonist, also blocks taurine effects on cerebral cortex (11). Therefore, the possibility that taurine may be acting through glycine or GABA receptors cannot be ruled out.

Snyder and his coworkers (18,44) have designed a biochemical method for the characterization of the interaction of neurotransmitters with specific postsynaptic receptors. They have measured the specific binding of labeled neurotransmitter amino acids of high specific activity to membrane preparations and have characterized a sodium-independent, temperature insensitive binding, showing a high affinity and saturation at very low concentration of the ligand. This binding is usually inhibited by antagonists of the physiological effects of the amino acids, such as strychinine and bicuculline for glycine and GABA binding, respectively. Using this technique, Snyder et al. (18,44) have identified postsynaptic receptors in different regions of the central nervous system for GABA and glycine, whose distribtuion closely follows that of the neurons sensitive to these amino acids. We have followed this procedure in an attempt to identify taurine receptors in various nervous regions: retina (34), cerebral cortex, hypothalamus and olfactory bulb. However, no specific binding for ^{3}H-taurine to membrane preparations, measured in frozen membranes and in a sodium-free medium, could be detected. All the studies were carried out in parallel with GABA and taurine, and whereas ^{3}H-GABA binding was always observed, taurine binding to the same preparation was undetectable. These results cannot exclude the existence of postsynaptic taurine receptors, localized at taurinergic synapses; however, we can conclude that if such receptors exist in any of the regions examined, they are not quantitatively significant. These results suggest that taurine is not playing a major role as an inhibitory neurotransmitter in the areas studied.

TAURINE RELEASE

The release from presynaptic nerve terminals in response to nerve excitation has been considered as a requirement to be fulfilled by putative neurotransmitters. The release of taurine in vitro and in vivo from various brain areas, in association with stimulation of neuronal activity, is well documented (7,24,26,39). This release does not seem to be calcium-dependent, which argues against a neuronal release and suggests a release from glia, which is calcium-independent. The release of taurine from isolated nerve terminals is stimulated by depolarizing concentrations of KCl. It is much lower than that observed for GABA and similar to that observed in vivo, it is not reduced when superfusion of synaptosomes is carried out with a calcium-free medium. On the contrary, the potassium-stimulated release of taurine is significantly increased in the absence of calcium (43). In order to obtain more information concerning the effect of depolarizing agents on taurine release from nerve terminals, we have measured the effect of high potassium concentrations, veratrine and ouabain on the simultaneous release of ^{3}H-GABA and ^{35}S-taurine. In this way the effect of depolarizing conditions on the release of the two amino acids can be compared in the same preparation. Fig. 1 shows that perfusion with a high potassium medium produces a 9-fold increase in the efflux of ^{3}H-GABA. The release of ^{35}S-taurine under these conditions is only 0.9 times the prestimulation value. The other depolarizing agents used,

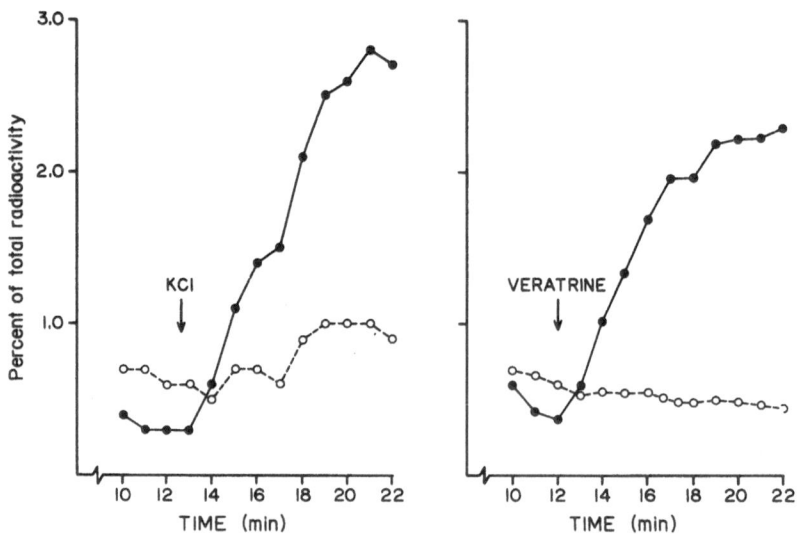

Fig. 1. Effect of KCl (56 mM) and of veratrine (12 µg/ml) on the
release of ^{3}H-GABA and ^{35}S-taurine from a crude synaptosomal
fraction of rat brain. A P_2 fraction was incubated for 10 min with
^{3}H-GABA (50 µM) and ^{35}S-taurine (50 µM). Aliquots of the suspension
were collected on Millipore filters and superfused with warmed
Krebs-bicarbonate medium. After an equilibration period of 12 min,
the superfusion medium was replaced with new medium containing
56 mM KCl, or veratrine, and superfusion continued for 10 min.
Superfusate fractions were collected every minute and the
radioactivity in the fractions and that remaining on the filters
was measured by liquid scintillation spectrometry. Each curve is
the average of 4-6 separate experiments. (●), ^{3}H-GABA; (0), ^{35}S-
taurine.

veratrine or ouabain, also increase the release of ^{3}H-GABA without
affecting that of ^{35}S-taurine. These results, together with the
absence of taurine receptors as measured by binding techniques,
do not favor a major neurotransmitter role for taurine in the
central nervous system. However, it should be pointed out that
a specific taurine uptake, exhibiting high-affinity constants, has
been described in nervous tissue slices as well as in synaptosomes
(21,40). This suggests that taurine is playing a role at the
synapse, though probably not as a neurotransmitter.

PHARMACOLOGICAL ACTIONS OF TAURINE

Some reports have been published describing certain effects

of taurine administration on pain, muscular tone and body temperature, which suggest that taurine might be involved in central pain mechanisms as well as in the mechanisms of thermoregulation and muscle tone control (33,41,42). One of the pharmacological effects of taurine which has raised much interest in recent years is its anticonvulsant action. The observation of Van Gelder (46) in 1972 on decreased taurine levels in human and cobalt-induced epileptogenic foci led to tests on the possible anticonvulsant action of taurine in several models of experimental epilepsy, both acute and chronic, with positive results (23,37,47). The protective effect of taurine against a wide variety of experimental models was part of the rationale for giving taurine a clinical trial as antiepileptic (4,5). The question raised by the experiments on the anticonvulsant action of taurine is whether its effects are related to its natural function in nervous system or represent only a pharmacological effect. In this respect, it is important to establish a causal relationship between a taurine depletion and the occurrence of the electro-encephalographic signs of epilepsy. Experiments of Durelli et al. (17) have shown that a penicillin focus is still developed when taurine levels in the cerebral cortex are 2-3 times higher than normal. This could be considered as an evidence against a direct causal involvement of taurine depletion in the development of epileptic spiking. However, a number of observations indicate an effect of taurine in preventing spread of seizure activity from the focus (17,37; thus, taurine may be exerting its anticonvulsant action by an antispreading effects.

The basic mechanisms underlying taurine effects as anticonvulsant are still unknown. A direct effect of taurine on excitable membranes, a taurine-induced modification of ionic permeability or an effect mediated through changes in the amino acid content of the epileptogenic tissue, tending to normalize their altered profile in brain, are some of the possibilities which have been considered. The results of the present work will be discussed later on in reference to these several possible actions of taurine.

TAURINE EFFECTS ON CALCIUM TRANSPORT

As mentioned previously, taurine is not an exclusive constituent of the nervous tissue. It is present at very high concentrations in contractile tissues - 40 mM in the rat heart (27) - or even higher in secretory tissue, - 60-110 mM in rat pineal or pituitary gland (8). The question raised by these observations is whether taurine is playing the same role in the different animal tissues or is involved in different functions at each organ. If the first proposal is true, taurine should be involved in a basic mechanism underlying both stimulus-secretion and stimulus-contraction processes.

Observations of Dolara et al. (16) have related some of the

cardiac actions of taurine, like its antiarrhythmic effects on
epinephrine and digoxine-induced arrhythmias, as well as its
inotropic effects, to modifications in calcium kinetics in heart
cells. Increased accumulation or retention of calcium caused by
taurine, might enhance the amount of calcium available for contraction
and explain the observed effects of taurine. In order to investigate
if taurine actions in the central nervous system might also be related
to modifications in calcium transport, we undertook a study of taurine
effects on calcium accumulation by rat brain synaptosomes.

Rat brain synaptosomes were prepared following the procedure
of Hajós (19). Sucrose homogenates of adult rat brain were
centrifuged at 1000 g for 10 min; the pellet was washed once and
the combined supernatants were centrifuged at 10,000 g for 20 min.
The crude synaptosomal fraction obtained in the pellet was
resuspended in 0.3 M sucrose and layered over 10 ml of 0.8 M sucrose.
The tubes were centrifuged at 10,000 g for 20 min. Myelin banded at
the sucrose interphase, the free mitochondria were pelleted at the
bottom and the synaptosomes were localized throughout the 0.8 M
sucrose layer. The layer containing the synaptosomes was diluted
with water to a final sucrose concentration of 0.4 M and the
synaptosomes were sedimented at 10,000 g for 15 min. The synaptosomal
pellet was resuspended in 0.32 M glucose or in a Krebs-bicarbonate
medium. Calcium uptake by the isolated nerve endings was studied by
measuring the accumulation of ^{45}Ca by synaptosomes incubated in a
Krebs-bicarbonate medium (118 mM NaCl; 4.7 mM KCl, 1.17 mM $MgSO_4$;
1.2 mM KH_2PO_4; 2.5 mM $CaCl_2$, 25 mM $NaHCO_3$ and 5.6 mM glucose), pH
7.4. Incubations were carried out at 37°C or at 25°C, directly in
Microfuge tubes containing 0.28 ml of Krebs-bicarbonate medium and
varying concentrations of $CaCl_2$ or taurine in a final volume of
0.3 ml. After 5 min of incubation, the tubes were centrifuged, the
pellet was washed superficially and solubilized with NCS (tissue
solubilizer). The radioactivity accumulated was measured after the
addition of Tritosol. Values of blanks incubated at 4°C or without
tissue were subtracted at each experiment. Synaptosomes, incubated
in a Krebs-bicarbonate medium containing 2.5 mM $CaCl_2$, accumulate
^{45}Ca very rapidly, reaching saturation values of 1.5–2.0 nmoles/mg
protein after 1-2 min of incubation. The addition of taurine to the
incubation medium causes a reduction in ^{45}Ca accumulation, which
varies from 30% to 50% depending on taurine concentration. The
effect is observed from the first minute of incubation (Fig. 2).
The effect of taurine is observed at pH values ranging from 6.8 to
8.2, being maximal at pH 7.4. The presence of bicarbonate and
phosphate is require to observe the action of taurine on ^{45}Ca
accumulation. When bicarbonate is replaced by other buffers like
TRIS, HEPES or imidazole, taurine only slightly inhibits ^{45}Ca
accumulation. Omission of phosphate from the Krebs-bicarbonate
medium considerably reduces ^{45}Ca accumulation; taurine has no
effects on the remaining ^{45}Ca accumulation (Fig. 3). Taurine action
on ^{45}Ca transport seems to be specific. Other amino acids like

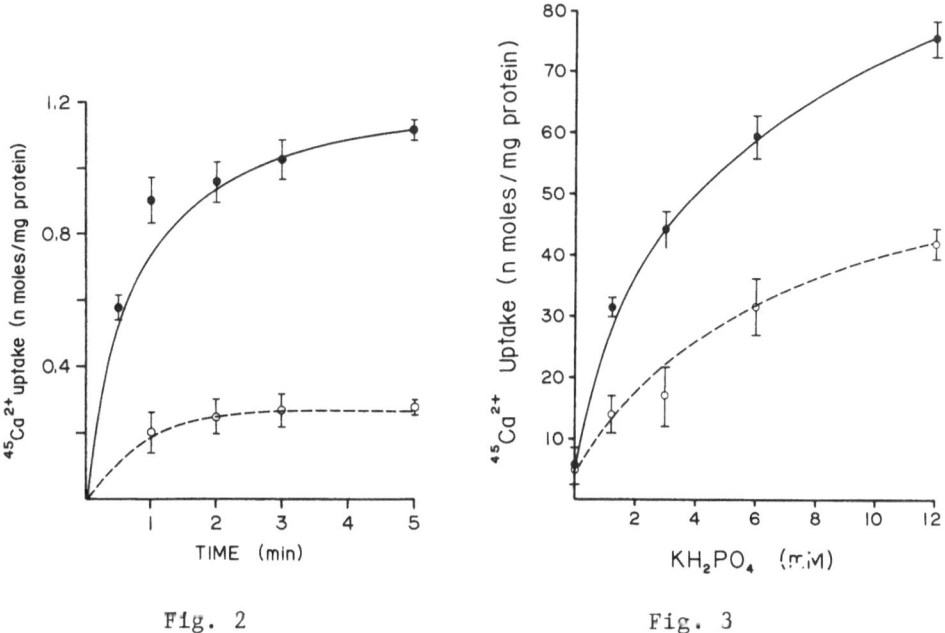

Fig. 2   Fig. 3

Fig. 2. The effect of taurine (25 mM) on ^{45}Ca accumulation by rat
brain synaptosomes. Synaptosomes were incubated in a Krebs-
bicarbonate medium containing 2.5 mM $CaCl_2$ and 2 μCi of ^{45}Ca. At
the indicated times, aliquots were withdrawn and centrifuged in a
Microfugue. The pellet was washed, solubilized and the radioactivity
accumulated was measured by liquid scintillation spectrometry.
Results are the means of 5-6 separate experiments. ($\bullet$), Control;
(0), 25 mM taurine.

Fig. 3. The effect of taurine (25 mM) on ^{45}Ca accumulation by rat
brain synaptosomes in the presence of varying concentrations of
KH_2PO_4. Synaptosomes were incubated during 5 min in a Krebs-
bicarbonate medium containing 2 μCi of ^{45}Ca and the indicated
concentrations of phosphate. The radioactivity accumulated was
measured as described in Fig. 2. Values are the means $\pm$ S.E.M. of
8 experiments. ($\bullet$) Control; (0), 25 mM Taurine.

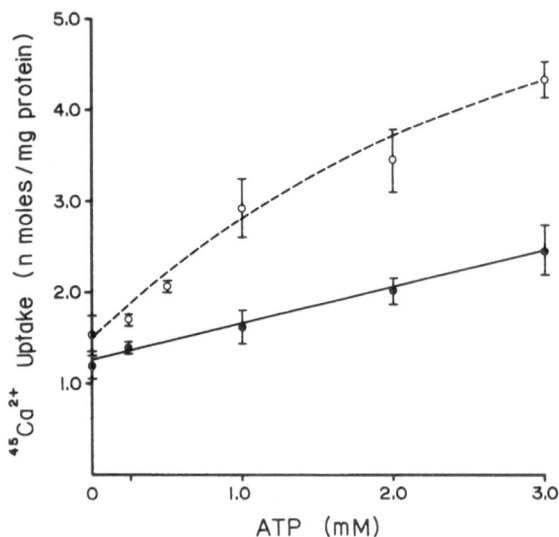

Fig. 4.   The effect of taurine (25 mM) on ^{45}Ca accumulation by rat
brain synaptosomes in the presence of low external calcium
concentration (40 μM), and of varying concentrations of ATP.
Incubation of synaptosomes and measurement of ^{45}Ca accumulation
were carried out as described in Fig. 2.   Values are the means
± S.E.M. of 4-8 separate experiments.   (●), Control; (0), 25 mM
taurine.

GABA, glycine, glutamic acid or leucine do not affect ^{45}Ca
accumulation by synaptosomes. ^{45}Ca accumulation by the synaptosomes
is not affected by ouabain or by ruthenium red (10 μM), and it is
not modified by depolarizing concentrations of KCl or by ATP.   It
is also unaffected by oligomycin or dinitrophenol, but it is
considerably reduced by low temperature.

     When the CaCl$_2$ concentration in the incubation medium is 20-40
μM, ^{45}Ca accumulation by synaptosomes is stimulated by ATP.
Increasing ATP concentration in the medium produces a concomitant
enhancement of ^{45}Ca accumulation (Fig. 4): 2 mM ATP produces an
increase of more than 65% on ^{45}Ca accumulation over control values
and 3 mM ATP increases ^{45}Ca uptake by 75% (Fig. 4).   Since all the
experiments were carried out in the presence of oligomycin,
dinitrophenol and ruthenium red, the possibility of ^{45}Ca accumulation
by mitochondria seems unlikely.   The presence of taurine in the
incubation medium produces an increase in the ATP-stimulated ^{45}Ca
accumulation.   This effect of taurine is specific and dose-dependent;
it is more evident in the presence of bicarbonate, but it is also
observed when imidazole or TRIS replace bicarbonate for maintaining
pH.   These results suggest that taurine may be involved in the

processes of calcium transport in nervous tissue. The precise steps
or mechanisms through which taurine may be acting remain to be
elucidated. Calcium regulation in synaptic terminals appears to be
a complex process. Calcium may enter the synaptosomes through at
least three different processes: a) voltage sensitive channels,
b) sodium-calcium exchange and c) difussion. An extremely rapid
and efficient control of intracellular calcium levels is critical
in nerve terminals, since they play a crucial role in the release
of neurotransmitters. Accordingly, the maintainance of the low
intracellular calcium levels appears to involve several intraterminal
mechanisms. Evidence presented by Blaustein et al. (6) suggests the
existence of at least two different mechanisms which store calcium
within the presynaptic terminal. One type of calcium storage site
has all the properties normally associated with mitochondrial calcium
accumulation: it is stimulated by ATP, and is blocked by the
mitochondrial poisons dinitrophenol, oligomycin and ruthenium red.
A second site of calcium sequestration, with properties different
from those of the mitochondrial site, apparently involves membrane-
bound organelles. From the various intraterminal structures which
might be considered as possible candidates for calcium sequestration
sites, biochemical and morphological evidence seems to favor the
smooth endoplasmic reticulum as the main site of calcium storage at
the synaptic terminal.

Our observations suggest that taurine may be related to the
mechanisms regulating calcium fluxes in synaptosomes. Much work is
necessary to identify the site or sites of taurine action as well as
the intimate mechanism of its effect; however, the present results
might explain some of the pharmacological actions of taurine and
could also be related to its "neurotransmitter-like" properties.
There is evidence that the gating of sodium and potassium channels
in nerve membranes is influenced by fixed negative charges in the
surface of the membrane. Normally, the divalent cations calcium and
magnesium reduce the electrostatic effect of fixed charges on the
outside of the membrane, either by forming a double ionic layer near
the charged sites or by binding to the membrane in the vicinity of
the charged sites. Solutions with low divalent cation concentration
are thought to increase excitability by reducing the availability
of cations to fixed negative charges on the outside surface of the
nerve membrane. Taking these statements into account we can
speculate on whether the widespread effects of taurine as
anticonvulsant are due to an specific action on nervous membranes
by blocking calcium entry into the cell and then increasing its
concentration in the vicinity of the membranes. In support of this
are the results of Izumi et al. (22) showing that the protective
effects of taurine against metrazol are prevented when taurine is
injected simultaneously with EDTA, a calcium chelator. This
observation provides experimental support for the hypothesis of
taurine actions on nervous excitability being mediated through
calcium interactions.

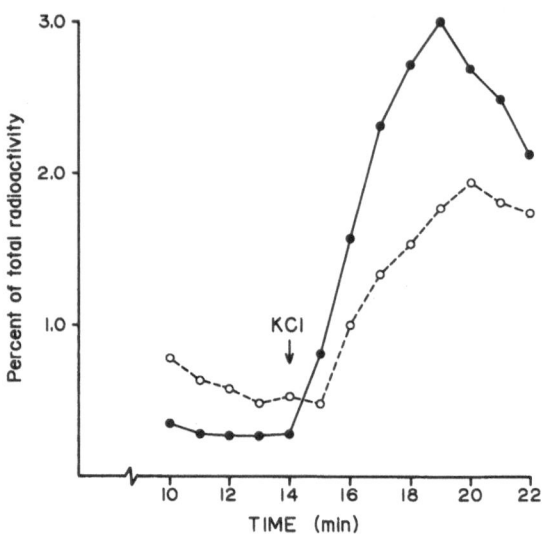

Fig. 5. The effect of taurine (25 mM) on the K⁺-stimulated release
of ^{3}H-GABA from a crude synaptosomal fraction of rat brain. Loading
and perfusion of synaptosomes were carried out as described in
Fig. 1. The curves are from a representative experiment. The
average increased release in controls was 7.8 times ± 0.83 (n=7)
and in the presence of taurine 4.1 times ± 0.58 (n=7). (●), Control;
(0), 25 mM taurine.

An alternate possibility relating taurine effects on calcium
transport to its suggested neuromodulatory action could be a possible
effect of taurine on neurotransmitter release. Kuriyama et al. have
shown that the addition of taurine (30 mM) to the superfusion medium
significantly reduces the potassium-evoked release of ^{14}C-acetyl-
choline from superior cervical ganglia without having any effect
on its spontaneous release (29). Also taurine significantly inhibits
the potassium-stimulated release of ^{3}H-norepinephrine from cerebral
cortex slices (29).

GABA appears to be a major inhibitory neurotransmitter in the
central nervous system, and it seems to be directly related to the
control of nervous excitability. Therefore, we have studied the
effect of taurine on the release of ^{3}H-GABA from rat brain
synaptosomes. ^{3}H-GABA loaded synaptosomes were perfused with a
Krebs-bicarbonate medium, pH 7.4, and fractions of the perfusion
effluent were collected every minute. After the baseline is attained,
the superfusion medium is changed to one containing 68 mM KCl. This
produces a marked increase in the release of ^{3}H-GABA of more than
10 times the prestimulation (9) value (Fig. 5). The addition of

25 mM taurine to the perfusion media significantly enhanced the spontaneous release of ^{3}H-GABA (Fig. 5); an increase of more than 90% is observed in the presence of taurine. Simultaneously, the potassium-evoked release of ^{3}H-GABA is markedly decreased from the synaptosomes perfused with a taurine containing medium. Fig. 5 shows a representative experiment of the effects of taurine on GABA release. The mean increase in ^{3}H-GABA efflux evoked by high potassium was 7.8 times $\pm$ 0.83 (n=7). In the presence of taurine this figure was 4.1 times $\pm$ 0.58 (n=7). Taurine did not affect ^{3}H-GABA uptake.

CONCLUSION

Although many reports are available concerning the neuro-pharmacological effects of taurine, nothing is clear about the mechanism underlying these effects, and more important, about the physiological role of taurine in nervous tissue. Some evidence is consistent with a neurotransmitter role for taurine, namely its uneven distribution in the central nervous system and high affinity uptake in synaptosomes. Other evidence, however, does not support such a role, namely, its Ca^{2+} independent release, the properties of its receptors and its depressant actions on neurons. As argued in this paper, the evidence does suggest a possible involvement in synaptic function, possibly acting on membrane-Ca^{2+} interactions.

Acknowledgements. The original work reported in the present chapter was supported in part by grant No. 5R01 EY02540-02 from the National Eye Institute and by grant No. PCCBNAL 790219, Consejo Nacional de Ciencia y Tecnología.

REFERENCES

1. Agrawal, H. C., Davies, S. M. and Himwich, W. A. Postnatal changes in free amino acid pool of rat brain. J. Neurochem. 13 (1966) 607-615.
2. Agrawal, H. C., Davies, J. M. and Himwich, W. A. Postnatal changes in free amino acid pool of rabbit brain. Brain Res. 3 (1966) 374-380.
3. Agrawal, H. C., Davison, A. N. and Kaczmarek, L. K. Subcellular distribution of taurine and cysteine sulfinate decarboxylase in developing rat brain. Biochem. J. 122 (1971) 759-763.
4. Barbeau, A. Zinc, taurine and epilepsy. Arch. Neurol. 30 (1974) 52-58.
5. Bergamini, L., Mutani, R., Delsedime, M. and Durelli, L. First clinical experience on the antiepileptic action of taurine. Eur. Neurol, 11 (1974) 261-269.
6. Blaustein, M. P., Ratzlaff, R. W., Kendrick, N. C. and Schweitzer, E. S. Calcium buffering in presynaptic nerve

terminals. I. Evidence for involvement of a nonmitochondrial ATP-dependent sequestration mechanism. J. Gen. Physiol., 72 (1977) 15–41.

7. Clark, R. M. and Collins, G. G. S. The release of endogenous amino acids from the mammalian visual cortex. J. Physiol. (Lond.) 246 (1975) 16P.

8. Crabai, F., Sitzia, A. and Pepen, G. Taurine concentration in the neurohypophysis of different animal species. J. Neurochem. 23 (1974) 1091–1092.

9. Cutler, R. W. P. and Dudzinski, D. S. Regional changes in amino acid content in developing rat brain. J. Neurochem. 23 (1974) 1005–1009.

10. Curtis, D. R., Duggan, A. W., Felix, D., Johnston, G. A. R. and McLennan, H. Antagonism between bicuculline and GABA in the cat brain. Brain Res. 33 (1971) 57–73.

11. Curtis, D. R., Hosli, L. and Johnston, G. A. R. A pharmacological study of the depression of spinal neurons by glycine and related amino acids. Exp. Brain Res. 6 (1968) 1–18.

12. Curtis, D. R. and Johnston, G. A. R. Amino acid transmitters in mammalian nervous system. Ergebn. Physiol. 69 (1974) 97–188.

13. Curtis, D. R. and Tebécis, A. K. Bicuculline and thalamic inhibition. Exp. Brain Res., 16 (1972) 210–218.

14. Curtis, D. R. and Watkins, J. C. The excitation and depression of spinal neurons by structurally related amino acids. J. Neurochem. 6 (1960) 117–141.

15. De Belleroche, J. J. and Bradford, H. F. Amino acids in synaptic vesicles from mammalian cerebral cortex: a reappraisal. J. Neurochem. 21 (1973) 441–451.

16. Dolara, P., Agresti, A., Giotti, A. and Pasquini, G. Effect of taurine on calcium kinetics of guinea-pig heart. Eur. J. Pharmacol. 24 (1973) 352–358.

17. Durelli, L., Mutani, R., Delsedime, M., Quattrocolo, G., Buffa, C., Mazzarion, M. and Fumero, S. Electroencephalographic and biochemical study of the antiepileptic action of taurine administered by cortical superfusion. Exp. Neurol. 52 (1976) 30–39.

18. Enna, S. J., Kuhar, M. and Snyder, S. H. Regional distribution of postsynaptic receptor binding for -aminobutyric acid (GABA) in monkey brain. Brain Res. 93 (1975) 168–175.

19. Hajós, F. An improved method for the preparation of synaptosomal fractions of high purity, Brain Res. 93 (1975) 485–489.

20. Hope, D. B. The persistence of taurine in the brains of pyridoxine-deficient rats. J. Neurochem. 1 (1957) 364–369.

21. Hruska, R., Huxtable, R., Bressler, J. and Yamamura, H. R. Sodium-dependent high affinity transport of taurine into rat brain synaptosomes. Proc. West. Pharmacol. Soc. 19 (1976) 152–156.

22. Izumi, K., Igisu, H. and Fukuda, T. Effects of edetate on seizure supressing actions of taurine and GABA, Brain Res. 88 (1975) 576–579.

23. Izumi, K., Igisu, H. and Fukuda, T.   Supression of seizures by taurine –specific or nonspecific? Brain Res. 76 (1974) 171-173.
24. Jasper, H. H. and Koyama, I.   Rate of release of amino acids from the cerebral cortex in the cat as affected by brain stem and thalamic stimulation.   Can. J. Physiol. Pharmacol. 47 (1969) 889-905.
25. Jacobsen, J. G. and Smith, H. L.   Biochemistry and physiology of taurine and taurine derivatives.   Physiol. Rev. 48 (1968) 424-511.
26. Kaczmarek, L. K. and Davison, A. N.   Uptake and release of taurine from rat brain slices.   J. Neurochem. 19 (1972) 2355-2362.
27. Kocsis, J. J., Kostos, U. J. and Baskin, S. I.   Taurine levels in the heart tissues of various species.   In: R. Huxtable and A. Barbeau (Eds.), Taurine, Raven Press, New York, 1976, pp. 145-153.
28. Kontro, P., Marnela, K. M. and Oja, S. S.   Free amino acids in the synaptosome and synaptic vesicle fractions of different bovine brain areas.   Brain Res. 184 (1980) 129-141.
29. Kuriyama, K., Muramatsu, M., Nakagawa, K. and Kakita, K.   Modulating role of taurine on release of neurotransmitters and calcium transport in excitable tissues.   In: A. Barbeau and R. Huxtable   (Eds.)   Taurine and Neurological Disorders, Raven Press, New York, 1978, pp. 201-216.
30. Kuriyama, K., Yoneda, Y. and Kurihara, E.   Microdistribution of taurine and cysteine sulfinate decarboxylase activity in rat spinal cord and thalamus: comparison with γ-aminobutyric acid and L-glutamic acid decarboxylase.   In: A. Barbeau and R. Huxtable (Eds.) Taurine and Neurological Disorders, Raven Press, New York, (1978) pp. 35-48.
31. Lahdesmaki, P., Karppinen, A., Saarni, H. and Winter, R.   Amino acids in the synaptic vesicle fraction from calf brain: content, uptake and metabolism.   Brain Res. 138 (1977) 295-308.
32. LeFauconnier, J. M., Urban, P. F. and Mandel, P. Taurine transport into the brain in rat.   Biochimie 60 (1978) 381-387.
33. Lipton, J. M. and Tickner, C. B. Central effect of taurine and its analogues on fever caused by intravenous leukocytic pyrogen in the rabbit.   J. Physiol. (Lond.) 287 (1979) 535-543.
34. López-Colomé, A. M. and Pasantes-Morales, H.   Taurine interactions with chick retinal membranes.   J. Neurochem. 34 (1980) 1047-1053.
35. Mandel, P. and Pasantes-Morales, H.   Taurine in the nervous tissue In: S. Ehrenpreis and I. Kopin (Eds.) Reviews in Neurosciences. Raven Press, New York. 1978, Vol. 3, pp. 157-194.
36. McBride, W. J. and Frederickson, R. C. A.   Neurochemical and neurophysiological evidence for a role of taurine as an inhibitory neurotransmitter in the cerebellum of the rat. In: A. Barbeau and R. Huxtable (Eds.), Taurine and Neurological Disorders, Raven Press, New York, 1978, pp. 415-428.
37. Mutani, R., Bergamini, L., Fariello, R. and Delsedime, H.   Effects of taurine on cortical acute epileptic foci, Brain Res.,

70 (1974) 170–173.

38. Pasantes-Morales, H., Bonaventure, N., Wioland, N. and Mandel, P. Effect of intravitreal injections of taurine and GABA on chicken ERG. Int. J. Neurosci. 5 (1973) 235–241.

39. Pasantes-Morales, H., Klethi, J., Urban, P. F. and Mandel, P. The effect of electrical stimulation, light and amino acids on the efflux of ^{35}S-taurine from the retina of the domestic fowl. Exp. Brain Res., 19 (1974) 131–142.

40. Schmid, R., Sieghart, W. and Karobath, M. Taurine uptake in synaptosomal fractions of rat cerebral cortex. J. Neurochem. 25 (1975) 5–9.

41. Sgaragli, G. P. and Pavan, F. Effects of neutral amino acids injected into cerebrospinal fluid space on glucose metabolism in the rat brain. Neuropharmacology 12 (1973) 653–661.

42. Sgaragli, G. P., Magnani, M., Carla, U. and Giotti, A. Muscle relaxation induced in the rabbit by intracerebroventricular taurine injection: a supraspinal effect. Naunyn Schmied. Arch. Pharmacol. 295 (1976) 95–97.

43. Sieghart, W. and Heckl, K. Potassium-evoked release of taurine from synaptosomal fractions of rat cerebral cortex. Brain Res. 116 (1976) 538–543.

44. Snyder, S. H. The glycine synaptic receptor in the mammalian central nervous system. Br. J. Pharmacol. 53 (1975) 475–484.

45. Sturman, J. A. Taurine pool sizes in the rat: effects of vitamin-B6 deficiency and high taurine diet. J. Nutr. 103 (1973) 1566–1580.

46. Van Gelder, N. M., Sherwin, A. L. and Rasmussen, T. Amino acid content of epileptogenic human brain: focal versus surrounding regions. Brain Res. 40 (1972) 385–393.

47. Van Gelder, N. M. Antagonism by taurine of cobalt induced epilepsy in cat and mouse. Brain Res. 47 (1972) 157–168.

A PRESYNAPTIC VOLTAGE CLAMP STUDY IN THE SQUID STELLATE GANGLION

R. Llinás and K. Walton

Department of Physiology and Biophysics
New York University Medical Center
550 First Avenue
New York, N.Y. 10016

INTRODUCTION

The importance of extracellular calcium in synaptic transmission was first recognized by Locke (29) in 1894. Working with a frog nerve-muscle preparation, he observed that if calcium was removed from the bathing medium, stimulation of the nerve did not elicit muscle contraction, although the muscle did contract when stimulated directly. Addition of calcium to the bathing medium restored the effectiveness of nerve stimulation. More recently, further work with the neuromuscular junction has elucidated the role of calcium in synaptic transmission (cf. 9). This work established the effective location (calcium must be present at the site of the junction itself) (7,10,11) and the timing (calcium must be present immediately before or during nerve stimulation) (13) of the calcium action.

More detailed studies followed in which tetrodotoxin was used to block nerve conduction. Under these conditions, graded transmitter release can be elicited by applying focal depolarizations to the nerve ending (14,15). Results of these studies led Katz and Miledi to propose what became known as the "calcium hypothesis", namely that the first step leading to the release of transmitter was the entry of calcium ions or a calcium complex into the presynaptic terminal.

The squid giant synapse preparation, first introduced by J. Z. Young in 1939 (35), provided an opportunity to study the presynaptic aspects of transmission (especially the role of calcium) more closely. A major advantage of this preparation is the accessibility and size of its pre- and post-fibers which allow simultaneous

microelectrode penetration of both elements without significant
damage to synaptic transmission (6,31,34). In this manner the
presynaptic electrical events leading to transmitter release can be
recorded and polarizing current applied directly across the membrane
of this terminal, allowing a direct study of depolarization-release
coupling. Early work with this preparation found that: (1) Calcium
is not only necessary but sufficient, for transmitter release. When
the sodium and potassium conductances are blocked, graded depolarization of the presynaptic terminal leads to graded transmitter release
(5,12,19,20). (2) There is a level of depolarization, the
"suppression potential", which was assumed to correspond to the
equilibrium potential for calcium or a calcium complex. At this
potential there is no transmitter release during the pulse (12,20,24).
However, after the end of the current pulse a postsynaptic response
is again recorded. These results were interpreted to indicate that
depolarization increases the presynaptic membrane conductance to
a cation or cation complex which enters the terminal until a
potential is reached at which there is no electrochemical gradient.
When the current pulse was terminated, the potential was assumed to
fall below the equilibrium potential and thus the cations would
again enter the terminal and transmitter would be released. (3) The
presynaptic terminal is capable of generating calcium spikes (16,17,
28). The size and duration of this regenerative activity varied
directly with the external calcium concentration and could be
recorded only from the presynaptic fiber and only in the region of
the synapse (16,26). Together, these findings provided indirect
evidence for a voltage-dependent calcium conductance in the
presynaptic terminal.

The first direct proof that a voltage-dependent increase in
intracellular calcium triggered transmitter release was provided by
Llinás and his colleagues (23,24), using the bioluminescent protein
aequorin. Aequorin reacts with very small quantities of calcium
(as low as 10^{-8} M) to produce light (4). Following presynaptic
intracellular injection of this protein, depolarization of the
terminal leads to transmitter release which is accompanied by an
increase in luminescence and, thus, intracellular free calcium.
Further, in the absence of extracellular calcium or the presence
of calcium blocking agents, light emission and synaptic release
were simultaneously blocked. In addition, following the aequorin
injection into the presynaptic terminal in a preparation treated
with tetraethylammonium (TEA) and tetrodotoxin (TTX), graded
depolarization led to a graded synaptic release and light emission
only to about 80 mV depolarization. Beyond this level, light and
transmitter output decreased gradually until the suppression
potential was reached. At the suppression potential both
transmitter release and light emission disappeared simultaneously,
only to reappear at the end of the pulse. These experiments
provided a direct demonstration that, as theorized by Katz and Miledi
(12), the suppression potential is indeed produced as V_m approach

E_{Ca}. Direct evidence for the triggering of transmitter release by an increase in intracellular calcium was provided by Miledi (30). He found that injection of calcium into the presynaptic terminal produced a small but significant postsynaptic depolarization having membrane noise characteristics typical of synaptic transmission.

Thus, a picture emerged of the crucial role played by calcium in depolarization-release coupling. Specifically, $[Ca^{2+}]$ i is increased by the activation of a voltage-dependent calcium conductance in the presynaptic terminal membrane and this increased calcium is instrumental in triggering transmitter release. Recently, in applying techniques of voltage clamping to the presynaptic terminal of the squid giant synapse we have studied this calcium conductance in some detail (22,25,26,27), providing a quantitative picture of the calcium current and its role in synaptic transmission. The findings of these studies provide the main subject of this paper. A model for the calcium channel, and for synaptic transmission itself, based on these experimental results, is also briefly discussed.

VOLTAGE CLAMP STUDIES

The large size of the presynaptic terminal of this squid synapse has allowed the use of voltage clamp techniques to determine the voltage and time dependence of the presynaptic ionic calcium currents during transmission (22,25,26) while simultaneously monitoring transmitter release (27). In preparations where sodium and potassium conductances are blocked respectively by TTX (32) and a 3-amino-pyridine (3-AmP) (28,33) and TEA (1) combination, rapid depolarization steps using voltage clamp techniques may be accomplished at the presynaptic terminal. Following a step depolarization, a calcium-dependent inward current may be observed which is blocked by manganese (50 mM) or cadmium chloride (1 mM) (22) and is absent if calcium is removed from the extracellular medium.

Following the onset of a sustained presynaptic depolarization, the membrane permeability to calcium increases rather slowly compared to that of sodium (8) and reaches a peak whose amplitude varies with the level of depolarization (Fig. 1A, B). At low levels of depolarization (Fig. 1A) a small inward current (lower record) is seen which begins very slowly and shows a close-to-linear increase with time. At the end of this pulse, a fast tail current is observed. The postsynaptic response generated by this current is illustrated in the middle trace. Note that as with the current, the response during the pulse ("on" response) has a close-to-linear rate of rise and that the EPSP which follows the break of the voltage clamp pulse ("off" response) is associated with the inward tail current.

The calcium current (I_{Ca}) increases in amplitude with increasing

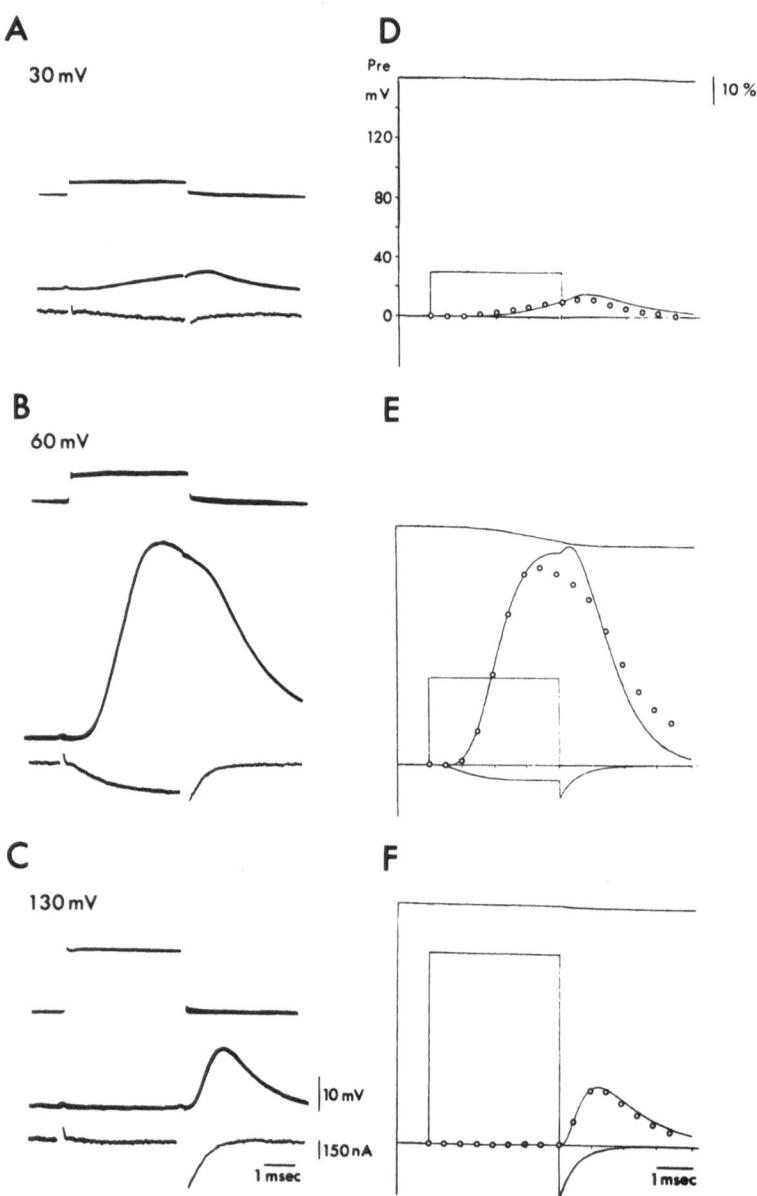

Fig. 1. Synaptic transmission during voltage clamp of presynaptic
terminal. A,B,C: Experimental data. Top trace, presynaptic voltage;
middle trace, postsynaptic response; lower trace, calcium current.
The S-shape of the current onset can be seen at 60 mV depolarization.
Note the fast tail current and the "on" and "off" EPSP. D,E,F:
Numerical solution to mathematical model. Top trace, vesicle
depletion; other traces as in A-C. Open circles, recorded EPSP.
$[Ca^{2+}]_o = 10$ mM.

levels of depolarization reaching a maximum in most synapses, at
60 mV. At this level the sigmoidal character of the current onset
is evident (Fig. 1B, bottom trace), following which the current
plateaus indicating the absence of inactivation, at least for pulses
of up to 30 msec. At the end of the pulse a tail current is, again,
apparent. Accompanying this increase in I_{Ca}, a change is seen in the
shape, rate of rise and amplitude of the EPSP. At 60 mV presynaptic
depolarization the EPSP reaches a peak and begins to decline during
the pulse. At the end of the pulse a secondary transmitter release,
the "off" response (Fig. 1B, middle trace) is seen.

As the presynaptic voltage is increased beyond 60 mV, the steady-
state presynaptic current decreases in amplitude while the tail
current increases to reach a maximum near 110 mV depolarization. At
the same time, the rate of rise and amplitude of the "on" EPSP
decreases while the "off" response increases. The steady-state
current continues to decrease beyond 110 mV until the suppression
potential (12) is reached at close to 130 mV depolarization (Fig. 1C,
lower trace). At this point no current is seen during the voltage
pulse, but a large tail current is observed following the pulse-break.
Similarly, no synaptic transmission is observed for the duration of
the pulse and a sharp, short-latency "off" postsynaptic response is
generated by the tail current. A clear correlation can thus be seen
between the characteristics of presynaptic I_{Ca} and the postsynaptic
response, the "on" EPSP being related to the I_{Ca} during the pulse
and the "off" EPSP to the tail I_{Ca} current.

The relationship between the amplitude of the presynaptic
voltage clamp depolarization and the amplitude of the inward steady-
state calcium current ($I_{Ca}\infty$) and tail calcium current is shown in
Fig. 2A. As seen in that figure, $I_{Ca}\infty$ follows a bell-shaped curve
with a rather rapid rise, reaching a peak near 60 mV. The decrease
in current amplitude at higher levels of depolarization is asymptotic,
making the estimation of a reversal potential by extrapolation very
difficult. The tail currents increase sharply for increasing levels
of depolarization up to about 80 mV, at which point they began to
level off reaching a maximum near 110 mV.

The quantitative relationship between the amplitude of the
presynaptic voltage clamp depolarization and the postsynaptic
potential is shown in Fig. 2B where the closed circles represent
the "on" and the crosses the "off" postsynaptic response. The peak
amplitude for the postsynaptic potential occurs, as does the peak
$I_{Ca}\infty$, near presynaptic depolarization of 60 mV and the suppression
potential in the vicinity of 120 to 130 mV. The amplitude of the
"on" EPSP was measured as the maximum level of postsynaptic
membrane potential prior to the termination of the voltage clamp
step. The amplitude of the "off" response was taken as the
difference between the "on" EPSP and the maximum postsynaptic
potential after the termination of the clamp.

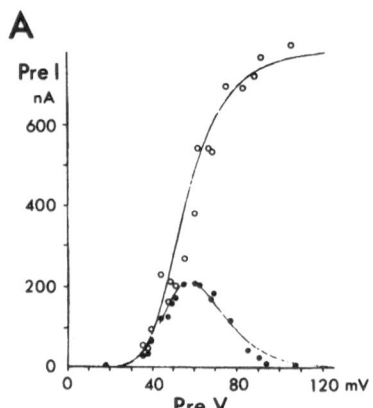

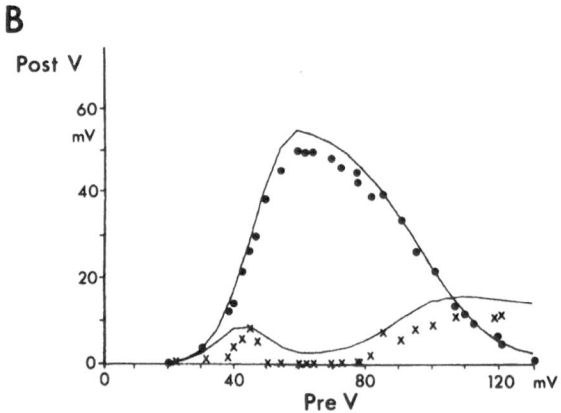

Fig. 2.   Dependence of calcium current (A) and EPSP (B) on presynaptic
depolarization.   In A: Dots, steady-state I_{Ca}; open circles, tail
current.   In B: Dots, "on" EPSP; crosses, "off" EPSP.   Continuous
lines in A and B, solution of mathematical model. $[Ca^{2+}]_o$ = 10 mM.

As can be seen in Fig. 2, the voltage dependence of the calcium
current and the postsynaptic response is very similar, suggesting
a close-to-linear relationship between these two variables.   In
fact, when I_{Ca}^{∞} is plotted against the "on" EPSP amplitude using
double-logarithmic coordinates, a mean of 1.36 ($\pm$ 0.13, n = 8)
is obtained.   Similarly, a mean of 1.11 ($\pm$ 0.22, n = 4) is obtained
when tail current amplitude and "off" EPSP amplitudes are plotted
using double-logarithmic coordinates (27).

Other results obtained from these voltage clamp studies relate
to the nature of synaptic delay.   As discussed in previous papers
(22, 25), the temporal relationship between depolarization and

A

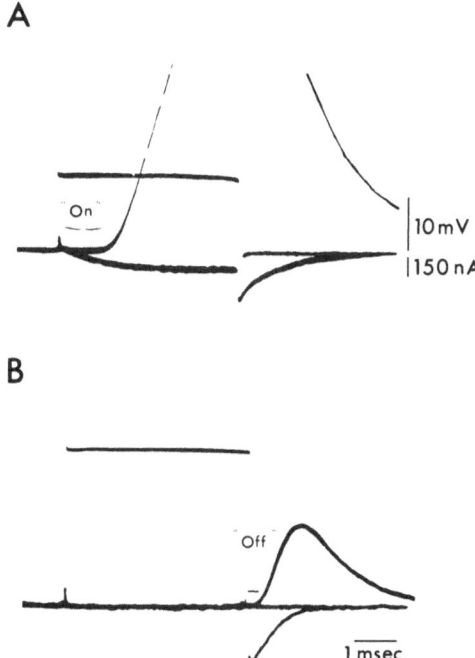

On

10 mV

|150 nA

B

Off

1 msec

Fig. 3. Synaptic delay for the "on" and "off" EPSP. A, "On" response
is shown for a 60 mV presynaptic voltage clamp; latency 733 usec.
B, The "off" response is seen following a voltage clamp to the
suppression potential (130 mV); 200 μsec latency.

transmitter release may be divided into two parts which we have
called a and b (22). The first component of this delay (a) is related
to the time required for opening of the calcium channels while
the b portion relates to the time between entry of calcium and the
onset of the postsynaptic potential. Examples of these two
components of delay can be obtained by comparing the "on" and "off"
release as shown in Fig. 3A and B. Here in record A a 60 mV
presynaptic voltage clamp step (from -70 mV holding potential)
demonstrates a characteristic 1 msec delay for the "on" response
(comprised of components a and b). The record in Fig. 3B was taken
at the suppression potential and illustrates the tail calcium
current and the latency (about 200 μsec; comprised of the b
component alone) for the "off" synaptic potential generated by
this short current pulse. The actual synaptic delay may, in fact,
be slightly shorter since the falling phase of the pulse is not
instantaneous. A direct measurement for these two components was
obtained (22) by comparing the time of onset of the voltage
clamp pulse with the first sign of inward calcium current and then

the delay between the current onset and the initiation of the
postsynaptic potential.

A MODEL FOR SYNAPTIC TRANSMISSION

    Based on the above findings a kinetic model was developed with
I. Steinberg for the relation between presynaptic depolarization
and calcium current and between calcium current and transmitter
release (25,26,27). The model for the calcium current indicates
that 1) the calcium conductance change follows fifth order
Michaelis-Menten kinetics and 2) the opening of the channel is
probably accompanied by a gating current (2,18) of five charges,
assuming that the fifth order kinetics corresponds to 5
noncooperative conformational changes occurring prior to the channel
opening (n = 5). These conformational changes, which are voltage
and time-dependent, are thought to be governed, as in the case of
the Hodgkin and Huxley model (8) by forward and backward rate
constants between the active and inactive states of the subunits in
the channels. Based on these considerations, I_{Ca} is given as the
product of the number of open gates ([G]) and the current flow per
open gate per unit time (j) (Equation 15 in ref. 26).

    The solution of this equation for presynaptic clamp voltages
of 30, 60 and 130 mV is given in Fig. 1 D-F (lower trace). An
expression for the steady-state current, I_{Ca}^{∞} , is obtained by
simplifying the equation to eliminate the time-dependent component.
The solution to the equation for I_{Ca}^{∞} for various values of V is
given by the continuous lines in Fig. 2A. As can be seen in that
figure, the agreement between the model and the data is quite good.

    Several factors must be considered when modeling the remaining
steps in synaptic transmission: 1) The high specificity of calcium
ions in promoting vesicle fusion and transmission release (9)
strongly suggests that a specific binding entity for calcium,probably
a protein, is involved in the process (3). Such a factor is
included in this model and will be referred to as the fusion-
promoting factor (fpf). 2) As fusion is initiated, we assumed that
vesicles are depleted from the immediate vicinity of the plasma
membrane (the immediately available store) and that this may become
a factor in the transmission process. 3) The time required for
the diffusion of the transmitter across the synaptic gap is expected
to be very short (less than a microsecond for a gap width of 200 A
and a diffusion coefficient of 10^{-5} cm^2/sec) and thus is not a
significant consideration. 4) The opening and closing time
constants of the gating of the postsynaptic receptor transmitter
complex have been assumed to be proportional to the rate of
transmitter release, and it is also assumed that the postsynaptic
receptors are far from saturation. 5) Finally, the electrical time
constants of the postsynaptic terminal membrane are included.

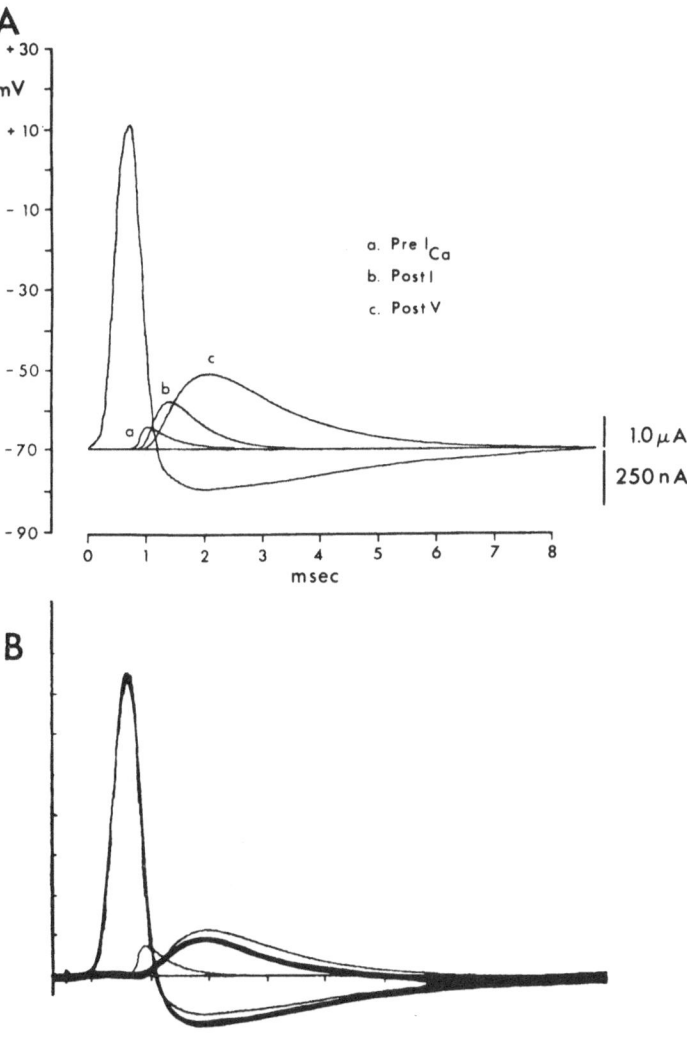

Fig. 4. Theoretical solution for propagating action potential in the presynaptic terminal and related steps in synaptic transmission. A, reconstruction of events during synaptic transmission based on Equations 10, 11, 14 and 18 of Llinás et al (submitted A). a, calcium current; b, postsynaptic current; c, postsynaptic potential. B, comparison of numerical solution model and experimentally observed synaptic transmission. Fine lines: presynaptic action potential plus calculated I_{Ca} and EPSP. Heavy lines: actual recording of pre- and postsynaptic response from synapse. Note that the calcium current has a late onset and a rather prolonged time course.

Following the above considerations, a 3-compartment scheme was developed to model transmitter release and postsynaptic current, each compartment having forward and backward first-order rate constants (Llinás, Steinberg and Walton, submitted B). The first compartment consists of a calcium protein reaction involving fpf. The second relates to a fusing factor-synaptic vesicle reaction leading to transmitter release, and the third component concerns synaptic transmitter triggering the postsynaptic conductance which generates the EPSP.

The computation of EPSP amplitude and time course for three levels of presynaptic voltage clamp are illustrated in Fig. 1 D-F (solid lines). Values obtained experimentally are given by open circles; the values obtained from the model for the peak postsynaptic "on" and "off" EPSP are given by the continuous lines.

As can be seen in Figs. 1 and 2, the present model of presynaptic calcium current, transmitter release and postsynaptic response agrees reasonably well with the voltage clamp data. The description of the relation between calcium current and postsynaptic potential provided by the model is of further interest if it can predict the events triggered by an actual presynaptic action potential and it can give a quantitative description of the time course and magnitude of these predicted events. Thus, as illustrated in Fig. 4A, an experimentally obtained action potential was utilized to obtain (via Equations 10, 11, 14 and 18 from ref. 26) the time courses of (a) the calcium current, (b) the postsynaptic current, and (c) the postsynaptic potential. The characteristics of these variables are amenable to comparison with actual records and resemble quite closely the typical experimental results (Fig. 4B).

One interesting detail which became clear from this model is that transmitter release by an action potential is mainly an "off" release since, given the time course of the action potential, I_{Ca} occurs normally at the falling phase of the spike. This time sequence has the interesting effect that the onset of g_{Ca} occurs at a time when the membrane potential is returning towards its resting level and thus the ionic conductance change occurs at a high calcium e.m.f. The present data and the model derived from them have provided a working hypothesis regarding the steps in the chain of events underlying synaptic transmission.

Acknowledgements. This research was supported by United States Public Health Service grants NS-14014 and NS-13742 from the National Institute of Neurological and Communicative Disorders and Stroke.

REFERENCES

1. Armstrong, C. M. and Binstock, L., Anomalous rectification in
    the squid giant axon injected with tetraethylammonium chloride.
    J. gen. Physiol., 48 (1965) 859-872.
2. Armstrong, C. M. and Bezanilla, F., Currents related to movement
    of the gating particles of the sodium channels. Nature, 242
    (1973) 459-461.
3. Baker, P. F. and Schaepfer, W. W., Uptake and binding of calcium
    by axoplasm isolated from giant axons of Loligo and Myxicola.
    J. Physiol. (Lond.), 276 (1978) 103-125.
4. Blinks, J. R., Prendergast, F. G. and Allen, D. G. Photoproteins
    as biological calcium indicators. Pharm. Rev. 28 (1976) 1-93.
5. Bloedel, J. R., Gage, P. W., Llinas, R. and Quastel, D. M. J.,
    Transmitter release at the squid giant synapse in the presence
    of tetrodotoxin. Nature 212 (1966) 49-50
6. Bullock, T. H. and Hagiwara, S., Intracellular recording from
    the giant synapse of the squid. J. gen. Physiol., 40 (1957)
    565-577.
7. Castillo, J. del and Katz, B., Changes in end-plate activity
    produced by presynaptic polarization. J. Physiol. (Lond.).
    124 (1954) 586-604.
8. Hodgkin, A. L. and Huxley, A. F., A quantitative description of
    membrane current and its application to conduction and excitation
    in nerve. J. Physiol. (Lond.) 117 (1952) 500-544.
9. Katz, B., The Release of Neural Transmitter Substances
    (Sherrington Lectures X), Charles C Thomas, Springfield, Ill.,
    1969.
10. Katz, B. and Miledi, R., Localization of calcium action at the
    nerve muscle junction. J. Physiol. (Lond.), 171 (1964) 10-12P.
11. Katz, B. and Miledi, R., The effect of calcium on acetylcholine
    release from motor nerve endings. Proc. R. Soc. (Lond.) B,
    161 (1965) 406-503.
12. Katz, B. and Miledi, R., A study of synaptic transmission in
    the absence of nerve impulses. J. Physiol. (Lond.), 192 (1967)
    407-436.
13. Katz, B. and Miledi, R., The timing of calcium action during
    neuromuscular transmission. J. Physiol. (Lond.), 189 (1967)
    535-544.
14. Katz, B. and Miledi, R., Tetrodotoxin and neuromuscular
    transmission. Proc. R. Soc. (Lond.) B, 167 (1967) 8-22.
15. Katz, B. and Miledi, R., The release of acetylcholine from nerve
    endings by graded electric pulses. Proc. R. Soc. (Lond.) B,
    167 (1967) 23-28.
16. Katz, B. and Miledi, R., Tetrodotoxin-resistant electric activity
    in presynaptic terminals. J. Physiol. (Lond.), 203 (1969) 459-
    487.
17. Katz, B. and Miledi, R., The effect of prolonged depolarization
    on synaptic transfer in the stellate ganglion of the squid.
    J. Physiol. (Lond.) 216 (1971) 503-512.

18. Keynes, R. D. and Rojas, E., Kinetics and steady state
    properties of the charged system controlling sodium conductance
    in the squid giant axon. J. Physiol. (Lond.), 239 (1974) 393–434.
19. Kusano, K., Further study of the relationship between pre- and
    postsynaptic potentials in the squid giant synapse. J. gen.
    Physiol., 52 (1968) 326–345.
20. Kusano, K., Influence of ionic environment on the relationship
    between pre- and postsynaptic potentials. J. Neurobiol., 1
    (1970) 437–457.
21. Kusano, K., Livengood, D. R. and Werman, R., Correlation of
    transmitter release with membrane properties of the presynaptic
    fiber of the squid giant synapse. J. gen. Physiol., 50 (1970)
    2579–2601.
22. Llinás, R. R. Calcium and transmitter release in squid synapse.
    In W. M. Cowan and J. A. Ferrendelli (Eds.) Society for
    Neuroscience Symposia, vol. 2, Society for Neuroscience,
    Bethesda, MD, 1977, pp. 139–160.
23. Llinás, R., Blinks, J. R. and Nicholson, C. Calcium transient
    in presynaptic terminal of squid giant synapse: Detection with
    aequorin. Science, 176 (1972) 1127–1129.
24. Llinás, R. and Nicholson, C., Calcium role in depolarization-
    release coupling: An aequorin study in squid giant synapse.
    Proc. Natl. Acad. Sci. (USA) 72 (1975) 187–190.
25. Llinás, R., Steinberg, I. Z. and Walton, K., Presynaptic calcium
    currents and their relation to synaptic transmission: Voltage
    clamp study in squid giant synapse and theoretical model for
    the calcium gate. Proc. Natl. Acad. Sci. (USA), 73 (1976)
    2918–2922.
26. Llinás, R., Steinberg, I. Z. and Walton, K., Presynaptic
    calcium currents in squid stellate ganglion: A voltage clamp
    study. Submitted for publication (A).
27. Llinás, R., Steinberg, I. Z. and Walton, K., Relationship
    between presynaptic calcium current and postsynaptic potential
    in squid giant synapse. Submitted for publication (B).
28. Llinás, R., Walton, K. and Bohr, V. (1976) Synaptic transmission
    in squid giant synapse after potassium conductance blockage with
    external 3- and 4-aminopyridine. Biophys. J., 16 (1976) 83–86.
29. Locke, F. S. Notiz ueber den Einfluss physiologischer
    Kochsalzloesung auf die elektrische Erregbarkeit von Muskel
    und Nerve. Zentrabl. Physiol., 8 (1894) 166–167.
30. Miledi, R. Transmitter release induced by injection of calcium
    ions into nerve terminals. Proc. R. Soc. (Lond.) B, 183 (1973)
    421–425.
31. Miledi, R. and Slater, C. R. (1966) The action of calcium on
    neuronal synapses in the squid. J. Physiol. (Lond.) 184 (1966)
    473–498.
32. Narahashi, T., Moore, J. W. and Scott, W. R. Tetrodotoxin
    blockage of sodium conductance increase on lobster giant axons.
    J. gen. Physiol., 47 (1964) 965–974.

33. Pelhate, M. and Pichon, Y. Selective-inhibition of potassium current in giant-axon of cockroach. J. Physiol. (Lond.), 242 (1974) P90-91.
34. Takeuchi, A. and Takeuchi, N. Electrical changes in pre- and postsynaptic axons of the giant synapse of Loligo. J. gen. Physiol., 45 (1962) 1181-1193.
35. Young, J. Z. Fused neurons and synaptic contacts in the giant nerve fibres of cephalopods. Phil. Trans. R. Soc. (Lond.) B, 229 (1939) 465-503.

CALCIUM TRANSPORT AND THE RELEASE OF NEUROTRANSMITTERS:

EFFECTS OF DRUGS IN VIVO AND IN VITRO

Ricardo Tapia and Clorinda Arias

Departamento de Neurociencias
Centro de Investigaciones en Fisiología Celular
Universidad Nacional Autónoma de México
México 20, D.F.

INTRODUCTION

The role of calcium ions in neurotransmitter release is now amply recognized. External calcium appears to enter the presynaptic terminal upon depolarization, and the increase in its intraterminal cytoplasmic concentration somehow induces the release of the neurotransmitter (21,25,26,39,40,51). In vertebrates this phenomenon occurs both at neuromuscular junctions, where the release of acetylcholine (ACh) occurs probably from synaptic vesicles, and in central synapses, where the role of vesicles is less clear. Thus, in several preparations from mammalian central nervous system, such as tissue slices or synaptosomes, it has been shown that the depolarization-induced release of catecholamines (6,10), ACh (8,45,64) or neurotransmitter amino acids such as GABA (10,16,27,44,56), glutamate (9,53) or glycine (27,44) is a phenomenon dependent on the presence of external Ca^{2+}. Because of this necessity for Ca^{2+} transport across the presynaptic membrane in order to induce transmitter release, drugs capable of modifying Ca^{2+} transport should affect the release of neuro-transmitters. This approach has been followed in the present communication in two different ways: 1) drugs which block Ca^{2+} transport or apparently increase its cytoplasmic concentration at the nerve terminals have been injected to experimental animals, and 2) the effect of such drugs on neurotransmitter release has been studied in synaptosomes prepared from brain tissue.

Among the drugs capable of increasing Ca^{2+} transport in biological membranes, it is possible to distinguish those which promote Ca^{2+} movements through a direct action on the membrane, such as the calcium ionophores X537A or A23187 (49), and those which increase Ca^{2+} transport probably secondarily to an effect on the membrane permeability to other ions, such as ouabain (2,13), veratridine (4,5,24), tetraethylammonium (20,21), guanidine (12, 17,18,32,36,47) or 4-aminopyridine (4-AP) (15,29,32,41,42). Furthermore, there are drugs, such as uncouplers of oxidative phosphorylation (7,52) or quinidine (3,52), which increase cytoplasmic Ca^{2+} concentration apparently by blocking its intraterminal sequestration or by facilitation of its release from intraterminal stores.

Drugs that diminish Ca^{2+} transport include those which are able to chelate this cation and those which bind to biological membranes in such a way that Ca^{2+} movements across the membrane are blocked. Ruthenium red (19,34,43) and the lanthanides (14,63) seem to act through this mechanism.

EFFECTS IN VIVO OF DRUGS AFFECTING Ca^{2+} TRANSPORT

The most simple way for studying the consequences of modifying Ca^{2+} transport on animal behavior is to alter its extra-cellular concentration; this can theoretically be accomplished by the administration of a calcium chelating agent, such as EDTA. As shown in Table 1, when we injected EDTA intraperitoneally (i.p.) to mice, the animals showed excitation followed by flaccid paralysis, which lasted for several minutes, and then the animals died. Although EDTA might exert several direct effects on nervous excitability, the fact that the paralyzing action of EDTA was antagonized by the simultaneous administration of $CaCl_2$ (Table 1) suggests that this action is related to a decrease of calcium concentration.

Ruthenium red (RuR) is an inorganic dye used for electron microscopy, which is capable of blocking the transport of Ca^{2+} in a variety of biological preparations (19,22,34,43,50). When this dye was administered i.p. to mice, at a dose much smaller than that of EDTA, the animals showed complete flaccid paralysis, which lasted for 3-4 hr (Table 1); in contrast to the effects of EDTA, the animals did not show any excitation period, and they did not die but recovered slowly. These differences might indicate that RuR does not have central actions after systemic administration, whereas EDTA could possess some central effects. As with EDTA, the injection of $CaCl_2$ prevented the paralyzing effect of RuR.

In contrast to the flaccid paralysis observed after the i.p. injection of EDTA or RuR, their intracisternal injection produced convulsions (Table 2). Again the effects of EDTA were very rapid, most of the mice died almost immediately in tonic generalized convulsion, whereas RuR produced tonic-clonic convulsions of both

Table 1. Effect of intraperitoneal administration of EDTA and RuR
in mice, and antagonic action of calcium*

Dose: theoretical concentration (μM) of drug in the body	Effect (Number of animals in parentheses)
EDTA, 400	Initial excitation; then flaccid paralysis and death (15)
EDTA, 400 + CaCl$_2$, 400	Depression; no flaccid paralysis (10)
RuR, 40	Total flaccid paralysis for 90–180 min (30)
RuR, 40 + CaCl$_2$, 400	Depression; no flaccid paralysis (10)

*Data from Ref. (61).

Table 2. Effect of intracisternal administration of EDTA and RuR
in mice, and antagonic action of calcium*

Dose: theoretical concentration (μM) of drug in the brain	Effect (Number of animals in parentheses)
EDTA, 200 or 400	Tonic convulsion and death immediately after injection or within 3 min (25)
EDTA, 400 + CaCl$_2$, 400	Depression and respiratory difficulties; 4 out of 10 animals died but convulsions were not observed
RuR, 20 or 40	Tonic-clonic convulsions for several hours, or death in convulsions (36)
RuR, 40 + CaCl$_2$, 600	No convulsions; depression and overreaction when touched (10)

*Data from Ref. (61)

forelimbs and hindlimbs, with extension of the head and loss of the
righting reflex; this status epilepticus lasted several hours when
the animals did not die. As in the case of the flaccid paralysis
observed after i.p. administration, the intracisternal injection of
$CaCl_2$ prevented the convulsions produced by RuR and by EDTA,
although the animals looked depressed and overeacted when touched
(Table 2).

RuR was also injected to cats permanently implanted with
electrodes in the cortex, in the hippocampus and in the neck
muscles (61). After i.p. administration of RuR (8 mg/kg) the cats
were clearly hypotonic, and this was evident from the electromyogram,
but no changes were observed in the electrical activity of the brain.
The systemic administration of carbachol (100 µg) to one of the
cats under this flaccid paralysis produced prominent myoclonic
responses, indicating that the paralyzing effect of RuR was not
due to a postsynaptic blocking action (61). When RuR was injected
into the lateral ventricle or the hippocampus (8-100 µg) through
permanently implanted cannulas, the cats showed generalized
seizures, evident both in the electrical recordings and clinically.
The convulsions lasted for several hours (61).

The results of the above experiments, taken together, suggest
that after i.p. administration both EDTA and RuR do not cross the
blood-brain barrier and block the Ca^{2+}-dependent release of ACh
at neuromuscular junctions, thus causing flaccid paralysis. In fact,
experiments in vitro (48) have shown that RuR greatly decreases
the frequency of miniature end-plate potentials (mepps) in the
frog neuromuscular junction, that this effect is prevented by
increasing external Ca^{2+} concentration and that it cannot be
explained by the relatively small postsynaptic action of RuR
described previously (1). In addition to this action on the
spontaneous release of ACh in neuromuscular junctions, RuR also
prevented its release induced by electrical stimulation (1) or by
the ionophore X537A in the presence of Ca^{2+} (48).

On the other hand, the convulsions observed after the
intracranial injections of RuR and EDTA could be accounted for
by at least two mechanisms: their blocking action on the Ca^{2+}-
dependent neurotransmitter release is a) restricted to an inhibitory
neurotransmitter involved in the regulation of cerebral excitability
or b) general for all neurotransmitters but the behavioral effect
reflects the deficit in the synaptic function of an inhibitory
transmitter which clearly predominates over the physiological
action of excitatory neurotransmitters.

The first alternative does not seem probable, since Ca^{2+} is
required for the release of all neurotransmitters. Since there
is evidence (57-59,62,65) that GABA is an inhibitory transmitter
with a tonic regulatory action on cerebral excitability, whereas

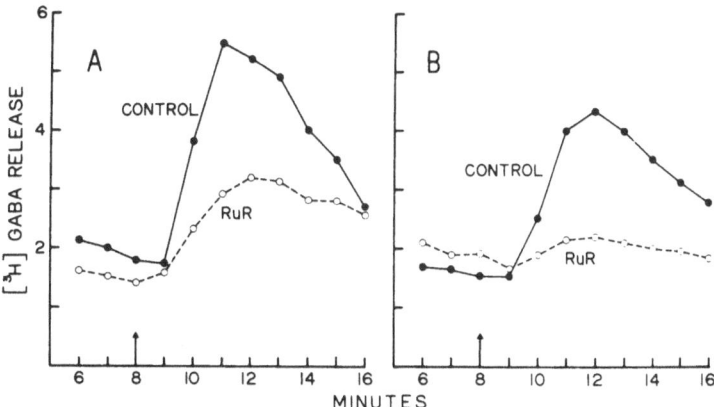

Fig. 1.   Inhibitory effect of RuR on the Ca^{2+}-dependent release of ^3H-GABA in mouse synaptosomes. A, synaptosomes from mice injected intracisternally with saline (control) or with RuR (8.8 µg); the latter animals were sacrificed at the time of tonic convulsion. B, synaptosomes preincubated without (control) or with RuR (8 µM). After loading with $[^3H]$ GABA, synaptosomes were superfused with Krebs-bicarbonate medium containing 48 mM KCl but no Ca^{2+}. At 8 min (arrow) the medium was substituted by a 48 mM KCl, 2 mM $CaCl_2$-containing medium. $[^3H]$ GABA release is expressed as percentage of total radioactivity (fractions plus filter). Data from Refs. (38) and (60).

other inhibitory transmitters are not related in this way to motoneurons firing, we favor the second alternative. Thus, we postulate that convulsions induced by the intracranial injection of RuR are due to a decrease in the Ca^{2+}-dependent release of GABA. In order to obtain evidence that could support this hypothesis, we studied the Ca^{2+}-dependent, depolarization-induced release of $[^3H]$ GABA from synaptosomal fractions prepared from brains of mice injected intracisternally with RuR and sacrificed during the status epilepticus described above. The results, shown in Fig. 1A, indicate that there was a 65% inhibition of the Ca^{2+}-dependent GABA release. Since RuR binds rapidly and irreversibly to membranes, including synaptosomal membranes (28,34; see below), it seems reasonable to conclude that RuR binds to presynaptic endings after its intracisternal administration, thus allowing the detection of its inhibitory action on GABA release in synaptosomes isolated from the treated animals.

Reversal of the Paralyzing Effect of the Intraperitoneal
Administration of RuR

Although there are possibly some differences in their mode of
action, both guanidine (17,18,32,36,47) and 4-AP (15,29,32,41,42)
produce an increase of neurotransmitter release probably through an
enhancement of Ca^{2+} entry into the presynaptic ending as a
consequence of depolarization. Both drugs reverse the paralyzing
effects of botulinum toxin when tested in neuromuscular preparations
from toxin-treated animals, and 4-AP is also effective when
administered in vivo (30,31). We therefore investigated whether
these two drugs were capable of antagonizing the paralyzing effect
of RuR described above. When guanidine or 4-AP were injected to
mice 30 min after RuR, during complete flaccid paralysis, the
animals started moving in just a few seconds after the injection,
and in 1 or 2 min they began to walk. It must be emphasized that
these effects were observed 30 min after RuR injection, whereas
the control mice did not start to recover until 90 min later. At
the doses of guanidine (1.25 mg/kg) and 4-AP (1-3 mg/kg) used,
these drugs by themselves did not have any noticeable effect, except
perhaps a slight excitation with 4-AP; higher doses of 4-AP (5-10
mg/kg) produced convulsions.

This antagonic effect of guanidine and 4-AP on RuR-induced
flaccid paralysis provides further evidence in favor of an
involvement of presynaptic Ca^{2+} mechanisms in the paralyzing action
of RuR. In fact, as already mentioned, both drugs seem to act
presynaptically by increasing the cytoplasmic concentration of
Ca^{2+} and consequently increasing the release of ACh at neuromuscular
junctions, an effect which would be expected to antagonize the
above discussed blocking action of RuR on Ca^{2+} transport.
Furthermore, these results suggest that guanidine and 4-AP
stimulate ACh release through a mechanism which probably involves
Ca^{2+} channels different from those presumably blocked by RuR, or
that these drugs induce the release of intraterminal Ca^{2+} from
intracellular stores.

Lanthanum is known as a Ca^{2+} antagonist in many biological
systems (63). La^{3+} competes with Ca^{2+} for its membrane transport
sites, and has a dual effect on neuromuscular transmission. It
increases by two or three orders of magnitude the frequency of
mepps and it also blocks the depolarization-induced, Ca^{2+}-
dependent release of ACh (14). It was therefore of interest
to study the action of the i.p. injection of La^{3+} to animals, as
well as to compare its effects with those of guanidine and 4-AP
regarding the antagonism with RuR. In contrast to RuR, $LaCl_3$ did
not have any apparent effect on the behavior of the mice, even at
doses theoretically equivalent to the extracellular Ca^{2+}
concentration. La^{3+} also did not antagonize at all the effect of
RuR when administered 30 min after the dye, in sharp contrast with

the effect of guanidine and 4-AP. However, when La^{3+} was administered 30 min prior to RuR, it completely prevented the paralyzing effect of RuR. These results indicate that the notable stimulation of the spontaneous release of ACh produced by La^{3+} in vitro (14) does not occur in vivo, or that this effect is not enough to overcome the action of RuR. On the other hand, the effectiveness of La^{3+} when injected prior to RuR might indicate that La^{3+} binds to the membrane in such a way that blocks the binding of RuR, but it is not capable of displacing it once the dye is bound. Experiments in vitro to be described in the next section were designed to study this possibility.

CORRELATION OF THE EFFECT OF DRUGS IN VIVO AND IN VITRO

Several attempts were made to study whether the mechanisms postulated to be functioning in vivo after the injection of the drugs used could be supported by experiments in vitro. Since we are dealing mainly with mechanisms of presynaptic release, the experimental model chosen for these experiments was the synaptosomal fraction prepared from mouse brain.

One of the important questions to be answered is whether RuR blocks the transport of Ca^{2+}. Some reports in the literature indicate that in fact RuR possesses this effect in synaptosomes; however, these experiments were carried out at relatively large RuR concentrations, or during long time periods (13). We therefore studied the effect of RuR on the spontaneous and K^{+}-stimulated uptake of ^{45}Ca by synaptosomes. Fig. 2 shows that RuR at a concentration similar to that used in vivo did in fact block the uptake of ^{45}Ca stimulated by 50 mM K^{+}. This result is well correlated not only with the effects in vivo of RuR but also with its inhibitory action on GABA release after its intracisternal injection (Fig. 1A) and after its incubation in vitro with synaptosomes (Fig. 1B).

The finding that the i.p. administration of La^{3+} prior to RuR, but not after RuR, prevented the paralyzing action of the latter, prompted us to study whether La^{3+} was affecting the binding of RuR to presynaptic membranes. For this purpose synaptosomes were incubated with RuR and its binding was studied spectrophotometrically after sedimenting the synaptosomes by centrifugation. As can be seen in Fig. 3, in the absence of ions (sucrose-tris medium) RuR binds to the synaptosomal membrane about 3 times more than in a Krebs-tris medium. Preincubation with La^{3+} at 25-100 μM concentrations strongly inhibited the binding of RuR in Krebs-tris medium, whereas in sucrose-tris medium only a 100 μM concentration of La^{3+} had a notable inhibitory effect. When La^{3+} was added to synaptosomes after RuR binding, it was incapable of displacing it (data not shown). If these results are extrapolated in order to

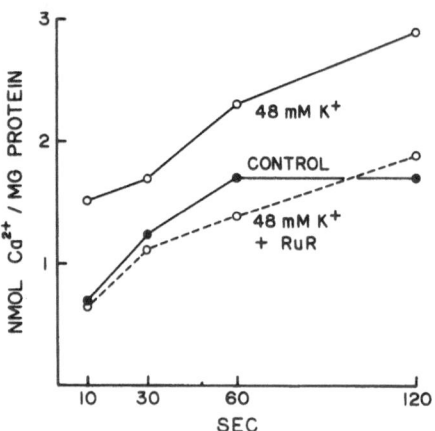

Fig. 2. Uptake of ^{45}Ca by mouse synaptosomes, and effect of RuR. Synaptosomes were incubated at 37° in 4.8 mM KCl Krebs-bicarbonate medium or in 48 mM KCl medium and the uptake was terminated at the times indicated by rapid filtration on 0.65 μm Millipore filters. The filters were rapidly washed with medium and the radioactivity was counted. When present, RuR (10 μM) was preincubated with the synaptosomes for 10 min. Mean values of 4-6 experiments.

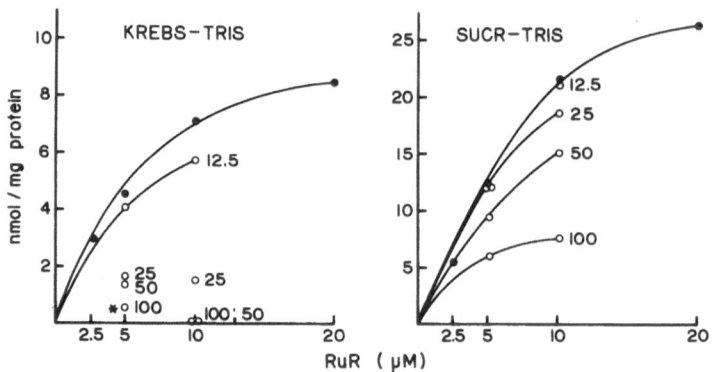

Fig. 3. Binding of RuR to mouse synaptosomes and effect of La^{3+}. Synaptosomes were incubated at room temperature for 10 min in Krebs-Tris or 0.32 M sucrose-25 mM Tris media, pH 7.0, with the indicated RuR concentrations, and the binding was stopped by centrifugation. RuR was measured spectrophotometrically at 540 nm in the supernatant. When present, LaCl$_3$ was preincubated with the synaptosomes for 10 min at the concentrations indicated by the numbers next to the curves or individual points. Mean values of 5-7 experiments. The asterisk indicate that at this La^{3+} and RuR concentrations 4 of 6 experiments gave a zero value (100% inhibition).

explain the effects of La^{3+} and RuR in vivo (i.p. administration), it may be concluded that La^{3+} blocks the paralyzing effect of RuR because it inhibits its binding to the motor nerve terminals in vivo, whereas it cannot reverse its action when injected after RuR because it does not displace it from the membrane sites.

From the above results it could also be concluded that RuR and La^{3+} bind to the same sites on the membrane. If this conclusion were correct, one should expect that RuR would block the binding of La^{3+} to synaptosomal membranes. We therefore studied the binding of La^{3+} to synaptosomes, using a double-beam spectrophometric technique with murexide as an indicator of the changes in La^{3+} concentration. It was observed that La^{3+} binds instantaneously to the synaptosomes, and that preincubation with RuR did not affect this binding (Fig. 4). Also interesting in this respect is the finding that La^{3+} did not produce any apparent effect when administered to mice, in spite of the fact that, as RuR, blocks the Ca^{2+}-dependent, depolarization-induced release of ACh (1, 14, 48; see below). These observations argue against a common site of binding of RuR and La^{3+} on the Ca^{2+}-transport membrane sites.

Since La^{3+} binds very rapidly to synaptosomal membranes (Fig. 4), inhibits the uptake of ^{45}Ca by synaptosomes (4) and, as mentioned above, produces a very large increase of the spontaneous release of ACh in neuromuscular junctions but inhibits its depolarization induced release (14), a series of experiments was carried out in order to test whether La^{3+} could affect the spontaneous and the depolarization-stimulated release of three neurotransmitters from synaptosomal fractions. In double-label experiments, synaptosomes were loaded simultaneously with [^{3}H] GABA and [^{14}C] glutamate, and their release was studied in a continuous superfusion system (see the legend to Fig. 1); in other experiments synaptosomes were loaded with [^{3}H] choline in order to study the release of labeled ACh. Fig. 5 shows that the release of both amino acids stimulated by 48 mM K^+ in the presence of 2 mM Ca^{2+} was almost completely inhibited by 2.5 mM or 1.0 mM La^{3+} and partially inhibited by 0.5 mM La^{3+}. In contrast, since La^{3+} was present from the beginning of the superfusion, it was possible to observe that during the pre-stimulation period of superfusion with normal, low K^+ medium, La^{3+} had a stimulatory action on this spontaneous release; this effect was more notable for glutamate than for GABA (Fig. 5).

When the synaptosomal fraction was superfused initially with a medium containing depolarizing K^+ concentrations but no Ca^{2+}, and then it was substituted by a high K^+, Ca^{2+}-containing medium, there was approximately a 3-fold stimulation of GABA and glutamate release. This stimulation was also almost completely inhibited by 1.0 mM or 2.5 mM La^{3+}, which was present only in the Ca^{2+}-containing medium (Fig. 6). In spite of the stimulatory effect of La^{3+} on the spontaneous release of GABA and glutamate (Fig. 5), when La^{3+} was

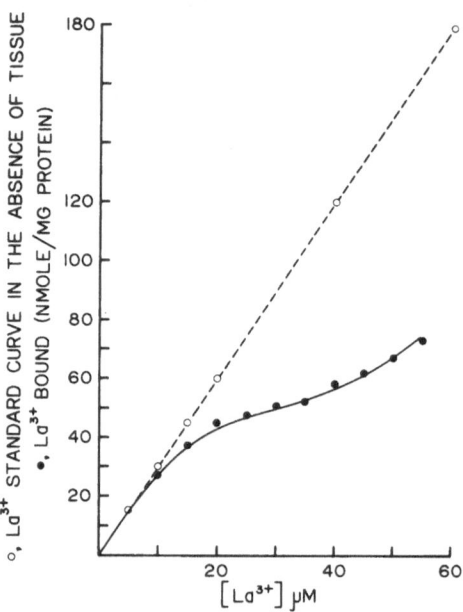

Fig. 4. Binding of La^{3+} to mouse synaptosomes. Soluble La^{3+} was determined in 0.32 M sucrose–25 mM Tris medium, pH 7.0, by the differential absorbance between 540 and 507 nm in the presence of 100 μM murexide as La^{3+} indicator, in a 3 ml volume, in a dual beam spectrophotometer. As shown by the standard curve in the absence of tissue, the detection of La^{3+} by this method is linear in the concentration range used. When synaptosomes were present in the spectrophotometer cell, the amount of free La^{3+} was calculated and the difference from the standard curve represented the amount of La^{3+} bound. The incubation was at 37°, with continuous recording of the changes. Values shown were obtained at one min incubation, but most of the binding, particularly at the lowest La^{3+} concentrations, was instantaneous. RuR added at 10 μM concentration did not affect the binding of La^{3+}. Mean values of 4 experiments.

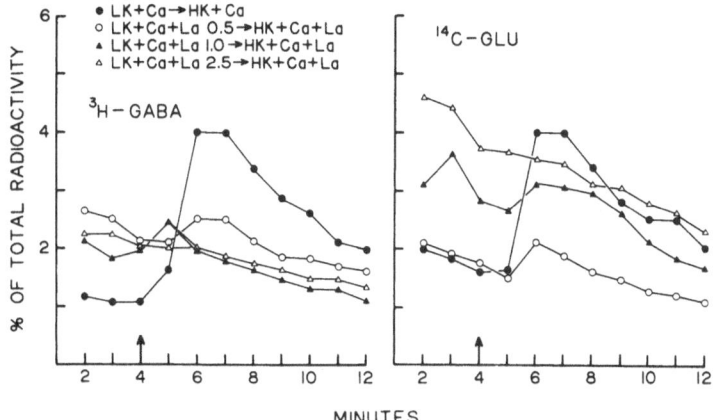

Fig. 5. Effect of La^{3+} on the release of labeled GABA and glutamate
in synaptosomes: stimulation by K$^+$-depolarization in the presence
of Ca^{2+} (2.5 mM). After loading with the amino acids, synaptosomes
were superfused (60) with Krebs-bicarbonate media containing 4.8 mM
KCl (low K$^+$, LK) or 48 mM KCl (high K$^+$, HK). At min 4 (vertical
arrow) the initial superfusing medium was substituted as indicated
by the horizontal arrows for each symbol; La^{3+} concentration (mM)
is shown. Mean values of 5-8 experiments.

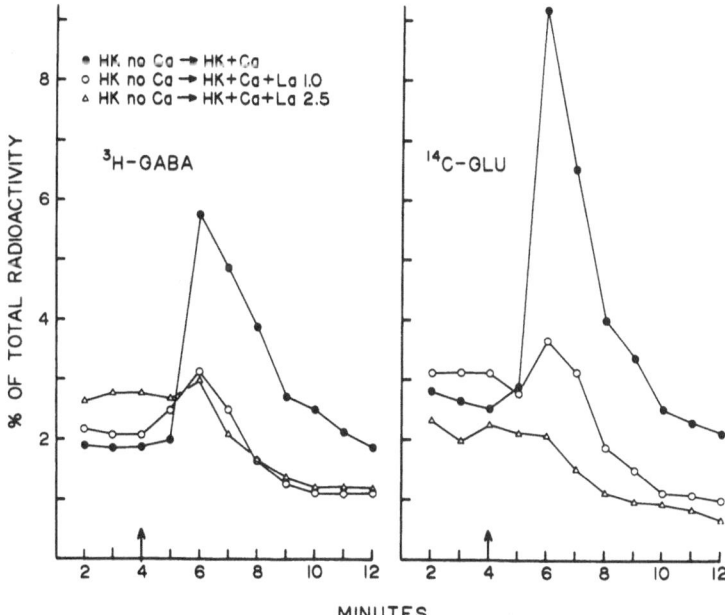

Fig. 6. Effect of La^{3+} on the release of labeled GABA and glutamate:
stimulation by Ca^{2+} in previously K$^+$-depolarized synaptosomes.
Details as for Fig. 5.

tested as a possible substitute for Ca^{2+} during the release
stimulated by high K^+ concentration, no effect of La^{3+} was observed
(Fig. 7). This result is at variance with a previous report with
synaptosomes, in which it was found that La^{3+} was capable of
substituting Ca^{2+} to induce GABA release with K^+-depolarization,
although the effect was very small as compared with that of Ca^{2+}
(23). Under different experimental conditions it had also been
described previously that La^{3+} stimulated the spontaneous release
of several endogenous amino acids, including GABA and glutamate (46).

Since in neuromuscular junctions the effect of La^{3+} on ACh
release has been relatively well studied, it was of interest to test
whether La^{3+} behaved similarly in synaptosomes with regard to this
transmitter. As shown in Fig. 8, after loading synaptosomes with
labeled choline, the release of radioactivity induced by K^+-
depolarization was dependent on the presence of Ca^{2+} in the medium,
which indicates that we were really measuring ACh release, since
the release of choline is not dependent on Ca^{2+} (8,64). La^{3+} at a
2.5 mM concentration abolished or greatly reduced the release of
ACh stimulated by K^+-depolarization in the presence of Ca^{2+}. However,
it is noteworthy that, in contrast to GABA and glutamate release
and to ACh release in neuromuscular junctions, La^{3+} did not modify
the spontaneous release of ACh in synaptosomes (Fig. 8).

GENERAL DISCUSSION

It is difficult to offer an explanation for the fact that La^{3+}
did not produce flaccid paralysis when injected i.p., since its
effects on Ca^{2+} transport and on Ca^{2+}-dependent neurotransmitter
release are similar to those of RuR. One possibility is that the
accessibility of La^{3+} to the presynaptic membrane is limited as
compared with that of RuR. However, the finding that the
administration of La^{3+} prior to that of RuR prevented the paralyzing
action of the latter, suggests that such a possibility is not
occurring. An alternative explanation for the lack of effect of
La^{3+} alone would be that the stimulatory effect of La^{3+} on the
spontaneous ACh release is enough for antagonizing its own blocking
action. This explanation implies that the spontaneous quantal
release of ACh, rather than the depolarization induced release,
is essential for the maintenance of muscular tone, and that this
spontaneous release is dependent on external Ca^{2+}. In agreement
with this possibility it has been shown in in vitro preparations
that small concentrations of RuR drastically reduce mepps frequency,
and that this effect is antagonized by increasing the concentration
of Ca^{2+} in the medium (48). An opposite effect of RuR, that is, a
small increase in mepps frequency, has also been reported and
ascribed to a blocking action of RuR on the intraterminal
mitochondrial uptake of Ca^{2+} (1). This explanation implies that
RuR is capable of penetrating the external membrane of the terminal

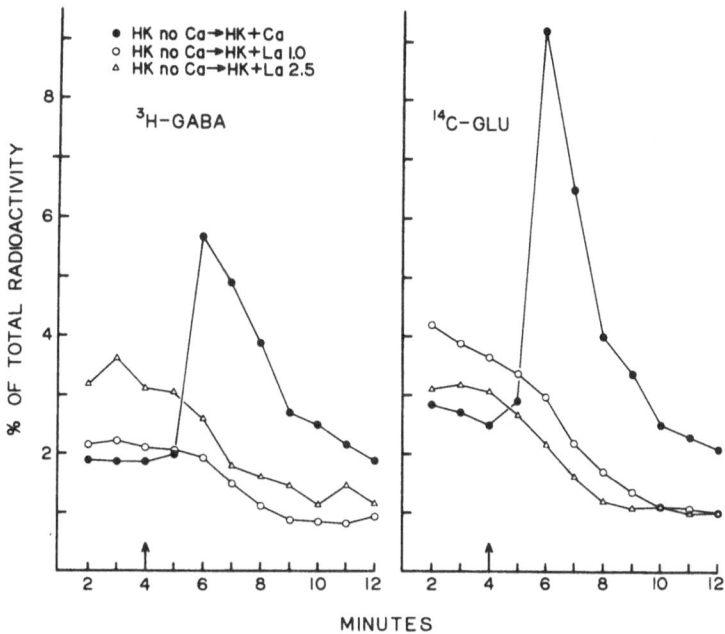

Fig. 7. Lack of stimulatory effect of La^{3+} on the release of labeled GABA and glutamate in previously K^+-depolarized synaptosomes. Details as for Fig. 5.

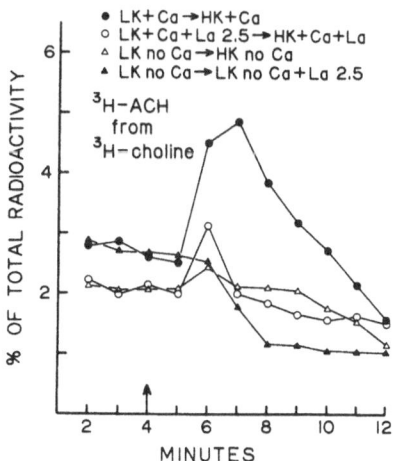

Fig. 8. Effect of La^{3+} on the release of labeled Ach derived from labeled choline: stimulation by K^+-depolarization in the presence and absence of Ca^{2+}. Details as for Fig. 5.

in order to reach the mitochondria. However, this is an improbable event, since morphological studies on the cellular location of RuR demonstrate that RuR does not penetrate biological membranes or does it very slowly (28,55). In addition, in unpublished electron microscopic observations, we have found that RuR stains the external synaptosomal membrane but does not stain intraterminal mitochondria or any other organelle. It might be argued that the concentration of RuR necessary for its visualization in the electron microscope is relatively high, and therefore it could be deposited on the intraterminal mitochondria in concentrations too low to detect but sufficient to block mitochondrial Ca^{2+} uptake. However, it must be considered also that the concentration of RuR necessary to inhibit the spontaneous release of ACh (48) is much lower than that used for the electron microscopic studies (28).

Our observations of the flaccid paralysis after RuR administration are certainly interpreted more easily as an inhibitory effect on Ca^{2+} entry at the external membrane of the terminal than by any effect on the intraterminal mitochondria. If the latter were occurring one would expect an increased release of ACh, which certainly could not result in flaccid paralysis.

Also in agreement with our conclusions of the previous paragraph, it has been reported that mepps frequency increases with increasing external Ca^{2+} concentration (11,33,35), and that the Ca^{2+} ionophores A23187 (11,54) or X537A (48), as well as oxidative phosphorylation uncouplers which act probably through the release of Ca^{2+} from intraterminal mitochondria (1,54), notably increase mepps frequency.

The results presented in this paper indicate that the administration into intact animals of drugs capable of modifying Ca^{2+} transport may be a useful and interesting approach to relate the movements of Ca^{2+} with the physiological role of the neuro-transmitters released at the specific synapses affected. This approach should be more valuable when combined with the pertinent in vitro experiments, which may provide more direct information on the synaptic mechanisms affected.

Acknowledgement. The work presented in this paper was supported in part by the Consejo Nacional de Ciencia y Tecnología, México, D. F. (Project PNCB-790214).

REFERENCES

1. Alnaes, E. and Rahamimoff, R., On the role of mitochondria in transmitter release from motor nerve terminals, J. Physiol. (Lond.), 248 (1975) 285-306.
2. Archibald, J. T. and White, T. D., Rapid reversal of internal Na^+ and K^+ of synaptosomes by ouabain, Nature, 255 (1974) 595-596.

3.   Batra, S., Mitochondrial calcium release as a mechanism for
     quinidine contracture in skeletal muscle, Biochem. Pharmacol.,
     25 (1976) 2631-2633.
4.   Blaustein, M. P., Effects of potassium, veratridine, and scorpion
     venom on calcium accumulation and transmitter release by nerve
     terminals in vitro, J. Physiol. (Lond.), 247 (1975) 617-655.
5.   Blaustein, M. P. and Goldring, J. M., Membrane potentials in
     pinched-off presynaptic terminals monitored with a fluorescent
     probe: evidence that synaptosomes have potassium diffusion
     potentials, J. Physiol. (Lond.), 247 (1975) 589-615.
6.   Blaustein, M. P., Johnson, E. M. and Needleman, P., Calcium-
     dependent norepinephrine release from presynaptic nerve endings
     in vitro, Proc. Nat. Acad. Sci. U.S.A., 69 (1972) 2237-2240.
7.   Carafoli, E. and Lehninger, A. L., A survey of the interaction
     of calcium ions with mitochondria from different tissues and
     species, Biochem. J., 122 (1971) 681-690.
8.   Carroll, P. T. and Goldberg, A. M., Relative importance of
     choline transport to spontaneous and potassium depolarized
     release of ACh, J. Neurochem. 25 (1975) 523-527.
9.   Cotman, C. W. and Hamberger, A., Glutamate as a CNS neuro-
     transmitter: properties of release, inactivation and
     biosynthesis. In F. Fonnum (Ed.), Amino Acids as Chemical
     Transmitters, Plenum, New York, 1978, pp. 379-412.
10. Cotman, C. W., Haycock, J. W. and White, W. F., Stimulus-
     secretion coupling processes in brain: analysis of nora-
     drenaline and gamma-aminobutyric acid release, J. Physiol.
     (Lond.), 254 (1976) 475-505.
11. Duncan, C. J. and Statham, H. E., Interacting effects of
     temperature and extracellular calcium on the spontaneous
     release of transmitter at the frog neuromuscular junction,
     J. Physiol. (Lond.), 268 (1977) 319-333.
12. Farley, J. M., Glavinović, M. I., Watanabe, S. and Narahashi,
     T., Stimulation of transmitter release by guanidine derivatives,
     Neuroscience, 4 (1979) 1511-1519.
13. Goddard, G. A. and Robinson, J. D., Uptake and release of
     calcium by rat brain synaptosomes, Brain Res., 110 (1976)
     331-350.
14. Heuser, J. and Miledi, R., Effect of lanthanum ions on
     function and structure of frog neuromuscular junctions, Proc.
     Roy. Soc. Lond. B., 179 (1971) 247-260.
15. Jacobs, R. S. and Burley, E. S., Nerve terminal facilitatory
     action of 4-aminopyridine: an analysis of the rising phase
     of the endplate potential, Neuropharmacology, 17 (1978) 439-444.
16. Johnston, G. A. R., Effects of calcium on the potassium-
     stimulated release of radioactive β-alanine and γ-aminobutyric
     acid from slices of rat cerebral cortex and spinal cord, Brain
     Res., 121 (1977) 179-181.
17. Kamenskaya, M. A., Elmqvist, D. and Thesleff, S., Guanidine
     and neuromuscular transmission. I. Effect on transmitter
     release occurring spontaneously and in response to single

nerve stimuli, Arch. Neurol., 32 (1975) 505–509.
18. Kamenskaya, M. A., Elmqvist, D. and Thesleff, S., Guanidine
and neuromuscular transmission. II. Effect on transmitter
release in response to repetitive nerve stimulation, Arch.
Neurol., 32 (1975) 510–518.
19. Kamino, K., Ogawa, M., Uyesaka, N. and Inouye, A., Calcium-
binding of synaptosomes isolated from rat brain cortex. IV.
Effects of ruthenium red on the co-operative nature of calcium-
binding, J. Membrane Biol., 26 (1976) 345–356.
20. Katz, B. and Miledi, R., A study of synaptic transmission in
the absence of nerve impulses, J. Physiol. (Lond.), 192 (1967)
407–436.
21. Katz, B. and Miledi, R., Tetrodotoxin-resistant electric
activity in presynaptic terminals, J. Physiol. (Lond.), 203
(1969) 459–487.
22. Kleineke, J. and Stratman, F. W., Calcium transport in isolated
rat hepatocytes, FEBS Letters, 43 (1974) 75–80.
23. Levy, W. B., Haycock, J. W. and Cotman, C. W., Effects of
polyvalent cations on stimulus-coupled secretion of $[^{14}C]$
-γ-aminobutyric acid from isolated brain synaptosomes, Mol.
Pharmacol., 10 (1974) 438–449.
24. Li, P. P. and White, T. D., Rapid effects of veratridine,
tetrodotoxin, gramicidin D, valinomycin and NaCN on the Na^+,
K^+ and ATP contents of synaptosomes, J. Neurochem., 28 (1977)
967–975.
25. Llinás, R. and Heuser, J. E., Depolarization-release coupling
systems in neurons, Neurosci. Res. Progr. Bull., 15 (1977)
557–687.
26. Llinás, R. and Nicholson, C., Calcium role in depolarization-
secretion coupling: an aequorin study in squid giant synapse,
Proc. Nat. Acad. Sci. U.S.A., 72 (1975) 187–190.
27. López-Colomé, A. M., Tapia, R., Salceda, R. and Pasantes-Morales, H., K^+-stimulated release of labeled γ-aminobutyrate,
glycine and taurine in slices of several regions of rat
central nervous system, Neuroscience, 3 (1978) 1069–1074.
28. Luft, J. H., Ruthenium red and violet. I. Fine structural
localization in animal tissues, Anat. Rec., 171 (1971)
369–416.
29. Lundh, H., Effects of 4-aminopyridine on neuromuscular
transmission, Brain Res., 153 (1978) 307–318.
30. Lundh, H., Cull-Candy, S. G., Leander, S. and Thesleff, S.,
Restoration of transmitter release in botulinum-poisoned
skeletal muscle, Brain Res., 110 (1976) 194–198.
31. Lundh, H., Leander S. and Thesleff, S., Antagonism of the
paralysis produced by botulinum toxin in the rat. The effects
of tetraethylammonium, guanidine and 4-aminopyridine, J. Neurol.
Sci., 32 (1977) 29–43.
32. Lundh, H. and Thesleff, S., The mode of action of 4-amino-
pyridine and guanidine on transmitter release from motor nerve
terminals, Eur. J. Pharmacol., 42 (1977) 411–412.

33. Madden, K. S. and Van der Kloot, W., Surface charges and the effects of calcium on the frequency of miniature end-plate potentials at the frog neuromuscular junction, J. Physiol. (Lond.), 276 (1978) 227-232.

34. Madeira, V. M. C. and Antunes-Madeira, M. C., Interaction of Ca^{2+} and Mg^{2+} with synaptic plasma membranes, Biochim. Biophys. Acta, 323 (1973) 396-407.

35. Matthews, G. and Wickelgren, W. O., Effects of guanidine on transmitter release and neuronal excitability, J. Physiol. (Lond.), 266 (1977) 69-89.

36. Matthews, G. and Wickelgren, W. O., On the effect of calcium on the frequency of miniature end-plate potentials at the frog neuromuscular junction, J. Physiol. (Lond.), 266 (1977) 91-101.

37. Meldrum, B. S., Epilepsy and γ-aminobutyric acid mediated inhibition, Intern. Rev. Neurobiol., 17 (1975) 1-36.

38. Meza-Ruiz, G. and Tapia, R., [³H] GABA release in synaptosomal fractions after intracranial administration of ruthenium red, Brain Res., 154 (1978) 163-166.

39. Miledi, R., Transmitter release induced by injection of calcium ions into nerve terminals, Proc. Roy. Soc. Lond. B., 183 (1973) 421-425.

40. Miledi, R. and Slater, C. R., The action of calcium on neuronal synapses in the squid, J. Physiol. (Lond.), 184 (1966) 473-498.

41. Molgo, J., Lemeignan, M. and Lechat, P., Effects of 4-amino-pyridine at the frog neuromuscular junction, J. Pharmacol. Exp. Ther., 203 (1977) 653-663.

42. Molgo, J., Lemeignan, M. and Lechat, P., Analysis of the action of 4-aminopyridine during repetitive stimulation at the neuromuscular junction, Eur. J. Pharmacol., 53 (1979) 307-311.

43. Moore, C. L., Specific inhibition of mitochondrial Ca^{++} transport by ruthenium red, Biochem. Biophys. Res. Comm., 42 (1971) 298-305.

44. Mulder, A. H. and Snyder, S. H., Potassium-induced release of amino acids from cerebral cortex and spinal cord slices of the rat, Brain Res., 76 (1974) 297-308.

45. Murrin, L. C., DeHaven, R. N. and Kuhar, M. J., On the relationship between [³H] choline uptake activation and [³H] acetylcholine release, J. Neurochem., 29 (1977) 681-687.

46. Osborne, R. H. and Bradford, H. F., The influence of sodium, potassium and lanthanum on amino acid release from spinal-medullary synaptosomes, J. Neurochem., 25 (1975) 35-41.

47. Otsuka, M. and Endo, M., The effect of guanidine on neuro-muscular transmission, J. Pharmacol. Exp. Ther., 128 (1960) 273-282.

48. Person, R. J. and Kuhn, J. A., Depression of spontaneous and ionophore-induced transmitter release by ruthenium red at the neuromuscular junction, Brain Res. Bull., 4 (1979) 669-674.

49. Pfeiffer, D. R., Taylor, R. W. and Lardy, H. A., Ionophore A23187: cation binding and transport properties, Ann. N. Y. Acad. Sci., 307 (1978) 402-423.

50. Prestipino, G., Ceccarelli, D., Conti, F. and Carafoli, E., Interactions of a mitochondrial Ca^{2+}-binding glycoprotein with lipid bilayer membranes, FEBS Letters, 45 (1974) 99–103.

51. Rubin, R. P., Calcium and the Secretory Process. Plenum, New York, 1974, 189 pp.

52. Sandoval, M. E., Studies on the relationship between Ca^{2+} efflux from mitochondria and the release of amino acid neurotransmitters, Brain Res., 181 (1980) 357–367.

53. Sandoval, M. E., Horch, P. and Cotman, C. W., Evaluation of glutamate as a hippocampal neurotransmitter: glutamate uptake and release from synaptosomes, Brain Res., 142 (1978) 285–299.

54. Shalton, P. M. and Wareham, A. C., Calcium ionophore A-23187 and spontaneous miniature end-plate potentials of mammalian skeletal muscle, Exp. Neurol., 63 (1979) 379–387.

55. Singer, M., Krishnan, N. and Fyfe, D. A., Penetration of ruthenium red into peripheral nerve fibers, Anat. Rec., 173 (1972) 375–390.

56. Srinivasan, V., Neal, M. J. and Mitchell, J. F., The effect of electrical stimulation and high potassium concentrations on the efflux of ^{3}H-γ-aminobutyric acid from brain slices, J. Neurochem. 16 (1969) 1235–1244.

57. Tapia, R. The role of γ-aminobutyric acid metabolism in the regulation of cerebral excitability. In R. D. Myers and R. R. Drucker-Colín (Eds.), Neurohumoral Coding of Brain Function, Plenum, New York, 1974, pp. 3–26.

58. Tapia, R. Biochemical pharmacology of GABA in CNS. In L. L. Iversen, S. D. Iversen and S. H. Snyder (Eds.), Handbook of Psychopharmacology, Vol. 4, Plenum, New York, 1975, pp. 1–58.

59. Tapia, R., Convulsions and the function of GABAergic synapses. In L. Battistin, G. Hashim and A. Lajtha (Eds.), Neurochemistry and Clinical Neurology, Alan R. Liss, New York, 1980, pp. 123–131.

60. Tapia, R. and Meza-Ruiz, G., Inhibition by ruthenium red of the calcium-dependent release of [^{3}H] GABA in synaptosomal fractions, Brain Res., 126 (1977) 160–166.

61. Tapia, R., Meza-Ruiz, G., Durán L. and Drucker-Colín, R. R., Convulsions or flaccid paralysis induced by ruthenium red depending on route of administration, Brain Res., 116 (1976) 101–109.

62. Tapia, R., Sandoval, M. E. and Contreras, P., Evidence for a role of glutamate decarboxylase activity as a regulatory mechanism of cerebral excitability., J. Neurochem., 24 (1975) 1283–1285.

63. Weiss, G. B., Cellular pharmacology of lanthanum, Ann. Rev. Pharmacol., 14 (1974) 343–354.

64. Wonnacott, S., Marchbanks, R. M. and Fiol, C., Ca^{2+} uptake by synaptosomes and its effect on the inhibition of acetylcholine release by botulinum toxin, J. Neurochem., 30 (1978) 1127–1134.

65. Wood, J. D., The role of γ-aminobutyric acid in the mechanism of seizures, Progr. Neurobiol., 5 (1975) 77–95.

ON THE ROLE OF MITOCHONDRIA IN NEUROTRANSMITTER RELEASE

María Elena Sandoval

Departamento de Neurociencias
Centro de Investigaciones en Fisiología Celular
Universidad Nacional Autónoma de México, México 20, D.F.

INTRODUCTION

Depolarization of the presynaptic membrane and the presence of calcium ions in the extracellular medium appear to be essential to evoke transmitter release under physiological conditions. Most probably depolarization of the nerve terminal causes an increase in the membrane Ca^{2+} conductance and Ca^{2+} flow inside the presynaptic nerve ending along their electrochemical gradient. However depolarization is not necessary to evoke transmitter liberation since procedures that elicit an increased intracellular calcium ($[Ca^{++}]i$) in the absence of depolarization lead to transmitter liberation. Thus, direct calcium injection into the squid axon and calcium entry facilitated by the calcium ionophore A23187 in synaptosomes, induce an increased transmitter release (22,31,40,46,48). This indicates that an elevated $[Ca^{++}]i$ is essential for transmitter release. At the frog neuromuscular junction spontaneous quantal release can be detected at rest and this rate of release is less dependent on extracellular calcium than evoked transmitter release (38,39). However, treatments that enhance $[Ca^{++}]i$ cause an increase in spontaneous transmitter liberation.

In the study of the stimulus-secretion coupling the depolarizing effects of treatments as high K^+ concentration, ouabain and veratridine have been largely used to mimic the stimulatory effect of action potentials. However, such treatments often fail to elicit secretion by a calcium-dependent mechanism. Depolarizing K^+ concentrations stimulate γ-aminobutyric acid (GABA), glutamate and norepinephrine release from brain slices or synaptosomes in EGTA-containing medium (29,46,48,56). Acetylcholine release from cortical slices has been demonstrated in the presence of ouabain and zero K^+

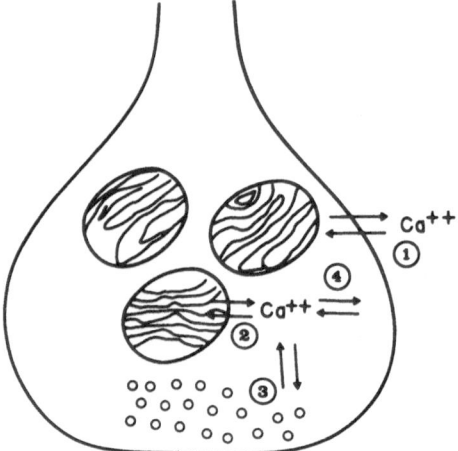

Fig. 1. Processes that may control $[Ca^{++}]i$: 1. Influx and extrusion
of calcium at rest and during activity. 2. Uptake and release of
calcium by mitochondria. 3. Calcium binding and release from the
endoplasmic reticulum-like system. 4. Calcium binding proteins.

medium in the absence of calcium (59). Recently Lowe et al. (37)
showed that insulin secretion can be induced in a Ca^{++}-free media
when isolated islets of Langerhans were incubated with veratridine.

    Several pieces of evidence suggest that spontaneous transmitter
liberation may be facilitated by changes in the monovalent cation
content of nerve terminals. At the frog neuromuscular junction the
miniature end-plate potential (mepp) frequency is greatly elevated if
the bathing medium contains digotoxin, ouabain or zero K^+, known
blockers of the Na^+-K^+ exchange pump (6,13,14,32). Similar results
have been reported when preparations were incubated with veratridine
or batrachotoxin, toxins known to increase Na^+ influx by opening the
Na^+-channels through a tetrodotoxin-sensitive process (32). This
increase in mepp frequency occurs also in calcium-free solutions with
EGTA and therefore does not seem to be due to an entry of calcium
into the motor nerve terminals (6,32). Thus, it has been suggested
that redistribution of the monovalent cation gradients within the
nerve terminal enhances transmitter liberation by releasing calcium
from intracellular stores (6,13,14,32). However, little is known
regarding the probable link between the processes that control
$[Ca^{++}]$ i and the mechanisms responsible for transmitter release.

    Several processes may control $[Ca^{++}]$ i at the nerve terminal
i.e.: a) the influx and extrusion of calcium at rest and during
activity at the surface of the synaptic membrane; b) the trapping of
Ca^{++} by mitochondria, which are known to accumulate calcium against

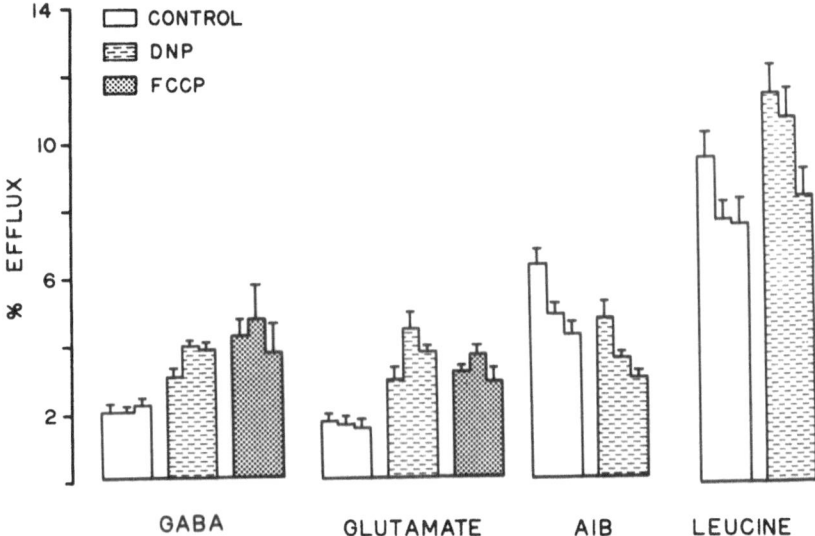

Fig. 2. Effect of DNP and FCCP on the spontaneous efflux of amino acids from synaptosomes. After loaded with [^{3}H] GABA, [^{14}C] glutamate, [^{3}H] leucine or [^{14}C] -α-aminoisobutyric acid, synaptosomes were washed with Krebs-Ringer buffer containing 1 mM EGTA in order to reach a baseline of spontaneous efflux. Then fractions received three consecutive 30 sec pulses with the same medium (control or medium supplemented with 100 µM DNP or 2 µM FCCP. Media were collected by filtration immediately after each pulse. Bars indicate efflux in each one of the experimental pulses. Efflux is expressed relative to the total radioactivity present on the filters immediately before the wash of interest by the formula: (dpm perfusate/dpm perfusate + dpm tissue) x 100 (22,48). Data represent the mean ± S.E.M. for 4-11 different observations.

a large electrochemical gradient; c) the trapping of Ca^{++} by the endoplasmic reticulum-like system recently reported by Blaustein et al. (16); and d) the activity of calcium-binding proteins (Fig. 1).

Pioneer studies by Alnaes and Rahamimoff showed that inhibition of the calcium storage by mitochondria in the frog neuromuscular junction leads to an increase in transmitter liberation. These authors suggest that mitochondria might modify transmitter release through the control of $[Ca^{++}]$ i (1,2).

In this paper I present evidence suggesting that mobilization of mitochondrial calcium may elicit transmitter release when Na^+ influx into isolated nerve endings is facilitated.

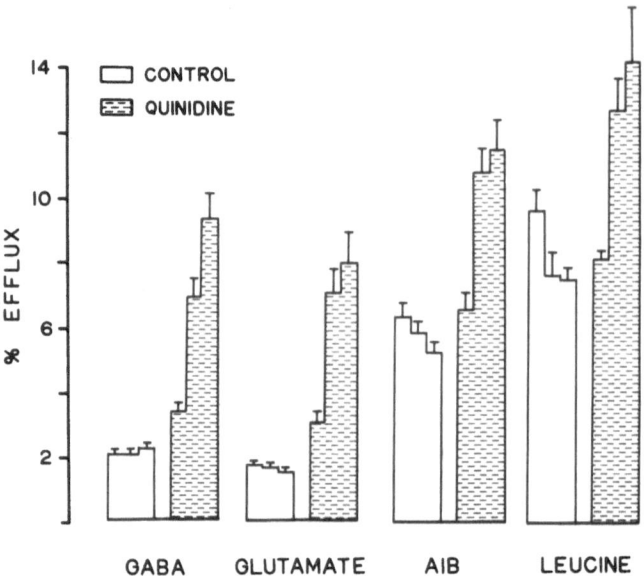

Fig. 3. Effect of quinidine on the spontaneous efflux of amino acids
from synaptosomes. Fractions were treated as described in Fig. 2 but
they received medium containing 2 mM quinidine during the experimental
pulses. Data represent the mean ± S.E.M. for 4-11 experiments.

INCREASED CALCIUM EFFLUX FROM MITOCHONDRIA AND TRANSMITTER RELEASE

It is well known that uncouplers of oxidative phosphorylation
elicit a rapid and remarkable loss of both, recently accumulated
$^{45}Ca^{++}$ and calcium ions stores within mitochondria isolated from
several tissues, including brain mitochondria (18,19). Previously
I tested the possibility that Ca^{++} released from mitochondria may
stimulate transmitter liberation (49). The effect of 2,4-dinitrophenol
(DNP) and carbonyl cyanide-p-trifluoromethoxy phenylhydrazone (FCCP),
known uncouplers of oxidative phosphorylation, on both $^{45}Ca^{++}$ loss
from isolated brain mitochondria and spontaneous efflux of transmitter
from isolated nerve terminals was studied. As shown in Fig. 2, both
DNP and FCCP elicit an augmented liberation of [^3H] GABA and [^{14}C]
glutamate, two suspected neurotransmitters, from synaptosomes. This
stimulatory effect is unlikely related to an increased influx of
external calcium since the experiments were carried out in EGTA-
containing Ca^{++}-free buffer.

Balzer et al. (8,9) and Andersson (3) have reported that the
alkaloid quinidine produces slowly developing contractures in several
muscle preparations even if muscle fibers are soaked in Ca^{++}-free
medium. These authors proposed that contractions are due to a direct

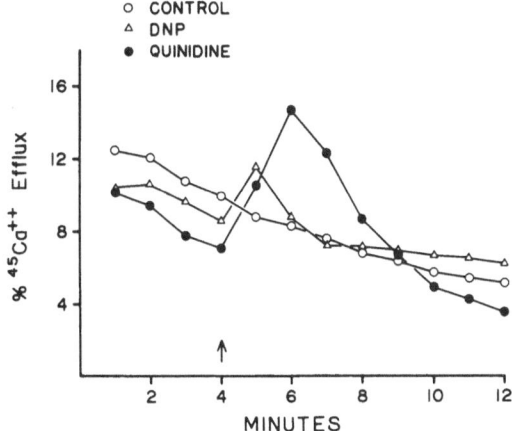

Fig. 4. Effect of DNP and quinidine on $^{45}Ca^{++}$ efflux from brain
mitochondria. After loading with $^{45}Ca^{++}$, mitochondria were washed
with sucrose, succinate and Tris medium in a superfusion filter system
(49). When indicated by the arrow media was replaced with the same
solution but supplemented with 100 μM DNP or 2 mM quinidine. A
typical experiment is shown.

effect of quinidine upon an intracellular calcium store that releases
calcium into the myofibrillar space. Data recently reported by Batra
(10,11) indicate the mitochondrial calcium stores as the most likely
system affected by quinidine.

The effect of quinidine on the spontaneous efflux of [^{3}H] GABA
and [^{14}C] glutamate from synaptosomes is shown in Fig. 3. The efflux
of both radiolabeled neurotransmitters was stimulated after three-
30 sec incubation periods with 2 mM quinidine.

In agreement with previously reported data (18,19) DNP produced
a rapid increase in the efflux of calcium from brain mitochondria. As
demonstrated in muscle mitochondria (10,11), quinidine elicited an
augmented calcium loss from isolated brain mitochondria (Fig. 4).

Non-transmitter amino acids as leucine and ∝-aminoisobutyric
acid (AIB) are poorly stored by synaptosomes and they are not
released from nerve endings in a calcium-dependent manner. In order
to know whether the effect of uncouplers and quinidine was restricted
to transmitter amino acids, the action of both DNP and quinidine on
the efflux of leucine and AIB was followed. Figs. 2 and 3 show that
DNP and quinidine fail to induce an increase of radiolabeled leucine
and AIB efflux from synaptosomes, or do it to a much lesser extent as
compared with GABA or glutamate.

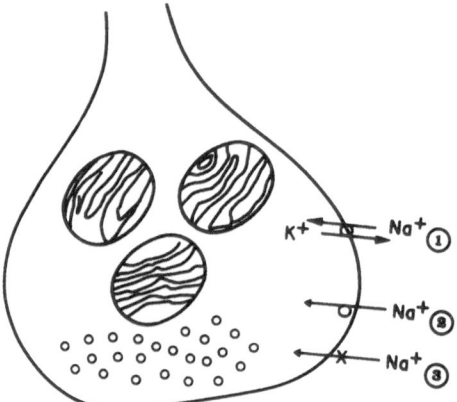

Fig. 5. Action of depolarizing agents: 1. Inhibitors of ATPase
activity: ouabain, K^+-free medium, digoxin, quinidine (?). 2. Openers
of the Na^+ channels: veratridine, batrachotoxin. 3. Na^+ (K)
ionophores gramicidin.

    Results show that DNP and FCCP as well as quinidine elicit a
rapid loss of Ca^{++} from brain mitochondria and stimulate the
spontaneous efflux of GABA and glutamate from synaptosomes, whereas
efflux of non-transmitter amino acids was much less affected. Thus,
we suggest that calcium ions released from mitochondria may trigger
transmitter liberation (49). This might be the explanation for the
neurotransmitter release induced by DNP and other uncouplers in
excitable tissues (21,24,26,27).

INCREASED $[Na^+]$ i AND TRANSMITTER RELEASE

    It is well known that several treatments that depolarize
synaptic membranes have the common effect of producing profound
changes in the Na^+ fluxes across the cell membrane, either by
inhibiting the $(Na^+-K^+-Mg^{++})$-activated ATPase (ouabain, zero K^+) or
by enhancing Na^+ entry into the nerve endings (veratridine, high
K^+-concentration (4,15,25,28,30,42,55,58) (Fig. 5). As mentioned
above, these same treatments often elicit secretion in the absence of
external calcium.

    Storage of calcium by mitochondria may be modified by several
pharmacological treatments. However, little is known regarding the
physiological control of mitochondrial calcium. Recently it has been
reported that sodium ions induce a rapid efflux of Ca^{++} from
mitochondria isolated from several excitable tissues, including
brain tissue (20,47,52). Most probably this effect is due to a
Na^+-Ca^{++} exchange mechanism present in the mitochondrial membrane

and it has been suggested as a likely process to control mitochondrial calcium (17,23). Calcium fluxes in mitochondria of non-excitable tissues are Na^+-insensitive (17).

Recently we studied the effect of several treatments known to enhance $[Na^+]$ i within the nerve terminal, such as zero K^+ concentration, gramicidin, ouabain, veratrine and high K^+ concentrations, on the spontaneous efflux of $[^3H]$ GABA from synaptosomes (50). In order to exclude the possibility that external Ca^{++} might be entering, all experiments were carried out in a Ca^{++}-free, EGTA-containing medium. Data are summarized in Fig. 6. When isolated nerve endings were incubated with an EGTA-supplemented medium containing ouabain, zero K^+, gramicidin, 56 mM KCl or veratrine, the efflux of $[^3H]$ GABA was stimulated. Zero K^+ buffer presented the lowest effect whereas veratrine showed the highest rate of stimulation. Fig. 5 also shows that all these treatments failed to increase $[^3H]$ GABA efflux or did it only slightly, when Na^+ was omitted from the incubation medium. This indicates that the stimulatory effect of the above treatments on transmitter release is Na^+-dependent. As indicated in Fig. 6 the increased $[^3H]$ GABA efflux induced by quinidine is also dependent on $[Na^+]$ i. Since quinidine inhibits ATPase activity in sarcoplasmic reticulum (12,43), this result suggests a similar action of this alkaloid on synaptosomal ATPase.

In order to know the specificity of increased $[Na^+]$ i to stimulate transmitter release we studied the effect of veratrine and gramicidin on the efflux of $[^{14}C]$ -aminoisobutyric acid; Fig. 6 shows that both treatments only poorly increased the efflux of this non-transmitter amino acid from synaptosomes.

These results show an augmented efflux of GABA from synaptosomes under conditions leading to an increased $[Na^+]$ i, in the absence of Ca^{++} in the medium.

Fig. 6. Effect of treatments that increase $[Na^+]$ i on the efflux of amino acids from synaptosomes. After loading with $[^3H]$ GABA or $[^{14}C]$ AIB fractions were washed with Krebs-Ringer buffer or with Na^+-free buffer solution to reach the baseline efflux; all media contained 1mM EGTA. During the experimental pulses (see the legend to Fig. 2) synaptosomes received the same medium containing the drug under study or zero K^+ medium. Drugs concentration were: ouabain, 4×10^{-4} M; gramicidin, 1 μg/ml; KCl, 56 mM; quinidine, 2 mM; veratrine, 60 μg/ml. Data represent the mean $\pm$ S.E.M. for 3-16 different experiments.

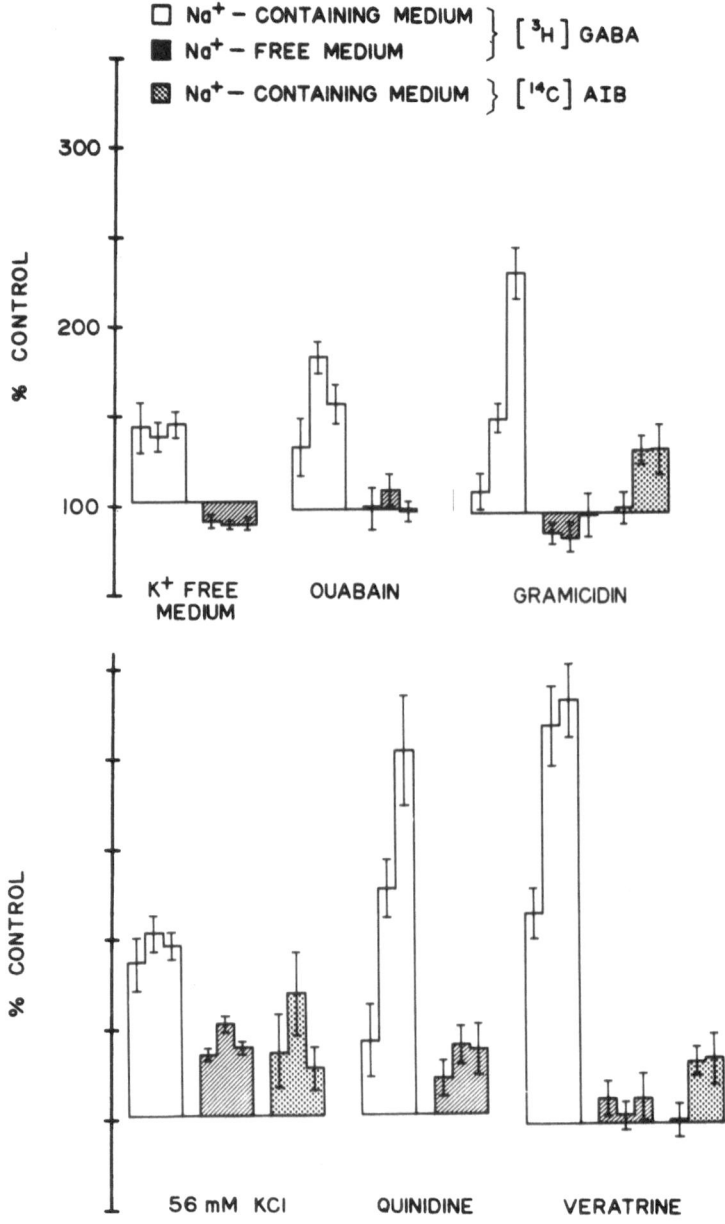

Fig. 6. See legend on the preceding page

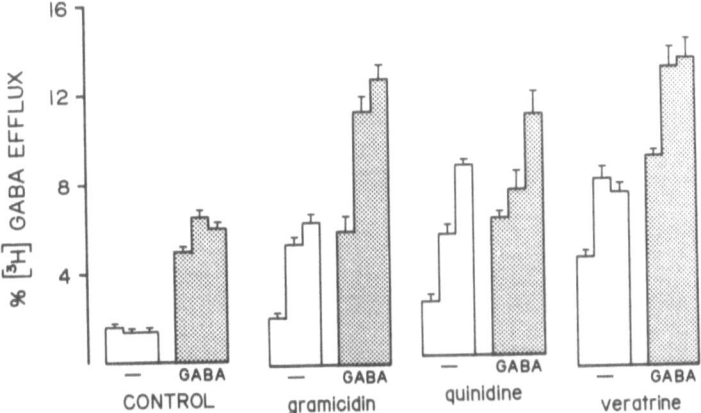

Fig. 7. Effect of external GABA on the increased [3H] GABA efflux induced by gramicidin, quinidine and veratrine. Synaptosomes were treated as indicated in Fig. 2 but during the experimental pulses they received Krebs-Ringer buffer containing 1 mM EGTA and supplemented with 30 µM GABA or 30 µM GABA plus either gramicidin, quinidine or veratrine. Drugs concentration as indicated in Fig. 6. Data represent the mean ± S.E.M. for 3-6 experiments.

Na$^+$-DEPENDENT GABA EFFLUX AND THE CARRIER-MEDIATED GABA TRANSPORT

Low GABA concentrations in the incubation medium stimulate an outward transport of intrasynaptosomal GABA through a Na$^+$-dependent homoexchange mechanism (33,54). Levi et al. (34,35) have reported that depolarization-induced, Ca^{++}-dependent release of GABA occurs simultaneously to the GABA stimulated GABA efflux; therefore they suggested that these processes are independent mechanisms for GABA transport. In the present experiments we studied the effect of GABA on the increased [3H] GABA efflux induced by gramicidin, quinidine and veratrine. Fig. 7 shows that 30 µM GABA augmented about three fold the baseline efflux of [3H] -GABA. When both GABA and either gramicidin, quinidine or veratrine were present simultaneously during the incubation period, an additive effect of both stimulations was observed (Fig. 7).

There is increasing evidence that treatment of isolated synaptic nerve terminals with 2,4-diaminobutyric acid (DABA) leads to a large inhibition of the Na$^+$-dependent, carrier-mediated GABA transport, since both GABA uptake and homoexchange are diminished (36,53). As shown in Fig. 8, preincubation of synaptosomes with DABA failed to modify the stimulatory effect on [3H] GABA efflux induced by conditions known to enhance [Na$^+$] i, whereas the GABA-induced [3H] GABA release was inhibited.

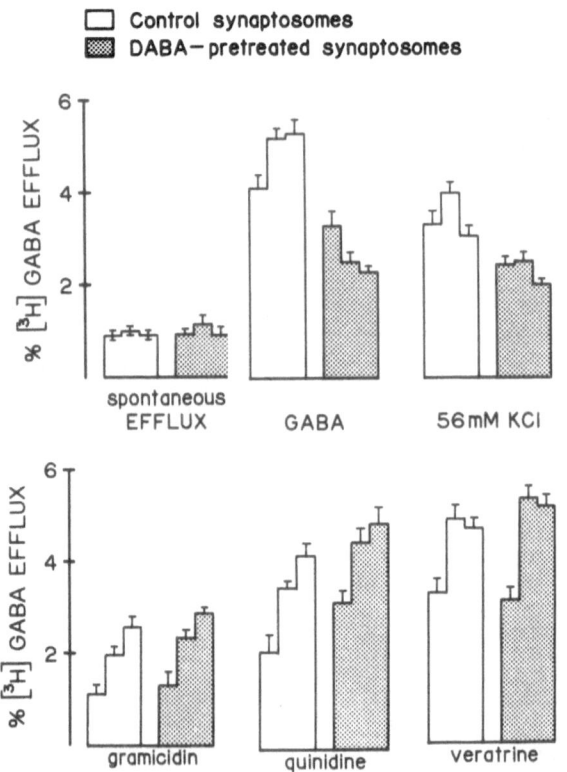

Fig. 8. Effect of DABA on the increased $[^3H]$ GABA efflux induced by GABA, KCl, gramicidin, quinidine and veratrine. Experimental conditions were as indicated in Fig. 6 but synaptosomes were loaded with $[^3H]$ GABA in the absence (control synaptosomes) or in the presence of 100 μM DABA (DABA-pretreated synaptosomes); $[^3H]$ GABA efflux was followed in a Krebs-Ringer solution containing EGTA. Data represent the mean ± S.E.M. for 4-6 experiments.

These data suggest that the activity of the outward carrier-mediated GABA transport is probably unrelated to the increased $[^3H]$ GABA efflux demonstrated under the above conditions of elevated $[Na^+]i$.

Taken together our data indicate that an increased entry of Na^+ into the nerve endings may elicit transmitter release. Since there is no external Ca^{++} present and the release of transmitter does not seem to be due to a carrier-mediated mechanism, we suggest mobilization of calcium from mitochondrial stores as the most likely mechanism of action of Na^+. Table 1 summarizes the effect of several treatments that elicit an elevated $[Na^+]i$ on distinct synaptic phenomena when preparations were bathed in a calcium-free, EGTA-

Table 1. Increased $[Na^+]_i$ and synaptic activity in the absence of external Ca^{++}

Treatment	Preparation	$[Na^+]_i$	Mepp frequency	Decay of long-term facilitation	Decay of post-tetanic potentiation
Batrachotoxin	rat diaphragm muscle (32)	↑	↑	—	↑
Digotoxin	frog neuromuscular junction (13,14)	↑	↑	—	—
	rat diaphragm muscle (32)	↑	↑	—	↑
Ouabain	frog neuromuscular junction (7,44)	↑	↑	—	↑
	opener and stretcher muscles in the walking legs of crayfish (5,57)	↑	—	↑	—
Veratridine	rat diaphragm muscle (32)	↑	↑	—	—
Zero K⁺	frog neuromuscular junction (7)	↑	↑	—	—
	crab stretcher muscle (5)	↑	—	↑	—

The numbers in parentheses indicate the references.

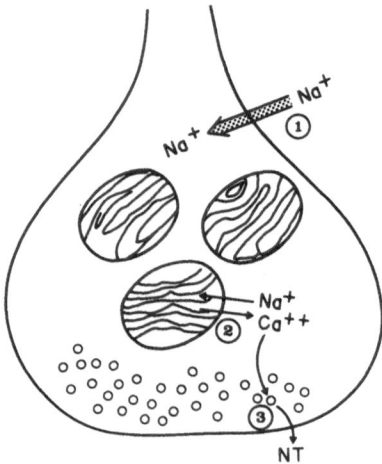

Fig. 9. Increased $[Na^+]$ i and neurotransmitter release. 1. Influx of
Na^+ into the nerve ending induced by: a) repetitive stimulation; b)
ATPase inhibitors; c) opening of the Na^+ channels; d) Na^+ ionophores.
2. Ca^{++} efflux from mitochondria elicited by increased $[Na^+]$ i,
3. Activation of the transmitter release processes induced by
elevated $[Ca^{++}]$ i.

containing media. All these treatments lead to an increased mepp
frequency at the vertebrate neuromuscular junction (6,7,13,14,32,44).
Since the rate of spontaneous efflux observed at rest is less
dependent on $[Ca^{++}]$ o than evoked transmitter release, it has been
suggested that it may be controled by internal calcium stores, namely
mitochondrial calcium (44).

The physiological significance of the above results is still
unclear. However, there is increasing evidence regarding the role
of monovalent cation content of the nerve terminal in several
processes of synaptic activity. At many chemical synapses,
transmitter release is facilitated by previous impulse activity in
the nerve terminal. At many crustacean neuromuscular junctions,
synaptic facilitation is maintained for several minutes and the
lasting phase is known as long-term facilitation. Sherman and
Atwood (51) proposed that this facilitation is linked to accumulation
of Na^+ by the nerve terminals during impulse activity. As shown in
Table 1, long term facilitation is enhanced by conditions which
promote accumulation of Na^+ in nerve endings (5,57).

Birks and Cohen (14) suggested that at vertebrate neuromuscular
junctions increased $[Na^+]$ i could be the initial cause of post-tetanic
potentiation (long-lasting increase in amplitude of the end-plate

potential following a number of closely spaced stimuli). As indicated in Table 1, it has been demonstrated that an increased Na^+ entry into the nerve terminal prolongs post-tetanic potentiation (6,32,44).

Altogether these results suggest that elevated $[Na^+]$ i may mobilize calcium from intraterminal stores and therefore enhance transmitter release (Fig. 9). The Na^+-sensitive, calcium efflux described in mitochondria of excitable tissues support mitochondrial calcium as the most likely calcium store affected by increased $[Na^+]$ i. Thus, it is suggested that mitochondria may modify transmitter liberation through the control of $[Ca^{++}]$ i. However, it remains to be known whether Na^+ can reach concentrations in the neighborhood of mitochondria high enough to induce this effect.

REFERENCES

1. Alnaes, E., Meiri, U., Rahamimoff, H., and Rahamimoff, R., Possible role of mitochondria in transmitter release, J. Physiol., 241 (1974) 30P.
2. Alnaes, E., and Rahamimoff, R., On the role of mitochondria in transmitter release from motor nerve terminals, J. Physiol., 248 (1975) 285-306.
3. Andersson, K.E., Effects of chloropromazine, imipramine and quinidine on the mechanical activity of single skeletal muscle fibers of the frog, Acta Physiol. Scand., 85 (1972) 532-546.
4. Archibald, J.T. and White T.D., Rapid reversal of internal Na^+ and K^+ contents of synaptosomes by ouabain, Nature, 252 (1974) 595-596.
5. Atwood, H.L., Swenarchuk, L.E. and Gruenwald, C.R., Long-term synaptic facilitation during sodium accumulation in nerve terminals, Brain Res., 100 (1975) 198-204.
6. Baker, P.F., Meves, H. and Ridgway, E.R., Calcium entry in response to maintained depolarization of squid axons, J. Physiol. (Lond.), 231 (1973) 527-548.
7. Baker, P.F., Crowford, A.C., A note on the mechanism by which inhibitors of the sodium pump accelerate spontaneous release of transmitter from motor nerve terminals, J. Physiol. (Lond.), 247 (1975) 209-226.
8. Balzer, H., and Hellenbrecht, D., Beeinflussung des Calcium-Austauchs und der Muskel-funktion des M. rectus and sartorius des Frosches durch Chloropromazin, Prenylamin, Imipramin und Reserpin, Naungn-Schmiedebergs Arch. Exp. Phath. Pharmak., 264 (1969) 129-146.
9. Balzer, H., Makinose, M., and Hasselbach, W., The inhibition of the sarcoplasmic calcium pump by prenylamine, reserpine, chloro-promazine and imipramine, Naunyn-Schmiedebergs Arch. Path. Pharmak., 260 (1968) 444-455.

10. Batra, S., The effects of drugs on calcium uptake and calcium release by mitochondria and sarcoplasmic reticulum of frog skeletal muscle, Biochem. Pharmacol., 23 (1974) 89–101.

11. Batra, S., Mitochondriasl calcium release as a mechanism for quinidine contracture in skeletal muscle., Biochem. Pharmacol., 25 (1976) 2631–2633.

12. Besch, H.R. Jr. and Watanabe, A.M., Binding and effect of tritiated quinidine on cardiac subcellular enzyme systems: sarcoplasmic reticulum vesicles, mitochondria and Na^+, K^+-adenosine triphosphatase, J. Pharmacol. Exp. Ther., 202 (1977) 354–364.

13. Birks, R.I. and Cohen, M.W., The action of sodium pump inhibitors on neuromuscular transmission, Proc. Roy. Soc. B. 170 (1968) 381–399.

14. Birks, R.I. and Cohen, M.W., The influence of internal sodium on the behavior of motor nerve endings, Proc. Roy. Soc. B, 170 (1968) 401–421.

15. Blaustein, M.P. and Goldring, J.M., Membrane potentials in pinched off presynaptic nerve terminals monitored with a fluorescent probe: evidence that synaptosomes have potassium diffusion potentials, J. Physiol. (Lond.), 247 (1975) 589–615.

16. Blaustein, MP., Ratzlaff, R.W., Kendrick N.C. and Schweitzer, E. S., Calcium buffering in presynaptic nerve terminals. I. Evidence for involvement of a non mitochondrial ATP-dependent sequestration mechanism, J. Gen. Physiol., 72 (1978) 15–41.

17. Carafoli, E. and Crompton, M., The regulation of intracellular calcium by mitochondria, Ann. N.Y. Acad. Sci., 307 (1978) 269–284.

18. Carafoli, E., and Lehninger, A.L., A survey of the interaction of calcium ions with mitochondria from different tissues and species, Biochem. J., 122 (1971) 681–690.

19. Carafoli, E., Malmstrom, K., Capano, M., Sigel, E., and Crompton, M., Mitochondria and the regulation of cell calcium. In E. Carafoli, E., Clementi, F., Drabikowski, W., and Margreth, A. (Eds.), Calcium Transport in Contraction and Secretion, Elsevier North Holland, 1975, pp. 53–64.

20. Carafoli, E., Tiozzo, R., Lugly, G., Crovetti, F. and Kratzing, C., The release of calcium from heart mitochondria by sodium, J. Molec. Cell Cardiol. 6 (1974) 361–371.

21. Chang, P., von Euler, U.S., and Lishajko, F., Effects of 2,4-dinitrophenol on release and uptake of noradrenaline in guinea pig heart, Acta Physiol. Scand., 85 (1972) 501–505.

22. Cotman, W.C., Haycock, J.W. and White, W.F., Stimulus-secretion coupling processes in brain: analysis of noradrenaline and gamma-aminobutyric acid release, J. Physiol. (Lond.), 254 (1976) 475–505.

23. Crompton, M., Capano, M. and Carafoli, E., The sodium-induced efflux of calcium from heart mitochondria. A possible mechanism for the regulation of mitochondrial calcium, Eur. J. Biochem., 69 (1976) 453–462.

24. Dembiec, D., and Cohen, G., Effect of carbonyl-binding agents and oxidative phosphorylation uncouplers on the release of [3H] norepinephrine from mouse heart, Biochem. Pharmacol., 25 (1976) 1369-1376.

25. Goddard, G.A. and Robinson, D., Uptake and release of calcium by rat brain synaptosomes, Brain Res., 110 (1976) 331-350.

26. Godfraind, J.M., Krnjevic, K., and Pumain, R., Unexpected features of the action of dinitrophenol on cortical neurons, Nature, 228 (1970) 562-564.

27. Godfraind, J.M., Kawamura, H., Krnjevic, K., and Pumain, R., Action of dinitrophenol and some other metabolic inhibitors on cortical neurons, J. Physiol. (Lond.) 215 (1971) 195-222.

28. Glynn, I.M., The action of cardiac glycosides on sodium and potassium movements in human red cells, J. Physiol. (Lond.) 136 (1957) 148-173.

29. Haycock, J.W., Levy, V.B., Denner, L. and Cotman, C.W., Effects of elevated K+ on the release of neurotransmitters from cortical synaptosomes: efflux or secretion?. J. Neurochem. 30 (1978) 1113-1125.

30. Hodgkin, A.L. and Keynes, R.D., Active transport of cations in giant axons from Sepia and Loligo., J. Physiol. (Lond.), 128 (1955) 28-60.

31. Holz, R.W., The release of dopamine from synaptosomes from rat striatum by the ionophores X-537 A and A 23187, Biochem. Biophys. Acta. 375 (1975) 138-152.

32. Jansson, S.E., Albuquerque, E.X. and Daly, J., The pharmacology of batrachotoxin VI. Effects on the mammalian motor nerve terminal, J. Pharmacol. Exp. Ther., 189 (1974) 525-537.

33. Levi, G. and Raiteri, H.M., Exchange of neurotransmitter amino acids at nerve endings can stimulate high affinity uptake, Nature, 253 (1974) 735-737.

34. Levi, G. and Raiteri, H.M., Modulation of γ-aminobutyric acid transport in nerve endings: Role of extracellular γ-aminobutyric acid and of cationic fluxes, Proc. Nat. Acad. Sci. U.S.A., 75 (1978) 2981-2985.

35. Levi, G., Roberts, P.J. and Raiteri, M., Release and exchange of neurotransmitters in synaptosomes: Effects of ionophore A 23187 and of ouabain, Neurochem. Res., 1 (1976) 409-416.

36. Levi, G., Rusca, G. and Raiteri, M., Diaminobutyric acid: a tool for discriminating between carrier-mediated and non-carrier mediated release of GABA from synaptosomes?, Neurochem. Res. 1 (1976) 581-590.

37. Lowe, D.A., Richardson, B.P., Taylor, P. and Donatsch, P., Increasing intracellular sodium triggers calcium release from bound pools, Nature, 260 (1976) 337-338.

38. Mambini, J. and Benoit, P.R., Action du calcium sur la jonction neuromusculaire chez le grenouille, C.R. Soc. Biol. (Paris) 158 (1964) 1454-1458.

39. Mathews, G. and Wickelegren, W.O., On the effect of calcium on frequency of miniature and plate potentials at the frog

neuromuscular junction, J. Physiol. (Lond.), 226 (1977) 91-101.

40. Miledi, R., Transmitter release induced by injection of calcium ions into nerve terminals, Proc. R. Soc. B., 183 (1973) 421-425.

41. Nadler, J.V., Vaca, K.W., White, W.F., Lynch, G.S. and Cotman C.W., Aspartate and glutamate as possible transmitters of excitatory hippocampal afferents, Nature, 260 (1976) 538-540.

42. Ohta, M., Narahashi, T. and Keeler, R.F., Effects of veratrum alkaloids on membrane potential and conductance of squid and crayfish giant axons, J. Pharmac. Exp. Ther., 184 (1973) 143-154.

43. Pang, C.D. and Briggs, N.F., Mechanism of quinidine and chloropromazine inhibition of sarcotubular ATPase activity, Biochem. Pharmacol., 25 (1976) 21-25.

44. Rahamimoff, R., Erulkar, S.D., Lev-Tov, A. and Meiri, H., Intracellular and extracellular ions in transmitter release at the neuromuscular synapse, Ann. N.Y. Acad. Sci., 307 (1978) 583-598.

45. Raiteri, M., Federico, R., Coletti, A. and Levi, G., Release and exchange studies relating to the synaptosomal uptake of GABA, J. Neurochem. 24 (1975) 1243-1250.

46. Redburn, D.A., Shelton, D. and Cotman, C.W., Calcium dependent release of exogenously loaded γ-amino $[U^{14}C]$ butyric acid from synaptosomes: time course stimulation by potassium, veratridine and calcium ionophore A23187, J. Neurochem., 26 (1976) 297-303.

47. Robinson, I.C.A.F., Russell, J.T. and Thorn, N.A., Calcium and stimulus-secretion coupling in the neurohypophysis. V. The effects of the Ca^{2+} ionophores A23187 and X537A on vasopressin release and $^{45}Ca^{2+}$ efflux; interactions with sodium and a verapamil analogue (D600), Acta Endocrinol., 83 (1976) 36-49.

48. Sandoval, M.E., Horch, P. and Cotman, C.W., Evaluation of glutamate as a hippocampal neurotransmitter: glutamate uptake and release from synaptosomes, Brain Res., 142 (1978) 285-289.

49. Sandoval, M.E., Studies on the relationship between Ca^{++}-efflux from mitochondria and the release of amino acid neurotransmitters, Brain Res. 181 (1980) 357-367.

50. Sandoval, M.E., Sodium-dependent efflux of $[^3H]$ GABA from synaptosomes probably related to mitochondrial calcium mobilization, J. Neurochem., 35 (1980) 915-921.

51. Sherman, R.G., and Atwood, H.L., Synaptic facilitation: long-term neuromuscular facilitation in crustaceans, Science, 171 (1971) 1248-1250.

52. Silbergeld, E.K., Na^+ regulates release of Ca^{++} sequestered in synaptosomal mitochondria, Biochem. Biophys. Res. Commun., 77 (1977) 464-469.

53. Simon, J.R. and Martin, D.L., The effects of 2,4-diaminobutyric acid on the uptake of gamma-aminobutyric acid by a synaptosomal fraction from rat brain, Arch. Biochem. Biophys., 157 (1973) 348-355.

54. Simon, J.R., Martin, D.L. and Kroll, M., Sodium-dependent efflux

and exchange of GABA in synaptosomes, J. Neurochem., 23 (1974) 981-991.

55. Skou, J.C., Further investigations on a Mg^{2+}-and Na^+-activated adenosine triphosphatase, possibly related to the active, linked transport of Na^+ and K^+ across the nerve membrane, Biochim. Biophys. Acta 42 (1960) 6-23.

56. Srinivasan, V., Neal, M.J. and Mitchell, J.F., The effect of electrical stimulation and high potassium concentrations on the efflux of $[^3H]$ aminobutyric acid from brain slices, J. Neuro-chem., 16 (1969) 1235-1244.

57. Swenarchuk, L.E. and Atwood, H.L., Long-term facilitation with minimal calcium entry, Brain Res., 100 (1975) 205-208.

58. Ulbricht, W., The effect of veratridine on excitable membranes of nerves and muscles, Ergbn. Physiol., 61 (1969) 18-71.

59. Vizi, E.S., Stimulation by inhibition of $(Na^+-K^+-Mg^{2+})$-Activated ATPase, of acetylcholine release in cortical slices from rat brain, J. Physiol. (Lond.), 226 (1972) 95-117.

CALCIUM, CALMODULIN, AND SYNAPTIC FUNCTION: MODULATION OF NEURO-
TRANSMITTER RELEASE, NERVE TERMINAL PROTEIN PHOSPHORYLATION, AND
SYNAPTIC VESICLE MORPHOLOGY BY CALCIUM AND CALMODULIN

Robert J. DeLorenzo

Yale University School of Medicine
Department of Neurology
333 Cedar Street, New Haven, CT 06510

INTRODUCTION

An understanding of the molecular mechanism underlying calcium-
dependent neurotransmitter release would greatly enhance our knowledge
of synaptic transmission and the action of specific neuropharmacologic
agents, and possibly provide new insights into human disease processes
involving synaptic modulation. Although calcium's role in neurotransmitter release from the presynaptic nerve terminal has been of
great interest, little is known about the molecular mechanisms of
Ca^{2+} in stimulating neurotransmitter release or its other
physiological functions in the nerve terminal. Ca^{2+} was shown to
stimulate the endogenous phosphorylation of whole brain (8-10,15) and
synaptosomal proteins (10,16) and it was suggested from these results
that the effects of calcium on protein phosphorylation might play a
role in mediating some of the effects of this ion on neuronal tissue.

This original hypothesis was further strengthened by studies with
the anticonvulsant phenytoin (diphenylhydantoin, DPH). Phenytoin has
been shown to inhibit calcium-dependent neurotransmitter release in
several preparations (27,28,39,40,41,51). If synaptic vesicle protein
phosphorylation mediates calcium-dependent release of neurotransmitter
from the nerve terminal, DPH would be expected to inhibit this
calcium-dependent protein phosphorylation. DPH in therapeutic
concentrations blocked calcium-dependent phosphorylation of specific
synaptosomal (8,10,15) and synaptic vesicle (18,19) proteins DPH-L
and DPH-M (specific synaptosomal fraction proteins with molecular
weights of approximately 63,000 and 53,000 respectively). These
results suggested that the inhibition of neurotransmitter release by
DPH may be caused by its inhibition of calcium-dependent synaptic
vesicle protein phosphorylation and suggested a role for these

phosphoproteins in neurotransmitter release (16).

To further elucidate the role of synaptic vesicle-associated phosphoproteins in neurotransmitter release, it would be important to demonstrate that calcium ions stimulate both neurotransmitter release and protein phosphorylation in the same preparation and to determine the precise molecular mechanism involved in the regulation of neurotransmitter release and protein phosphorylation by calcium. In this paper I will present evidence from my laboratory employing isolated synaptic vesicle and intact synaptosome preparations indicating that several of the effects of calcium on synaptic function, especially neurotransmitter release, are mediated by the calcium receptor protein, calmodulin and may be modulated throught calcium and calmodulin stimulated protein phosphorylation.

A "MORE PHYSIOLOGICAL" ISOLATION PROCEDURE FOR OBTAINING HIGHLY ENRICHED SYNAPTIC VESICLE PREPARATIONS

Synaptic vesicles are uniquely localized within the presynaptic nerve terminal and have been suggested to play an important role in synaptic transmission (38). Vesicles have been shown to contain certain neurotransmitter substances and have been implicated in neurotransmitter release, especially by the theory of exocytosis. Although there have been numerous studies and reviews on the role of synaptic vesicles in the release processes (4), little is known about the specific biochemical mechanisms involved in mediating the effects of calcium on synaptic vesicle function.

A major objective of this laboratory over the past four years has been to attempt to determine the biochemical mechanisms that mediate calcium's action on synaptic vesicle function. Because of these interests in synaptic vesicle function, we studied the effects of calcium on vesicle protein phosphorylation and neurotransmitter content (21). It quickly became apparent that synaptic vesicles isolated by standard procedures (26,49), did not show significant responses to calcium, and contained low concentrations of neuro-transmitter substances and showed only minimal calcium-dependent neurotransmitter release.

To study neurotransmitter release from synaptic vesicles, it was important to develop a vesicle preparation that was stable, as physiologically active as possible, and was highly enriched in synaptic vesicles. Experiments were initiated to prepare synaptic vesicles under numerous preparation conditions to determine if methods of preparation could affect the physiological viability of vesicles. The first set of experiments demonstrated that the nore-pinephrine content of synaptic vesicles isolated from whole brain was significantly influenced by methods of preparation (Table 1). Vesicles were isolated under isotonic and hypotonic conditions,

Table 1. Effects of Preparation Conditions on the Norepinephrine Content of Synaptic Vesicles

Preparation Conditions*	Norepinephrine Concentration[+]
KCl 160mM, NaCl 5mM, Tris-maleate 10mM	
pH 6.5, MgCl$_2$ 5mM (Iso-KCl media)	5.36±0.16
pH 6.5	5.03±0.11
pH 6.5, increased preparation time	3.64±0.14
pH 6.5, prolonged osmotic shock	1.72±0.23
pH 7.0	4.63±0.10
Sucrose 0.32 M, Tris-maleate 10mM	
pH 6.5	3.74±0.14
pH 6.5, increased preparation time	1.95±0.09
pH 6.5, prolonged osmotic shock	1.12±0.18
pH 7.0	3.11±0.13
KCl 100mM, Tris-maleate 10mM	
pH 6.5	2.98±0.12
Tris-maleate 10mM, KCl 10mM	
pH 6.5	2.23±0.11
pH 6.5, increased preparation time	1.01±0.23
pH 6.5, prolonged osmotic shock	0.61±0.31
pH 7.0	1.67±0.14

*Synaptic vesicles were prepared under standard conditions with the addition of the above reagents or conditions following the osmotic shock of the washed P$_2$ pellet (21). Preparation time was increased from less than 5 hours to 8 hours as indicated. The P$_2$ pellet was subjected to prolonged osmotic shock by homogenization in 5.5 times its volume of distilled water with 20 strokes at 500 rpm for 10 minutes (21).

[+]Each value is expressed as ng norepinephrine/mg protein and represents the mean values and ranges for 4 determinations. (From ref. 21).

at different pH values, and under different techniques of preparation.
The results indicate that hypotonic media, as well as vigorous or
prolonged preparation conditions affected the neurotransmitter content
of the vesicles.

Synaptic vesicles are traditionally isolated from synatosomes by
inducing osmotic shock of the synaptosome preparation and isolating
the vesicles by differential centrifugation. The hypotonic and non-
physiological condition that result during these procedures produce
synaptic vesicle preparations that have low neurotransmitter content
(Table 1).

Several support solutions were used to attempt to restore a more
physiological intracellular environment following osmotic shock and
release of synaptic vesicles. One of these physiological support media
developed in this investigation, designated Iso-KCl media (21)
produced the highest norepinephrine content. Isotonic sucrose was
not as good (Table 1). Hypotonic buffered media were also not as
good support solutions. Standard hypotonic methods for isolating
synaptic vesicles from brain caused a significant reduction in norepinephrine content of vesicles, and the vesicles were very unstable
in such solutions in the absence of added soluble protein (Table 1).
In addition to support media, it was shown that more vigorous
homogenization during osmotic shock of the P_2 pellet, prolongation
of the preparation time, and increasing or decreasing pH
significantly affected the neurotransmitter content of the vesicles
(Table 1).

These experiments developed an isolation procedure for synaptic
vesicles that produced significantly higher vesicle neurotransmitter
content. The preparation procedure is summarized in Figure 1. This
vesicle preparation was highly enriched in synaptic vesicles as
determined by electronmicroscopy (Fig. 2) and enzyme markers (21).
This vesicle preparation was then employed to study the possible
role of calcium-dependent phosphorylation in neurotransmitter release.

CALCIUM-DEPENDENT NEUROTRANSMITTER RELEASE AND PROTEIN PHOSPHORYLA-
TION IN SYNAPTIC VESICLE PREPARATIONS

The effect of calcium on the release of neurotransmitter
substances from synaptic vesicles was investigated (Table 2). Calcium
ions caused a marked decrease in the norepinephrine content of
vesicles prepared in Iso-KCl media, but had much less effect on the
release of neurotransmitter from vesicles prepared in sucrose or
hypotonic media (Table 2).

In our standard vesicle preparation (Iso-KCl) calcium caused
a significant decrease in norepinephrine content of synaptic vesicles
and a corresponding increase in the amount of norepinephrine

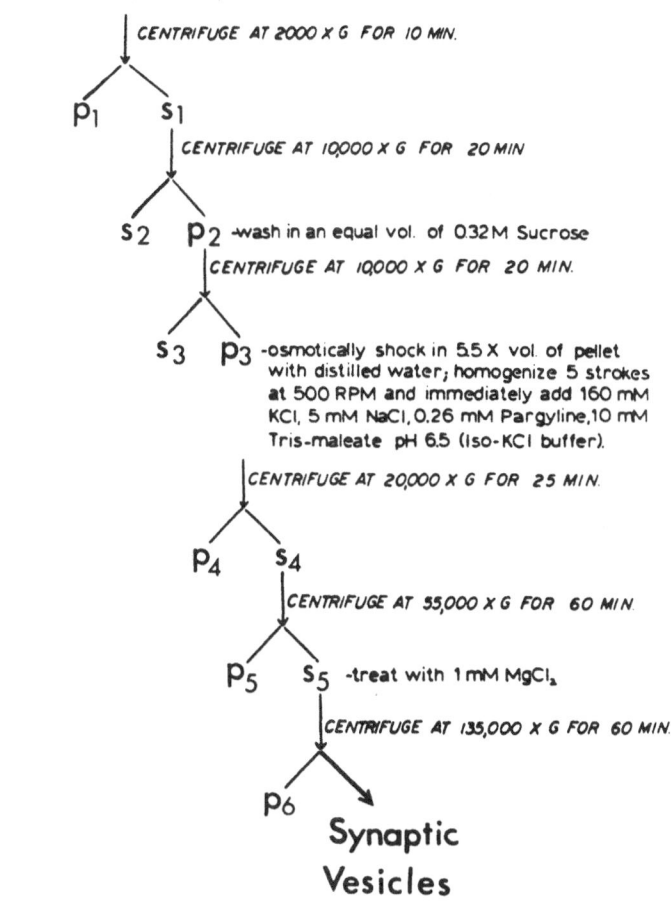

HOMOGENATE -homogenize rat cortex in 8 parts by weight of 0.32 M
Sucrose in a glass homogenizer at 4°C with a Teflon
pestle at approx. 500 RPM FOR 12 strokes

CENTRIFUGE AT 2000 X G FOR 10 MIN.

P_1 S_1

CENTRIFUGE AT 10,000 X G FOR 20 MIN

S_2 P_2 -wash in an equal vol. of 0.32M Sucrose

CENTRIFUGE AT 10,000 X G FOR 20 MIN.

S_3 P_3 -osmotically shock in 5.5 X vol. of pellet
with distilled water; homogenize 5 strokes
at 500 RPM and immediately add 160 mM
KCl, 5 mM NaCl, 0.26 mM Pargyline, 10 mM
Tris-maleate pH 6.5 (Iso-KCl buffer).

CENTRIFUGE AT 20,000 X G FOR 25 MIN.

P_4 S_4

CENTRIFUGE AT 55,000 X G FOR 60 MIN.

P_5 S_5 -treat with 1 mM $MgCl_2$

CENTRIFUGE AT 135,000 X G FOR 60 MIN.

P_6 Synaptic
Vesicles

Fig. 1. Isolation of synaptic vesicles

released from the vesicles. The action of calcium on neurotransmitter
release was dependent upon the presence of magnesium (Table 3). The
free calcium ion concentration required to produce a half-maximal
increase in norepinephrine release under standard conditions was
approximately 0.5 µM. The free calcium concentration was determined
by employing a calcium-EGTA buffer system and verified with a calcium
specific electrode (Orion).

Under the same conditions for neurotransmitter release calcium
ions simultaneously produced significant increases in the level of
endogenous phosphorylation of several synaptic vesicle proteins,
especially vesicle associated proteins DPH-L and DPH-M (Fig. 3).

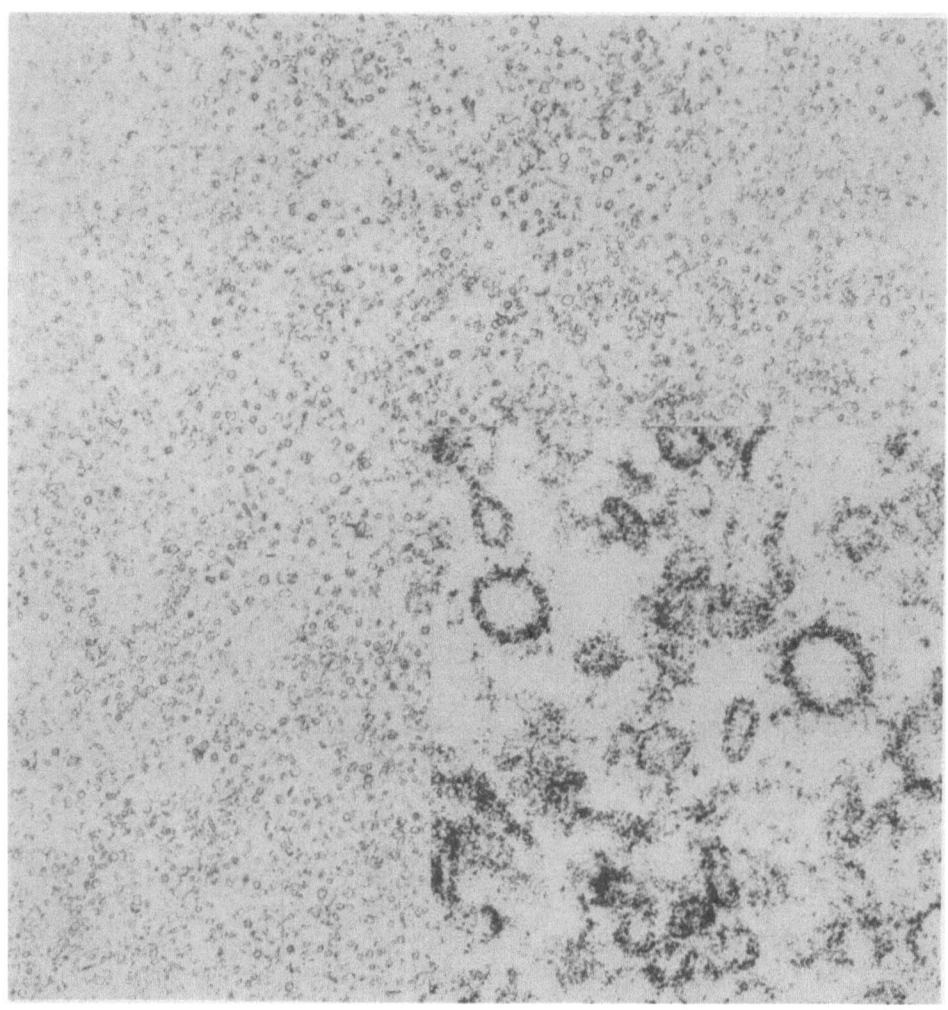

Fig. 2. Electronmicrograph of the synaptic vesicle preparation.
Synaptic vesicles were isolated by our newly developed procedure
(Fig. 1) and prepared for thin sectioning (21). The electron
micrographs demonstrate the high enrichment of synaptic vesicles
in this preparation (50,000X; insert 350,000X). The vesicle
preparation is mainly composed of plain synaptic vesicles, but some
coated vesicles are also seen. Reprinted from ref. (21).

Table 2. Effect of Calcium on the Norepinephrine Content of
          Synaptic Vesicles Prepared Under Different Conditions*

Preparation Conditions	Norepinephrine Concentration	
	Control	Calcium
Iso-KCl media	5.21 ± 0.12	3.27 ± 0.18
Sucrose 0.32M, Tris-maleate 10mM pH 6.5	3.66 ± 0.17	3.18 ± 0.09
Tris-maleate 10mM, KCl 10mM pH 6.5	2.31 ± 0.10	1.98 ± 0.21

*Synaptic vesicle plus soluble protein fractions obtained with the
addition of the above reagents following osmotic shock were
incubated under standard conditions in the presence or absence of
calcium ions. Synaptic vesicles were isolated by centrifugation and
assayed for norepinephrine (21). Each value is expressed as ng
(norepinephrine)/mg (protein) and represents the mean values and
ranges for 4 determinations. (Modified from ref. 21).

The stimulatory effects of calcium on protein phosphorylation in
vesicles isolated by our new procedure was also much more dramatic
and more easily reproduced than when vesicles isolated by standard
procedures were studied. Although calcium markedly stimulated the
phosphorylation of proteins DPH-L and DPH-M, the phosphorylation
of several other proteins were also stimulated by calcium. Protein
DPH-M was the main phosphoprotein in the vesicle preparation, when
incubated under standard conditions. Most of the quantitation of
protein phosphorylation in this report is of the incorporation of
32P-phosphate into protein DPH-M, since this protein was
representative of the phosphorylation patterns of the numerous other
bands. Considerable research is needed to clarify the precise function
and role of each phosphoprotein in these synaptic fractions.

The synaptic vesicle preparation developed in my laboratory
offers distinct advantages in studying the mechanisms of neurotransmitter release and "bridges" the gap between purely in vivo
and in vitro preparations. It is possible with this preparation
to simultaneously study the effects of calcium on not only biochemical
reactions such as protein phosphorylation, but also to investigate

Table 3. Effect of Calcium on Neurotransmitter Release and Protein Phosphorylation in a Highly Purified Preparation of Synaptic Vesicles*

| Condition | NOREPINEPHRINE RELEASE+ | | SYNAPTIC VESICLE PROTEIN PHOSPHORYLATION# | | | |
| | Synaptic Vesicle Concentration | % Release | Protein DPH-L | | Protein DPH-M | |
			Arbitrary Units	Percent	Arbitrary Units	Percent
Control	5.27	-	-	-	-	-
Mg^{2+}	5.25	3.3	7.8	13.3	10.5	11.8
Ca^{2+}	5.20	2.1	4.2	7.2	6.0	6.7
$Ca^{2+} + Mg^{2+}$	3.16	47.6	58.7	100.0	88.9	100.0

*Synaptic vesicles were incubated for 1min under standard conditions (21) in the presence of calcium (10 μM) and/or magnesium (5mM) with 25 μM ATP (norepinephrine release) or [γ-32P] ATP (protein phosphorylation). Vesicles were isolated by centrifugation at 4°C and assayed for norepinephrine and protein phosphorylation (21).

+Norepinephrine concentration is expressed as ng/mg protein. % Release is expressed as % release of norepinephrine as compared to control conditions. The recovery of released and bound norepinephrine was 96-98% in each condition. The data represent the mean value of four determinations and the largest ranges about the mean were 0.18 ng/mg and 1.4% and were thus omitted for clarity.

#Each arbitrary unit equals approximately 36.7cpm and percent represents the percent of the Ca + Mg condition. The data give the mean values for 6 experiments. The largest ranges about the mean were 1.98 arbitrary units and 2.4% and were omitted for clarity.

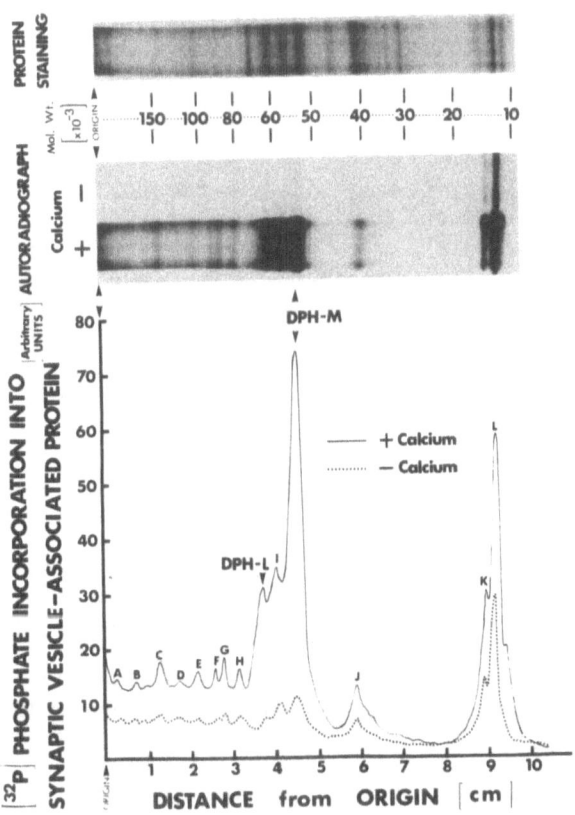

Fig. 3. Effect of calcium on the endogenous phosphorylation of synaptic vesicle-associated proteins. Synaptic vesicles were obtained and incubated under standard conditions in the presence of $MgCl_2$ (4mM) and $\gamma-^{32}P$ ATP (25µM) plus or minus $CaCl_2$ (10µM) under identical conditions to those described in Table 3 for norepinephrine release. Following the isolation of the vesicles after the reaction by centrifugation, the vesicle pellets were subjected to sodium-dodecyl sulfate (SDS) - polyacrylamide gel electrophoresis, protein staining, autoradiography, and quantitation, as described previously (8,10,21). The results shown are representative of 12 individual experiments. Each arbitrary unit equals approximately 38.6 cpm. Molecular weight determinations were performed as described previously (8,10).

physiological functions such as neurotransmitter release and turnover and morphological changes that might accompany the biochemical reactions or physiological functions.

The synaptic vesicle preparation is a useful model system to study in conjunction with intact synaptosome preparations. The ability to isolate synaptic vesicles that retain calcium-dependent functions may provide a powerful tool for studying neurotransmitter release and neuropharmacological agents.

With our assistance, several other laboratories interested in neurotransmitter release and storage have set up the Iso-KCl vesicle preparation technique in their laboratories. Very recently (44), it has been shown that vesicles isolated by our procedure contained serotonin, demonstrating for the first time that serotonin is present in synaptic vesicles. These results confirm our findings and demonstrate further that this "more physiological vesicle preparation" may be a powerful tool for studying synaptic function.

PHENYTOIN INHIBITION OF NEUROTRANSMITTER RELEASE AND PROTEIN PHOSPHORYLATION IN SYNAPTIC VESICLE PREPARATIONS

The calcium-dependent phosphorylation of synaptic vesicle-associated proteins was significantly inhibited by DPH. The effects of DPH and calcium on the phosphorylation of synaptic vesicle-associated proteins DPH-L and DPH-M were independent of ATP concentration over the range 0.5 to 50 μM. Proteins DPH-L and DPH-M are clearly present in highly purified synaptic vesicle preparations (18-20). Calcium and DPH have the same antagonistic effects on the level of these synaptic vesicle proteins as they do on proteins DPH-L and DPH-M in the synaptosome preparation. Since synaptic vesicles are localized within the presynaptic nerve terminal (1,38) and are not found in glial or other synaptosome fraction contaminations, these results demonstrate that the antagonistic actions of DPH and calcium on brain protein phosphorylation occur within the presynaptic nerve terminal in close association with the synaptic vesicles.

Studies of the relative specific distribution of proteins DPH-L and DPH-M also demonstrated that these phosphoproteins are enriched in synaptic vesicle fractions. The presence of phosphoproteins DPH-L and DPH-M within preparations of presynaptic nerve terminals indicates that the antagonistic effects of DPH and calcium on the level of phosphorylation on these presynaptic nerve terminal proteins could account for the opposing effects of DPH and calcium on the release of neurotransmitter from the presynaptic nerve ending.

CALMODULIN'S ROLE IN MEDIATING THE EFFECTS OF CALCIUM ON SYNAPTIC
FUNCTION

Recent studies have suggested that some of calcium's effects
on nerve function may be modulated by a heat-stable Ca^{2+}-dependent
regulator protein (calmodulin), since the Ca^{2+}-dependent activation
of several important enzyme systems in brain required calmodulin
(5,6,39,42). Calmodulin has been purified and characterized from
many sources and appears to be a Ca^{2+} receptor protein with a specific
and strong binding affinity for Ca^{2+} (34,45,50). These results plus
the presence of calmodulin-like proteins in a wide variety of
mammalian and invertebrate tissues (47) have suggested that many of
calcium's physiological functions may be mediated by Ca^{2+}-receptor
proteins such as calmodulin (32).

It would be important to determine if calmodulin or calmodulin-
like proteins play a role in mediating calcium's effect on
depolarization-dependent neurotransmitter release. However,
essentially no evidence has been presented to demonstrate a role
of calmodulin in neurotransmitter release, partially because
conventional techniques for studying the effects of calmodulin on
isolated enzyme systems cannot be easily applied to physiologically
active in vivo or in vitro systems without destroying the viability
of the preparation. My approach to this problem has been to study
the physiologically active preparation of synaptic vesicles that has
been developed in this laboratory, demonstrating both Ca^{2+}-dependent
neurotransmitter release and protein phosphorylation (21). This
vesicle preparation offers distinct advantages for studying the
effects of calmodulin on these Ca^{2+}-dependent vesicle functions and
offers new insights into the molecular mechanism of neurotransmitter
release from the intact nerve terminal.

The experimental evidence reviewed below provides strong evidence
that calmodulin is present in the nerve terminal and may modulate
several of the effects of calcium on synaptic activity. The effects
of calcium on synaptic vesicle neurotransmitter release and protein
phosphorylation were shown to be dependent on calmodulin. These
results provide the first evidence for the possible role of calmodulin
in calcium-dependent neurotransmitter release and synaptic modulation.

DISTINCTION BETWEEN CALCIUM AND CYCLIC AMP STIMULATED PROTEIN KINASES

It is important to demonstrate that the calcium and calmodulin-
stimulated protein phosphorylation is a distinct system from the well
described cyclic AMP protein kinase systems in brain (46,48) and that
the effects of calcium are not being mediated through cyclic AMP.
If cyclic AMP mediated the effects of calcium on vesicle protein
phosphorylation, it would be expected that cyclic AMP would stimulate
the endogenous phosphorylation of the same vesicle proteins. Cyclic

Table 4.   Effect of cyclic AMP on the endogenous phosphorylation of
           synaptosomal fraction proteins I and II and proteins DPH–L
           and DPH–M

Substrates	Protein Phosphorylation (pmol/mg protein)	
	Control	Cyclic AMP
Protein I	1.6 ± 0.3	21.8 ± 3.6*
Protein II	2.2 ± 0.4	13.5 ± 2.6*
Protein DPH–L	4.8 ± 1.1	3.0 ± 1.1
Protein DPH–M	4.3 ± 0.9	6.2 ± 2.1

Synaptosomal preparations were incubated under the conditions of
Ueda et al (46) in the presence or absence of 5×10^{-6}M cyclic AMP
and analyzed for incorporation of ^{32}P-phosphate into protein (46).
The positions of proteins I, II, DPH–L, and DPH–M were determined
by molecular weight. Cyclic AMP concentrations from 1×10^{-7} to
5×10^{-4} did not significantly stimulate the phosphorylation of
proteins DPH–L and DPH–M. The data give the means $\pm$ SEM for 6
determinations. $P < 0.001$ in comparing control to cyclic AMP
conditions.

AMP did not significantly stimulate the phosphorylation of proteins
DPH–L or DPH–M in synaptic vesicle preparations (Table 4), but it
did stimulate the phosphorylations of proteins I and II, described
by Greengard's group (46). Proteins DPH–L and DPH–M were clearly
distinct proteins from proteins I and II, as determined on SDS–polyacrylamide gel electrophoresis.

     To further demonstrate the distinction between the endogenous
cyclic AMP and calcium protein kinases, cyclic AMP protein kinase
inhibitor (43) was isolated from rat brain cerebellum (43) and
tested on synaptic vesicle and synaptosome protein phosphorylation.
This inhibitor blocked the cyclic AMP stimulation of protein phosphorylation, but had no significant effect on the calcium stimulated
phosphorylation of proteins DPH–L or DPH–M (5). These results
indicate the calcium dependent kinase system originally described
in this laboratory in whole brain (10,15), synaptosome (3,9,17),
and synaptic vesicle (21) fractions is a distinct endogenous kinase
system from the cyclic AMP stimulated kinases described in brain
(46). The protein substrates of these enzymes are different and the

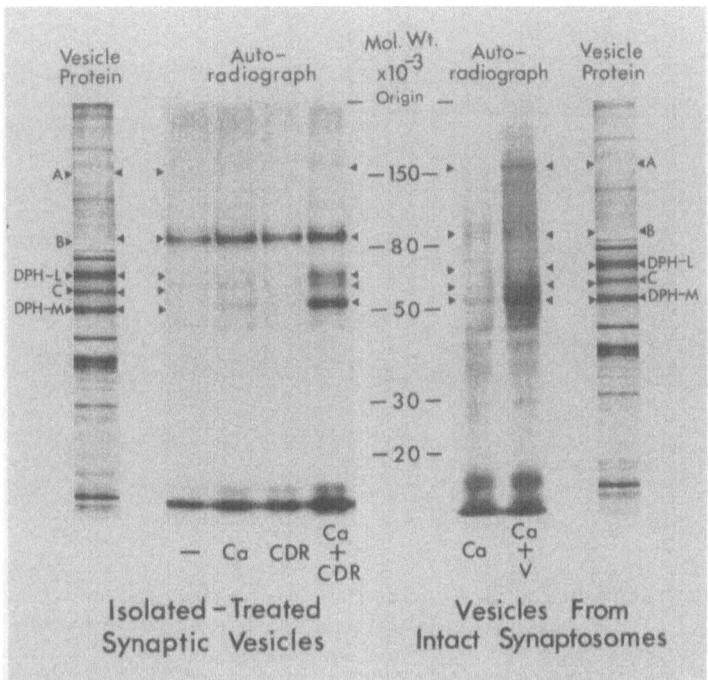

Fig. 4. Effects of Ca^{2+} and/or calmodulin (CDR) on protein phosphorylation in isolated treated synaptic vesicle preparations and of synaptosomal depolarization-dependent Ca^{2+}-uptake on protein phosphorylation of synaptic vesicle proteins isolated from ^{32}P-prelabeled intact synaptosome preparations. For experiments with isolated vesicles, $[\gamma-^{32}P]$ ATP was added directly to the reaction mixture containing 100 μg vesicle protein and incubated under standard conditions in the presence and/or absence of Ca^{2+} (10 μM) and rat brain CDR (5 μg). Qualitatively similar results were obtained with heat-treated SVE and bovine brain CDR. For experiments with intact nerve terminals, synaptosomes were first preincubated with radioactive inorganic phosphate and then incubated with Ca^{2+} alone (1.2 mM) or Ca^{2+} plus Veratridine (V, 75 μM) for 5 min. Following incubation, synaptic vesicles were rapidly isolated from each incubated synaptosome mixture and studied for synaptic vesicle protein phosphorylation. Experimental details were described previously (23) and in the legend to Table 8. Proteins DPH-L, DPH-M, A, B, and C are designated with arrows for clarity. (Reproduced from reference 23).

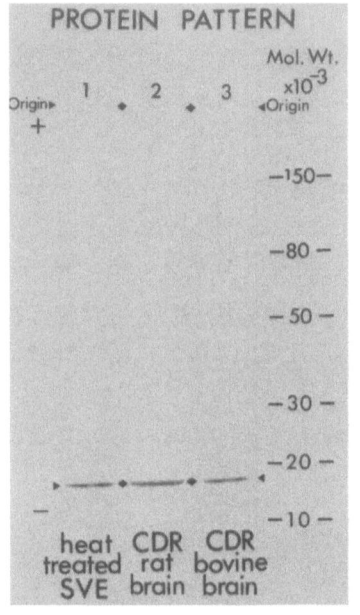

Fig. 5. Electrophoretic protein patterns on SDS-polyacrylamide gel of heat-treated SVE, rat brain Calmodulin (CDR), and bovine brain Calmodulin (CDR). Protein was stained with Coomassie brilliant blue.

conditions of activation are unique to each system.

REGULATION OF CALCIUM DEPENDENT NOREPINEPHRINE RELEASE AND PROTEIN PHOSPHORYLATION BY AN ENDOGENOUS, HEAT STABLE VESICLE PROTEIN FRACTION

The loss of calcium-dependent synaptic vesicle neurotransmitter release and protein phosphorylation under less physiological conditions of isolation (21), suggested that some of the enzymes or substrates required to mediate these calcium-dependent processes were removed from the synaptic vesicles during isolation. The nature of this activating factor was investigated (22,23).

Synaptic vesicles could be repeatedly washed in Iso-KCl media without significant change in protein pattern or loss of Ca^{2+}-dependent norepinephrine release or protein phosphorylation, demonstrating that the enzymes and substrates involved in these processes were tightly bound to the vesicles. However, treating the vesicles with EDTA and homogenization almost completely removed the ability of Ca^{2+} to stimulate release of norepinephrine and phosphorylation of proteins DPH-M, DPH-L and other minor bands, without significantly affecting vesicle norepinephrine concentration or vesicle morphology. Less than 5% of the total vesicle protein was removed in the synaptic vesicle extract (SVE).

Calcium's ability to stimulate release and phosphorylation was

restored to the treated vesicle preparation, when SVE or boiled SVE
was added back to the vesicle preparation. SVE did not significantly
stimulate calcium's effects on release and phosphorylation in plain
or washed synaptic vesicle preparations, indicating that these preparations were saturated with endogenously bound activator. Heat
stable SVE did not exhibit endogenous Ca^{2+} or Mg^{2+} -dependent protein
phosphorylation, nor did it show protein kinase activity with the
artificial substrates histone, casein, or protamine. Characterization of SVE by treatment with heat, dialysis, DNase, RNase, and trypsin,
and subsequent measurement of norepinephrine release and protein
phosphorylation after addition to treated vesicles showed that the
active compound in SVE was heat stable and protein in nature. These
results are summarized in Table 5 and Figure 4.

CALMODULIN AND CALCIUM-ACTIVATED NOREPINEPHRINE RELEASE AND PROTEIN
PHOSPHORYLATION IN SYNAPTIC VESICLE PREPARATIONS

     The heat stability and properties of the stimulating factor in
SVE suggested a possible relationship to calmodulin that modulates
adenylate cyclase and cyclic nucleotide-dependent phosphodiesterase
activities. Thus, calmodulin was purified to homogeneity from bovine

---

Legend to Table 5. Plain, washed or treated synaptic vesicles were
incubated for 1 min under standard conditions (23) to study neurotransmitter release (10 mg vesicle protein containing 4.3-5.6 ng
norepinephrine/mg protein) and protein phosphorylation (1 mg vesicle
protein) with or without the addition of various preparations of
SVE or calmodulin (50 µg of added fraction per mg vesicle protein)
in the presence or absence of Ca^{2+} (10 µM). The reactions were
terminated by centrifugation at 4°C and quantitated for vesicle
norepinephrine content and protein DPH-M phosphorylation (21).
Endogenous vesicle norepinephrine content was determined spectro-
photometricaly and is expressed as ng per reaction mixture. The
total recovery of released and bound norepinephrine was 95-100% in
each condition. Ca^{2+}-stimulated release is expressed as % of Ca^{2+}-
dependent norepinephrine release from the vesicles as compared to
control conditions. Protein DPH-M phosphorylation is expressed as
cpm of [32P] phosphate incorporated per 100 µg of reaction protein.
Ca^{2+}-stimulated phosphorylation is expressed as % of Ca^{2+}-dependent
protein DPH-M phosphorylation as compared to control conditions.
Results qualitatively similar to those for DPH-M were obtained for
protein DPH-L and other protein bands shown in Fig. 5. The data for
release and phosphorylation give the mean values of five determinations and are representative of four separate experiments. The
largest standard errors about the means for release, phosphorylation,
% release, and % phosphorylation were 5 ng, 143 cpm, 7% and 26%
respectively, and were omitted for clarity. Table 5 is in next page.

Table 5. Effects of synaptic vesicle extract (SVE) and calmodulin on Ca^{2+}-dependent synaptic vesicle neurotransmitter release and protein phosphorylation

	Neurotransmitter Release			Protein Phosphorylation		
	Synaptic Vesicle Norepinephrine Content (ng)		Ca^{2+}-Stimulated Release	Protein DPH-M Phosphorylation (cpm)		Ca^{2+} Stimulated Phosphorylation
Conditions	Control	Ca^{2+}	%	Control	Ca^{2+}	%
Plain Synaptic Vesicles	53	26	50.9	369	1588	330
+ calmodulin (rat)	50	24	52.0	340	1506	343
Washed Synaptic Vesicles	49	29	40.8	325	1406	333
+ calmodulin (rat)	47	24	48.9	307	1498	388
Treated Synaptic Vesicles	46	42	8.7	344	503	46
+ SVE	45	26	42.2	310	1393	349
+ Heat-treated SVE	44	23	47.7	358	1481	314
+ Trypsin-treated SVE	47	41	12.8	322	486	51
+ RNase-treated SVE	49	25	49.0	361	1461	305
+ DNase-treated SVE	43	20	53.5	293	1432	389
+ calmodulin (rat)	44	19	56.8	308	1516	392
+ calmodulin (bovine)	47	24	48.9	329	1503	357

The legend is in the previous page

Table 6. Stimulation of activator-depleted phosphodiesterase
          activity by SVE and calmodulin

Activator Added	Phosphodiesterase Activity A_{660}	Activation Fold
None	0.16 ± 0.08	--
SVE	0.83 ± 0.11	5.2
Boiled SVE	0.93 ± 0.13	5.8
Calmodulin	0.99 ± 0.09	6.2

The reaction mixture contained 50 µg of activator-depleted
phosphodiesterase and 0.1 mM $CaCl_2$ in the presence or absence of
5µg of SVE, boiled SVE, or calmodulin (rat brain) and was incubated
under the standard conditions of Lin, et al. (34). Cyclic AMP
(2mM) was added last to initiate the reactions and phosphodiesterase
activity was determined by converting its reaction product, 5'-AMP,
into adenosine and inorganic phosphate by a two stage reaction
procedure and then determining the amount of inorganic phosphate
released by a colorimetric assay that was quantitated spectrophotometrically at 660nM (34). Phosphodiesterase activity is
expressed as absorbance (A) at 660 nM and represents the mean $\pm$
S.E.M. of 5 determinations. The data are representative of three
separate experiments.

and rat brain preparations to test its effect on the vesicle
preparation. Bovine and rat preparations of calmodulin migrated as
single bands of essentially identical molecular weights on SDS-
polyacrylamide gel electrophoresis (Fig. 5). Both preparations of
calmodulin were as effective as SVE in restoring calcium's ability
to stimulate norepinephrine release and protein phosphorylation when
added to treated vesicle preparations (Table 5 and Fig. 4).
Calmodulin-depleted cyclic nucleotide-dependent phosphodiesterase
was prepared from rat brain by DEAE-cellulose column chromatography
to determine if SVE was as effective as calmodulin in activating
this enzyme system (Table 6). Both calmodulin and boiled SVE were
equally as effective in stimulating activator-deficient phosphodiesterase. These results indicate that SVE and calmodulin are
functionally equivalent in their abilities to stimulate calcium-
dependent phosphodiesterase activity, synaptic vesicle protein kinase
activity, and synaptic vesicle norepinephrine release.

Table 7.   Distribution of synaptosomes, norepinephrine, and calcium
           -stimulated phosphorylation of proteins DPH-L, and protein
           DPH-M, in P_2 subfractions

	P_2 subfractions		
	Myelin fraction (0.32–0.8M interface)	Synaptosome fraction (0.8–1.2M interface)	Mitochondrial fraction (pellet)
Nerve endings (synaptosomes)	0.38*	1.50	0.57
Norepinephrine (ng/mg protein)	0.98+	2.45	1.11
Protein DPH-L			
Isotonic	127#	362	213
Osmotic shock	192	672	312
Protein DPH-M			
Isotonic	229	483	298
Osmotic shock	351	1,235	567

*Data taken from the work of Blaustein et al. (3) showing relative
"specific" distribution of synaptosomes. Electron micrographs from
this laboratory confirm these results showing that approximately
50–70% of synaptosomes are recovered in 0.8–1.2 M sucrose interface.
+Norepinephrine levels expressed in ng/mg fraction protein as
described (13). Data give mean of three experiments. The largest
range about the mean was 0.22, and they were thus omitted for clarity.
#Counts per minute per 75 µg of subfraction protein. Reactions were
conducted under standard conditions in the presence of calcium ions
(10 µM). Each value represents mean of results obtained from five
experiments. The largest range about the mean was 46 counts/min, and
the ranges were thus omitted for clarity.

        Concentrations of calmodulin required to produce half-maximal
stimulation of norepinephrine release and protein DPH-M
phosphorylation were nearly identical, 0.55 and 0.40 µg respectively.
The free Ca^{2+} concentrations required to produce half-maximal
stimulation of norepinephrine release and protein phosphorylation were
0.7 and 0.5 µM, respectively. The dependence of calcium's effects
on both norepinephrine release and vesicle-associated protein
phosphorylation on SVE or calmodulin, and the similarities between

concentration curves for release and phosphorylation, further suggest
a possible relationship between presynaptic nerve terminal protein
phosphorylation and neurotransmitter release.

ISOLATION OF CALMODULIN FROM PRESYNAPTIC NERVE TERMINAL PREPARATIONS

The major protein component visualized on the electrophoretic
pattern of boiled SVE, representing approximately 65-80% of the
total protein in SVE, co-migrated identically with calmodulin,
molecular weight 17-19,000 (Fig. 5). The major protein staining
component of SVE, like calmodulin, was also extremely heat stable,
sensitive to trypsin, and resistant to dialysis, RNase, and DNase,
suggesting that the active fraction in SVE is a vesicle bound
calmodulin-like protein. Therefore, we have recently purified to
homogenity and characterized this major protein component in SVE
and demonstrated that this heat stable, vesicle bound protein was
the stimulating factor in SVE and is essentially identical to
calmodulin isolated from rat brain. Two-dimensional SDS-dyacrylamide
gel electrophoresis of both proteins showed that they comigrated as

---

Legend to Table 8. Synaptosomes were preincubated with Na + 5K
medium for 12 min at 30°C for all experiments except for phosphorylation studies, in which $^{32}P_i$ was added to the preincubation medium.
Additional aliquots of test solutions were then added to the
synaptosome suspensions, giving a final concentration of 1.2 mM
Ca^{2+}, 60 mM K^+, 75 µM veratridine, 12 µg of scorpion venom per ml,
and 0.4 µM TTX. Synaptosome mixtures were incubated for 5 min with
each addition; Ca^{2+} uptake, neurotransmitter release, and protein
phosphorylation were measured. Protein DPH-M was measured in both
whole synaptosomes and vesicles isolated from incubated synaptosomes.
The sum of $[Na^+]$ + $[K^+]$ in the medium was always 137 mM. Data are
means of four determinations, representative of three experiments,
and are expressed as percentage of the maximally stimulated conditions
(Ca^{2+}, K^+) so that the effects of these agents on Ca^{2+} uptake, neurotransmitter release, and protein phosphorylation can be easily
compared. The largest SEM was $\pm$ 6%; they were omitted for clarity.
The means of maximal stimulation (100%) for Ca^{2+} uptake, neurotransmitter release, and protein DPH-M phosphorylation in intact
synaptosomes and isolated vesicles were 14.3 nmol of Ca^{2+} per mg of
protein, 0.7 ng of norepinephrine released per mg of protein, 621
cpm/100 µg of protein, and 1561 cpm/100 µg of protein, respectively.
Percent of total synaptosomal norepinephrine released in 5 min under
maximal stimulation (Ca^{2+}, K^+) was 24-33%. Detailed methodology has
been described previously (23). Acetylcholine was assayed by
previously established procedures (49). Table 8 is in next page.

Table 8. Effects of high K+, veratridine (V), and scorpion venom (SV) in the absence or presence of TTX on synaptosomal Ca^{2+} uptake, norepinephrine release, and protein DPH-M phosphorylation

Condition	Ca^{2+} uptake %	Acetylcholine release %	Norepinephrine release %	Protein DPH-M phosphorylation, %	
				Intact synaptosomes	Isolated synaptic vesicles
Control	–	46	48	61	41
Ca^{2+}	53	48	50	65	46
Ca^{2+}, K+	100	100	100	100	100
Ca^{2+}, K+, TTX	97	96	98	95	98
Ca^{2+}, V	88	85	91	86	94
Ca^{2+}, V, TTX	64	49	58	69	50
Ca^{2+}, SV	89	88	92	86	94
Ca^{2+}, SV, TTX	55	52	47	67	47
K+	–	63	54	60	49
V	–	55	49	65	44
SV	–	60	51	58	39
TTX	–	49	55	66	43

The legend appears on the preceding page.

a single acidic protein spot with an isoelectric point of 8.9. Amino acid analysis of both proteins was also essentially identical.

These results demonstrate that calmodulin is bound to synaptic vesicles, isolated under "more physiological" conditions, and can be purified from vesicle preparations. Since synaptic vesicles are localized within the presynaptic nerve endings, these results provide evidence that this vesicle-associated calmodulin is present within the nerve terminal. This data (22,23) represents the first demonstration of a presynaptic localization for calmodulin in nervous tissue.

DEPOLARIZATION-DEPENDENT CALCIUM UPTAKE, NEUROTRANSMITTER RELEASE, AND PROTEIN PHOSPHORYLATION IN INTACT SYNAPTOSOME PREPARATIONS

The phosphorylation of proteins DPH-M and DPH-L in crude synaptosome fractions (P_2) was also present in more highly enriched synaptosome preparations (17). The calcium-stimulated phosphorylation of these proteins was enriched in the synaptosome fractions isolated from P_2 (Table 7). Thus, it was important to determine if the phosphorylation of these proteins observed in broken synaptosome fractions could be observed in intact synaptosome preparations.

Intact synaptosomes were used to simultaneously study depolarization-dependent Ca^{2+}-uptake, neurotransmitter release, and synaptic vesicle protein phosphorylation (23). When compared to the standard vesicle preparations, vesicles isolated from incubated intact synaptosomes manifested essentially identical protein patterns (Fig. 4) and vesicle morphology (as determined by electron microscopy). Agents that depolarize synaptosomes - high external K^+, veratridine, and scorpion venom (2,3) - stimulated synaptosomal $^{45}Ca^{2+}$ uptake and neurotransmitter release, confirming previous results (2), and simultaneously stimulated $[^{32}P]$-phosphate incorporation into protein DPH-M and other particular proteins in both synaptosomes and synaptic vesicles isolated from incubated intact synaptosomes (Table 8 and Fig. 4). TTX blocked the stimulatory effects of veratridine and scorpion venom, but not of increased external K^+, on $^{45}Ca^{2+}$ uptake, neurotransmitter release, and protein phosphorylation (Table 8), confirming previously demonstrated effects of TTX on this preparation (2). In the absence of external Ca^{2+}, the depolarizing effects of high external K^+, veratridine, and scorpion venom had no significant effect on neurotransmitter release and synaptic vesicle protein phosphorylation.

Both the protein staining and autoradiographic images of Ca^{2+} - and calmodulin-dependent phosphoprotein DPH-M from isolated vesicles and from vesicles obtained from prelabeled intact synaptosomes comigrated as a single band on Na-DodSO$_4$/polyacrylamide gel electrophoresis. Osmotic shock and incubation at 4°C of some synaptosome samples for the time interval required to isolate vesicles from the incubated synaptosomes did not significantly affect depolarization-dependent stimulation of $[^{32}P]$ phosphate incorporation into protein DPH-M when compared to regularly treated synaptosomes. Depolarization-dependent phosphorylation of several other vesicle proteins was also observed (Fig. 4). The phosphorylation of proteins A and C (Fig. 4) was more prominent in the intact synaptosomes when compared to isolated vesicles, but protein DPH-L appeared more prominent in the isolated vesicles.

The specific activity of depolarization-dependent $[^{32}P]$ - phosphate incorporation into protein DPH-M (per mg of reaction protein) was significantly higher in isolated vesicles when compared to whole synaptosome fractions (Table 8). Stimulation of the phosphorylation of protein DPH-M by depolarization-dependent Ca^{2+} uptake was also seen in crude synaptic membrane fractions from intact synaptosomes.

These results indicate that the levels of phosphorylation of several synaptosome and more specifically synaptic vesicle proteins are regulated by the depolarization-dependent uptake of calcium into the nerve terminal. Thus, the physiological variations of calcium concentrations in the nerve ending can influence the phosphorylation of synaptosomal fraction proteins (23,33) and also regulate the phosphorylation of synaptic vesicle associated protein within the nerve terminal (23). This data enhances the relevance of the synaptic vesicle phosphorylation model, since synaptic vesicle protein phosphorylation is regulated by depolarization-dependent calcium uptake in intact synaptosome systems under conditions that initiate the release of neurotransmitter substances.

PHENYTOIN INHIBITION OF CALCIUM- AND CALMODULIN- DEPENDENT NEURO-TRANSMITTER RELEASE AND PROTEIN PHOSPHORYLATION

Since phenytoin has been shown to inhibit the effects of calcium on protein phosphorylation and neurotransmitter release, the effects of phenytoin on calmodulin-dependent functions were studied. Phenytoin inhibited the effects of calcium on the calmodulin-dependent release of norepinephrine and phosphorylation of specific synaptic-vesicle-associated proteins (13). Therapeutic concentrations of phenytoin produced a 10 to 20% inhibition of calmodulin's effectiveness in this system, and toxic concentrations caused a 60 to 70% inhibition of release and phosphorylation. These results demonstrate that phenytoin inhibits calcium- and calmodulin-activated neurotransmitter release and protein phosphorylation in the synaptic vesicle preparation and suggest that phenytoin inhibits some of calcium's actions on neuronal tissue by interfering with the calmodulin- and calcium-induced activation of protein phosphorylation and neurotransmitter release. Experiments are presently being initiated to determine the precise mechanism of phenytoin's inhibition of activated calmodulin.

Our preliminary results suggest that phenytoin competitively inhibits
the effects of the calcium-calmodulin complex and not the binding
of calcium to calmodulin.

CALMODULIN AND CALCIUM STIMULATED ENDOGENOUS PHOSPHORYLATION OF POSTSYNAPTIC DENSITY PROTEINS

To further establish the role of Ca^{2+}-dependent protein
phosphorylation in synaptic function, Ca^{2+} and calmodulin-dependent
protein kinase activity was studied in highly enriched preparations
of postsynaptic densities (PSD, 12). PSD fractions were isolated by
Cotman's technique (7) incubated with 10 μM γ-^{32}P - ATP, 4mM $MgCl_2$,
and 50 mM PIPES buffer pH 7.0 in the presence or absence of Ca^{2+} and/
or calmodulin isolated from rat brain, and studied for endogenous
incorporation of   P-phosphate into specific PSD proteins by our
previously described procedures. In the absence of calmodulin, Ca^{2+}
did not stimulate phosphorylation of PSD protein. However, Ca^{2+}
significantly stimulated endogenous phosphorylation of several PSD
proteins in the presence of calmodulin. Proteins that comigrated
with vesicle proteins DPH-L and DPH-M were markedly phosphorylated
in the PSD fraction. The concentrations of free Ca^{2+} and calmodulin
required to produce half maximal levels of phosphorylation were
0.93 and 0.56 μM, respectively. The results demonstrate that PSD
fractions contain calcium and calmodulin stimulated protein kinases
that endogenously phosphorylate many major proteins of the PSD,
suggesting that this phosphorylating system may be of significance
in modulating some of the actions of Ca^{2+} on the PSD.

CALCIUM AND CALMODULIN-STIMULATED MORPHOLOGICAL CHANGES IN SYNAPTIC VESICLES

Recent experiments in this laboratory (12) have been directed
at determining the effects of calcium on the morphology of synaptic
vesicles. I wanted to determine if the synaptic vesicles changed
shape or size under the conditions that simultaneously induced
calcium-dependent neurotransmitter release and protein phosphorylation. To approach these questions, vesicles were treated under
various conditions and then subjected to examination by electron-
microscopy (7,36). Both negatively stained specimens and their
sections of embedded vesicle preparations were examined. The data
were analyzed using established techniques of stereology and
morphometry.

Negatively stained preparations of synaptic vesicles prepared
under the Iso-KCL standard conditions contained a population composed
of free and grouped vesicles as shown in the representative electronmicrographs reproduced in Figure 6. The vesicles appeared
predominantly round in shape and almost looked as if they surrounded

Table 9.   Effects of Calcium on Synaptic Vesicle Morphology

Vesicle Morphology	Conditions	
	Control	Calcium
Mean Vesicle Diameter (Å)	410	298
Percent Free Vesicles %	63	5

Vesicles were incubated under standard conditions in the presence or absence of calcium (10 μM) and subjected to negative staining procedures and examination by electron-microscopy. Mean vesicle diameters (A° units) and percent Free vesicles (%) were quantitated by standard techniques (36). The data for vesicle diameter represent the mean of 2000 determinations with the largest ± S.E.M. of 35A°. The data for percent free vesicles represent the mean of 200 determinations and the largest ± S.E.M. was 8%.

or embedded in an amorphous material, when viewed in stained or unstained preparations. Vesicle protein was evenly distributed over the grid surface.

Incubating the vesicles under standard conditions in the presence of calcium ions caused a striking change in vesicle morphology, as presented in the representative electron micrographs in Fig. 6. Calcium caused marked aggregation of vesicles into very electron dense aggregates. Quantitation of the distribution of vesicle protein demonstrated that calcium caused a significant decrease in the area of grid surface covered by vesicle protein and a marked decrease in the percentage of free vesicles.

The high electron density and sharp three-dimensional out-lines of the calcium-induced vesicle aggregates suggested that the vesicles had not just associated in large groups, but had actually contracted in size. Contraction of vesicle mass was shown by quantitation of mean vesicle diameters of vesicles that could be visualized near the periphery of the vesicle network. Calcium caused approximately a 30% decrease in the mean vesicle diameter (Table 9) and shifted the whole frequency distribution of vesicle size to the left. Vesicles appeared squeezed, elongated or contracted within these clusters.

Thin sections of control and calcium treated preparations confirmed our observations with negatively stained preparations

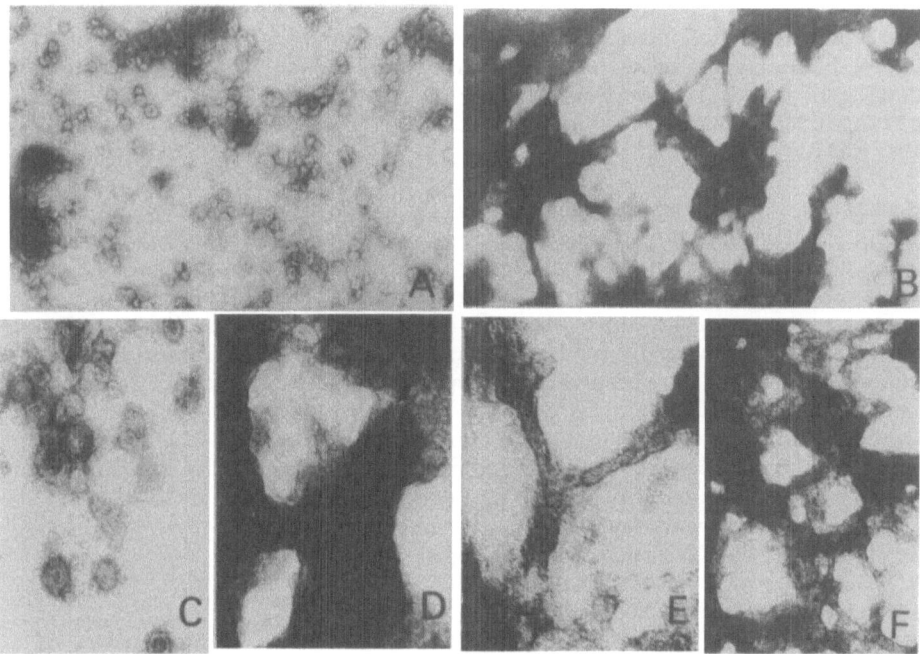

Fig. 6. Effect of calcium on synaptic vesicle morphology: negative
staining. Synaptic vesicle were incubated under standard conditions
in the presence (B, 40,000X; D-F, 150,000X) or absence (A, 40,000X;
C, 150,000X) of 10μM free calcium and prepared for negative staining.
Control conditions (A,C) show many plain vesicles that are round in
shape. Many free vesicles are also seen in the control conditions
(A and C). Calcium caused dramatical changes in the vesicle morphology
(B, D-F). Calcium caused marked vesicle aggregation with a three-
dimensional structure (B,D) and projections with stretching of the
vesicles (E), and net-like formation (F). Vesicle size was noted to
be smaller (decreased mean vesicle diameter) in the calcium treated
preparation, and the vesicles appeared very distorted in shape. The
effect of calcium on calmodulin-depleted vesicles was not as dramatic,
but by adding calmodulin to the vesicles the effect of calcium on
vesicle morphology was restored. The dependence of these changes on
calmodulin suggest that these effects of calcium are not due to
nonspecific changes. Further evaluation of this preparation by freeze
drying techniques may help to rule out drying artifacts. These
experiments are in progress.

as shown in the representative electronmicrographs presented in
Figure 7A. Embedded sections of pelleted control vesicles viewed at
low magnification revealed an even distribution of vesicle protein
across the field and numerous free vesicles were observed. The
distance between vesicles and the number of free vesicles in the
control preparation was inversely proportional to the force of
centrifugation, demonstrating that the vesicles were freely dispersed
in the media.

Calcium caused striking changes in vesicle morphology in thin
sections of fixed and pelleted vesicle samples that paralleled the
changes seen in negative staining. Vesicles were grouped in net-like
cluster or honeycomb formations (Fig. 7) that gave a three-dimensional
appearance, especially when viewed in thicker sections (2000 A°).
Comparison of low power fields demonstrated that calcium significantly
decreased the area of the field occupied by vesicle protein. Free
vesicles were rarely seen in calcium treated preparations. The
distance between vesicles in each vesicle cluster and the spaces
within the formations were independent of centrifugation force and
could not be collapsed by pelleting at forces as high as 130,000
x g. Distances between individual vesicle clusters however, did vary
inversely with the force of centrifugation.

Incubation with calcium ions caused a significant decrease in
the mean vesicle diameter and a shift of the whole distribution of
vesicle diameters to the left (smaller diameters). Calcium also caused
the vesicles to appear elliptical or distorted in shape rather than
round as in control preparations. When the ratio of the largest to
smallest vesicle diameters were determined for 2,000 vesicles in
control and calcium treated preparations more than 78.6% of the
control vesicles and only 18.4% of the calcium treated vesicles had
a ratio greater than 0.8. This result is representative of three
experiments. Within the calcium treated preparations apparent vesicle
fusion with the connection of vesicle membrane was observed. Vesicle

Fig. 7. Effect of calcium on synaptic vesicle morphology: thin
sectioning. Synaptic vesicles were incubated in the presence (B) or
absence (A) of calcium as described in Fig. 6 and subjected to
fixation, dehydration, and embedding for thin sectioning. The control
conditions (A) show round synaptic vesicles with numerous free
vesicles. The calcium treated preparation demonstrates that calcium
caused a marked aggregation of vesicles with a significant reduction
of free vesicles. The mean vesicle diameter in the calcium condition
was also noted to be approximately 40% less than in the control
condition (Table 9).

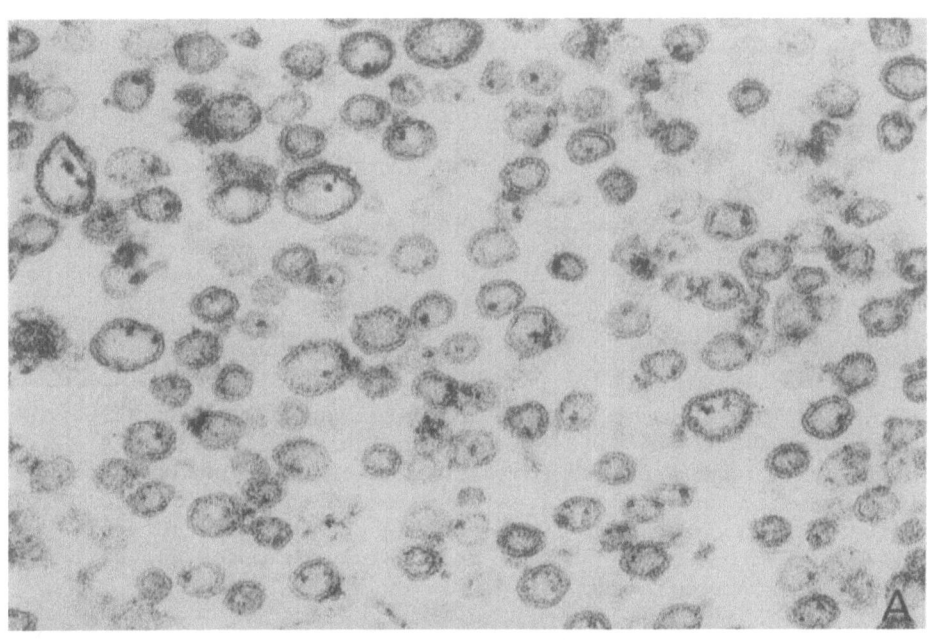

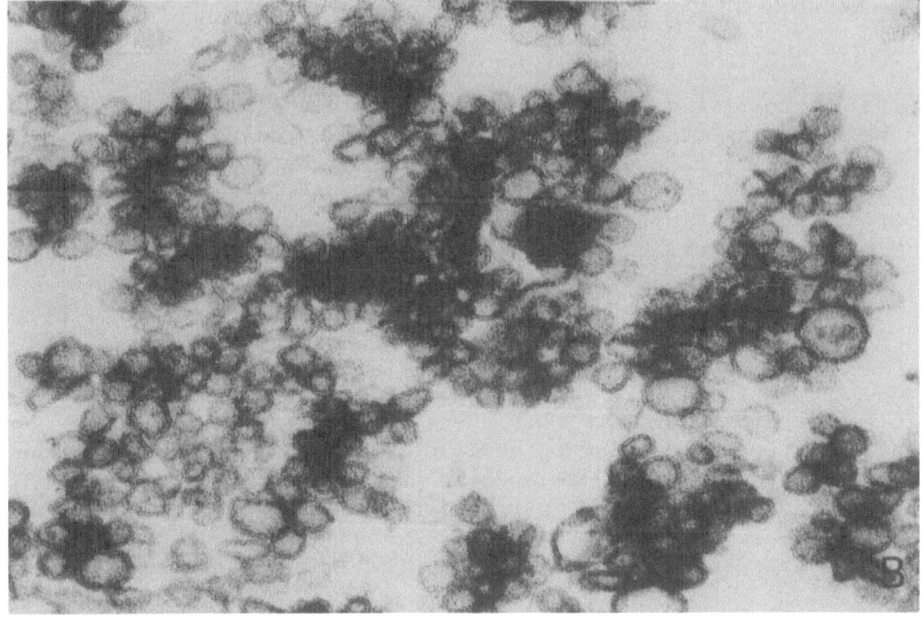

Table 10.   Effects of Calcium and Calmodulin on Synaptic Vesicles
            Diameter and Percent Free Vesicles

Condition	Vesicle Diameter (Å)	Vesicles (%)
Calmodulin-Depleted Synaptic Vesicles		
Control	410	46.2
Calcium	380	39.8
Calcium + Calmodulin	299*	10.4*
Calmodulin	405	47.6

Calmodulin-depleted vesicles were incubated under standard
conditions as described in the legend to Table 9, isolated by
centrifugation, and prepared for electronmicroscopy and morphology.
Vesicle diameter and percent free vesicles an each field were
quantitated. The data shown represent the means of 400 determinations.
The largest SEM was 30 Å and 7.6% for vesicle diameter and percent
free vesicles, respectively. *P<0.001.

fusion and compression was rarely seen in control preparations.

The effects of calcium on synaptic vesicle diameter and percent
free vesicles was also shown to be dependent on calmodulin (Table 10).
The dependency of calmodulin was demonstrated by employing calmodulin-
depleted synaptic vesicles. Calcium (10 μM) had little effect on
vesicle morphology in the absence of calmodulin. Addition of
calmodulin to the calmodulin-depleted vesicles restored calcium's
ability to affect vesicle morphology.

Although the functional significance of these vesicle
morphological changes remains to be determined, the fact that
calmodulin was involved in modulating the effects of calcium on
vesicle morphology suggests that these changes may represent an
important effect of calcium on vesicle function. It is possible that
the effects of calcium and calmodulin on vesicle morphology may
relate to the release of neurotransmitter substances from the
vesicles. A more detailed morphological study with isolated vesicles
and intact synaptosomes is being conducted in my laboratory.

Our preliminary studies demonstrate that it is possible to
simultaneously evaluate neurotransmitter release, protein

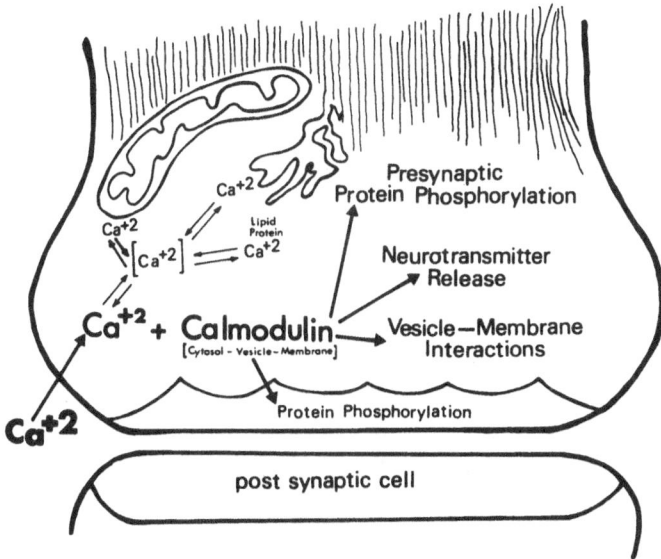

Fig. 8. Schematic model of the possible role of calmodulin in mediating several of the effects of Ca^{2+} on synaptic function. Calmodulin bound to synaptic vesicles serves as a Ca^{2+} receptor in close proximity to the synaptic junction. Calmodulin may also be present in the synaptosome cytosol and synaptic membrane. Following the depolarization-dependent entry of Ca^{2+} into the nerve terminal, Ca^{2+} is immediately bound to calmodulin near the membrane. The binding of Ca^{2+} to calmodulin in the presynaptic terminal could then initiate several processes: presynaptic protein phosphorylation, neurotransmitter release, vesicle membrane interactions, and other Ca^{2+} regulated synaptic functions. This model serves as framework for initiating further experimentation to more clearly delineate the full functional significance of calmodulin in neurotransmitter release and synaptic function. The transsynaptic localization of the Ca^{2+} and calmodulin protein kinase system suggests that this Ca^{2+} regulated biochemical process may play an important role in modulating synaptic function. (Modified from ref. 11).

phosphorylation and morphological change in the same preparation of synaptic vesicles. This preparation can be a powerful tool to investigate the functional role of protein phosphorylation in mediating neurotransmitter release. I plan to utilize this system to determine if the phosphorylation of specific vesicle proteins induces conformational or allosteric changes in these phosphoproteins that then initiate the morphological changes and release of neurotransmitter from the vesicles. The results of these experiments may not only provide insights into the molecular mechanism of

neurotransmitter release, but also may hopefully give insights into
the functional or structural role of specific phosphoproteins in
synaptic transmission.

PROPOSED MODEL FOR THE MODULATION OF NEUROTRANSMITTER RELEASE AND
SYNAPTIC FUNCTION BY CALCIUM AND CALMODULIN

The demonstration that calmodulin was required to mediate calcium
-stimulated neurotransmitter release from synaptic vesicles and
possibly intact nerve terminals (11,22,23) provides the first
evidence for the possible role of the calcium receptor protein,
calmodulin, in mediating the effects of calcium on neurotransmitter
release and synaptic function. These results offer a molecular
approach to investigating calcium-stimulated neurotransmitter
release and its modulation by various drugs. The schematic model
shown in Fig. 8 presents this hypothesis. The expanded model suggests
that depolarization-dependent calcium entry at the presynaptic nerve
terminal initiates neurotransmitter release by binding to the
synaptic vesicle bound calmodulin. When calcium binds to calmodulin
it stimulates the vesicle associated protein kinase system causing
the phosphorylation of specific proteins associated with the
vesicles and the synaptic junctional complex. The model suggests that
the calcium/calmodulin-stimulated protein phosphorylation of vesicle
proteins in some way may provide the motor force for initiating
neurotransmitter release, possibly by causing vesicle concentration
or other morphological changes.

This molecular model for the possible effect of calcium and
calmodulin on neurotransmitter release is supported by the following
evidence: 1) Calcium ions stimulate the phosphorylation of specific
proteins in broken and intact (8-10,17,23,33) preparations of
presynaptic nerve terminals, synaptosomes (especially proteins DPH-L
and DPH-M); 2) the calcium-dependent phosphorylation of synaptosomal
proteins DPH-L and DPH-M occurs within the presynaptic nerve terminal
in association with the synaptic vesicles (13,21,23); 3) the
depolarization-dependent entry of calcium into synaptosomes (23,33)
and calcium added to synaptic vesicles (21,23) stimulate the level
of phosphorylation of proteins with the same molecular weight as
proteins DPH-L and DPH-M under conditions which have been shown to
initiate the release of neurotransmitter from the synaptosome (2,13)
and the vesicles (21,23); 4) calcium simultaneously stimulates both
the phosphorylation of specific proteins and the release of norepinephrine, acetylcholine and other putative neurotransmitters in the
same preparation of synaptic vesicles incubated under identical
conditions (13,24); 5) the concentrations of calcium required to
produce half-maximal norepinephrine release and protein phosphorylation in vesicle preparations are almost identical (13,14); 6) high
concentrations of calcium inhibit both norepinephrine release and

protein phosphorylation (21); 7) calcium-dependent vesicle norepinephrine release and protein phosphorylation both require the presence
of magnesium and ATP (13,21,23); 8) the anticonvulsant phenytoin,
which inhibits calcium-dependent neurotransmitter release from the
presynaptic nerve terminal in several intact and broken preparations
(39-41,51), simultaneously inhibits both the calcium and calmodulin-
dependent release of norepinephrine and phosphorylation of proteins
DPH-L and DPH-M in vesicle preparations (10,13,21); 9) the calcium-
binding regulator protein (calmodulin) that stimulates vesicle
protein phosphorylation by activating the vesicle bound protein
kinase also mediates the effect of calcium on the release of vesicle-
bound norepinephrine (13,22,23); 10) the same concentration of added
calmodulin produces maximal calcium-dependent norepinephrine release
and protein phosphorylation in calmodulin-depleted vesicles (22,23);
11) calcium and calmodulin also stimulate the endogenous phosphorylation of synaptic junction proteins, suggesting a possible role of
phosphoproteins in mediating trans-synaptic modulation (13); 12)
calmodulin and calcium simultaneously stimulate synaptic vesicle
protein phosphorylation, neurotransmitter release, and morphological
changes (12).

Llinas et al. (35) has provided kinetic and electrophysiologic
evidence that only the vesicles or proteins immediately adjacent
to the membrane or actually touching it are able to respond rapidly
enough to calcium inflow to account for the short latency of release.
The extremely close proximity of synaptic vesicles to the synaptic
junction (38) and the evidence that the components of the phos-
phorylation-release system exist in the form of a calcium-activated
molecular complex bound to the synaptic vesicles (23) suggest that
the localization of this molecular system in the nerve terminal makes
it an attractive model for studying the biochemical basis of neurotransmitter release. Special rate studies must be conducted before it
can be determined if this calcium-dependent biochemical process can
mediate rapid calcium-dependent synaptic transmission or mediate
longer-lasting modulations of transmitter release.

The results from my laboratory provide evidence that calmodulin
is present in the presynaptic nerve terminal and that calcium and
calmodulin stimulate neurotransmitter release, protein phosphorylation, and morphological changes in various preparations of the
presynaptic nerve terminal. These observations indicate that
calmodulin plays an important role in modulating the effects of
calcium on synaptic function (14). Since calcium and calmodulin
simultaneously effect protein phosphorylation, neurotransmitter
release, and morphological changes in vesicle preparations is
intriguing to suggest that these three events are related. The data
provided in this report further suggests a relationship between
these calcium stimulated events, but determinations of their precise
relationship requires further investigation.

Acknowledgements. This investigation was supported by U.S. Public Health Service Grant NS 13632 and Research Career Development Award NSI-EA 1KO4 NS from the National Institute of Neurological and Communicative Disorders and Stroke, and by the Esther A. and Joseph Klingenstein Fund. The support and encouragement of Dr. G.H. Glaser are greatly appreciated.

REFERENCES

1. Barondes, H.B., Synaptic macromolecules: Identification and metabolism, Ann. Rev. Biochem., 43 (1974) 147–168.
2. Blaustein, M.P., Effects of potassium, veratridine, and scorpion venom on calcium accumulation and transmitter release by nerve terminals in vitro, J. Physiol. (London), 247 (1974) 617–655.
3. Blaustein, M.P. and Goldring, J.M., Membrane potentials in pinched-off presynaptic nerve terminals monitored with a fluorescent probe: evidence that synaptosomes have potassium diffusion potentials, J. Physiol. (London), 247 (1975) 589–615.
4. Boyne, A.F., Neurosecretion: integration of recent findings into the vesicle hypothesis, Life Sci., 22 (1978) 2057–2066.
5. Brostrom, C.O., Huang, Y.C., Breckenridge, B., and Wolff, D.J., Identification of a calcium-binding protein as a calcium-dependent regulator of brain adenylate cyclase, Proc. Natl. Acad. Sci., 72 (1975) 64–68.
6. Cheung, W.Y., Cyclic 3',5'-nucleotide phosphodiesterase: Demonstration of an activator, Biochem. Biophys. Res. Commun. 38 (1970) 533–538.
7. Cotman, C.W. and Flansburg, D.A., An analytical micro-method for electron microscopic study of the composition and sedimentation properties of subcellular fractions, Brain Res., 22 (1970) 152–156.
8. DeLorenzo, R.J., Antagonistic action of diphenylhydantoin and calcium on the endogenous phosphorylation of specific brain proteins, Neurology, 26 (1976) 386.
9. DeLorenzo, R.J., Possible role of phosphoproteins in mediating calcium-dependent neurotransmitter release, Neurosci. Abstracts, 2 (1976) 1001.
10. DeLorenzo, R.J., Antagonistic action of diphenylhydantoin and calcium on the level of phosphorylation of particular rat and human brain proteins, Brain Res., 134 (1977) 125–138.
11. DeLorenzo, R.J., Role of calcium dependent regulator proteins in neurotransmitter release, Trans. Am. Soc. Neurochem., 10 (1979) 100.
12. DeLorenzo, R.J., Calcium-dependent morphological changes in synaptic vesicles, Trans. Int. Soc. Neurochem., 7 (1979) 68.
13. DeLorenzo, R.J., Phenytoin: calcium- and calmodulin-dependent protein phosphorylation and neurotransmitter release. In Glaser G.H., Penry, J.K., and Woodbury, D.M., (eds.), Antiepileptic

Drugs: Mechanisms of Action, Raven, New York, 1980, pp. 399-414.

14. DeLorenzo, R.J., Role of calmodulin in neurotransmitter release and synaptic function, Ann. N.Y. Acad. Sci., 356 (1980) in press.

15. DeLorenzo, R.J., Emple, G.P., and Glaser, G.H., Regulation of the level of endogenous phosphorylation of specific brain proteins by diphenylhydantoin, J. Neurochem., 28 (1976) 21-30.

16. DeLorenzo, R.J., and Freedman, S.D., Possible role of calcium-dependent protein phosphorylation in mediating neurotransmitter release and anticonvulsant action, Epilepsia, 18 (1977) 357-365.

17. DeLorenzo, R.J., and Freedman, S.D., Calcium-dependent phosphorylation of specific synaptosomal fraction proteins: Possible role of phosphoproteins in mediating neurotransmitter release, Biochem. Biophys. Res. Commun., 71 (1977) 590-597.

18. DeLorenzo, R.J., and Freedman, S.D., Calcium-dependent phosphorylation of synaptic vesicle proteins and its possible role in mediating neurotransmitter release and vesicle function, Biochem. Biophys. Res. Commun. 77 (1977) 1036-1043.

19. DeLorenzo, R.J., and Freedman, S.D. Phenytoin inhibition of calcium-dependent protein phosphorylation in synaptic vesicles, Neurology 27 (1977) 375.

20. DeLorenzo, R.J., and Freedman, S.D. Possible role of synaptic vesicle-associated phosphoproteins in mediating calcium-dependent neurotransmitter release, Neurosci. Abstracts, 3 (1977) 513.

21. DeLorenzo, R.J., and Freedman, S.D., Calcium-dependent neuro-transmitter release and protein phosphorylation in synaptic vesicles, Biochem. Biophys. Res. Commun., 80 (1978) 183-192.

22. DeLorenzo, R.J., Freedman, S.D. and Yoho, W.B., Purification of a synaptic vesicle-bound protein mediating calcium dependent vesicle neurotransmitter release and protein phosphorylation, Soc. Neurosci. Abs., 4 (1978) 578.

23. DeLorenzo, R.J., Freedman, S.D., Yohe, W.B., and Maurer, S.C., Stimulation of Ca^{2+}-dependent neurotransmitter release and presynaptic nerve terminal protein phosphorylation by calmodulin and a calmodulin-like protein isolated from synaptic vesicles, Proc. Natl. Acad. Sci. U.S.A., 76 (1979) 1838-1842.

24. DeLorenzo, R.J., and Glaser, G.H., Regulation of endogenous phosphorylation of a specific protein from rat brain homogenates by diphenylhydantoin, Neurosci. Abstracts, 1 (1975) 342.

25. DeLorenzo, R.J., and Glaser, G.H., Effect of diphenylhydantoin on the endogenous phosphorylation of brain protein, Brain Res., 105 (1976) 381-386.

26. DeRobertis, E., and DeLores Arnaiz, G.R., Structural components of the synaptic region. In: A. Lajtha (ed.), Handbook of Neuro-chemistry, Plenum, New York, 1969, vol. 2, pp. 365-392.

27. Esplin, D.W., Effects of diphenylhydantoin on synaptic trans-mission in cat spinal cord and stellate ganglion, J. Pharmacol. Exp. Ther., 120 (1957) 301-323.

28. Esplin, D.W., Criteria for assessing effects of depressant drugs on spinal cord synaptic transmission, with examples of drug

selectivity, Arch. Int. Pharmacodyn. Ther., 143 (1963) 479–497.

29. Kakiuchi, S., and Yamazaki, R., Calcium dependent phospho-
    diesterase activity and its activating factor (PAF) from brain,
    Biochem. Biophys. Res. Commun., 41 (1970) 1104–1110.

30. Katz, B., and Miledi, R., Tetrodotoxin-resistant electric
    activity in presynaptic terminals, J. Physiol. (Lond.), 203 (1969)
    459–487.

31. Katz, B., and Miledi, R., Further study of the role of calcium in
    synaptic transmission, J. Physiol. (Lond.), 207 (1970) 789–801.

32. Kretsinger, R.H., in Carafoli, E., Clementi, F., Drabikowski, W.
    and Margreth, A., (eds.), Calcium Transport in Contraction and
    Secretion, North-Holland, Amsterdam, 1975, pp. 469–478.

33. Krueger, B.K., Forn, J., and Greengard, P., Depolarization-
    induced phosphorylation of specific proteins, mediated by calcium
    ion influx, in rat brain synaptosomes, J. Biol. Chem., 252 (1977)
    2764–2773.

34. Lin, Y.M., Lin, Y.P., and Cheung, W.Y., Cyclic 3' :5'-nucleotide
    phosphodiesterase: Purification, characterization and active
    forms of the protein activator from bovine brain, J. Biol. Chem.,
    249 (1974) 4943–4954.

35. Llinas, R., Steinberg, I.Z., and Walton, K., Presynaptic calcium
    currents and their relation to synaptic transmission: Voltage
    clamp study in squid giant synapse and theoretical model for the
    calcium gate, Proc. Natl. Acad. Sci. U.S.A., 73 (1976) 2918–2922.

36. Matthews, E.K. and Nordmann, J.S., The synaptic vesicle: calcium
    ion binding to the vesicle membrane and its modification by drug
    action, Mol. Pharm., 12 (1976) 778–788.

37. Miledi, R., Transmitter release induced by injection of calcium
    ions into nerve terminals, Proc. R. Soc. Lond. (Biol.), 183
    (1973) 421–425.

38. Pfenninger, K., The cytochemistry of synaptic densities, J.
    Ultrastruct. Res., 34 (1971) 103–118.

39. Pincus, J.H., and Lee, S.H., Diphenylhydantoin and calcium:
    Relation to norepinephrine release from brain slices, Arch.
    Neurol., 29 (1973) 239–244.

40. Raines, A., and Standaert, F.G., An effect of diphenylhydantoin
    on post-tetanic hyperpolarization of intramedullary nerve
    terminals, J. Pharmacol. Exp. Ther., 156 (1967) 591–597.

41. Raines, A., and Standaert, F.G., Pre- and post- junctional
    effects of diphenylhydantoin at the soleus neuromuscular junction,
    J. Pharmacol. Exp. Ther. 153 (1966) 361–366.

42. Schulman, H., and Greengard, P., Stimulation of brain membrane
    protein phosphorylation by calcium and an endogenous heat-stable
    protein, Nature (Lond.), 271 (1978) 478–479.

43. Szmigielski, A., Guidotti, A., and Costa, E., Endogenous protein
    kinase inhibitors. Purification, characterization and distribu-
    tion in different tissues, J. Biol. Chem., 252 (1977) 3848–3853.

44. Tamir, H., Rapport, M.M., and Gershon, M.D. Serotonin binding
    protein in synaptic vesicles from brain, Trans. Am. Soc.

Neurochem., 10 (1979) 184.
45. Teo, T.S., Wang, H., and Wang, J.H., Purification and properties of the protein activation of bovine heart cyclic adenosine 3',5'-monophosphate phosphodiesterase, J. Biol. Chem., 248 (1973) 588-595.
46. Ueda, T., Maeno, H., and Greengard, P., Regulation of endogenous phosphorylation of specific proteins in synaptic membrane fractions from rat brain adenosine 3': 5'-monophosphate, J. Biol. Chem., 248 (1973) 8295-8305.
47. Waisman, D., Stevens, F.C., and Wang, J.H., The distribution of the calcium-dependent protein activator of cyclic nucleotide phosphodiesterase in invertebrates, Biochem. Biophys. Res. Commun., 65 (1975) 975-982.
48. Weller, M., and Rodnight, R., Protein kinase activity in membrane preparations from ox brain: stimulation of intrinsic activity by adenosine 3': 5'-cyclic monophosphate, Biochem. J., 132 (1973) 483-492.
49. Whittaker, V.P., The synaptosome. In A. Lajtha (ed.), Handbook of Neurochemistry, Plenum, New York, 1969, Vol. 2, pp. 327-364.
50. Wolff, D.J., and Siegel, F.L., Purification of a calcium-binding phosphoprotein from pig brain, J. Biol. Chem., 247 (1972) 4180-4185.
51. Yaari, Y., Pincus, J.H., and Argov., Z., Depression of synaptic transmission by diphenylhydantoin, Ann. Neurol., 1 (1977) 334-338.

BRAIN CLATHRIN: A STUDY OF ITS PROPERTIES

Saul Puszkin, Kenan Haver, and William Schook

The Department of Pathology, Mount Sinai School of
Medicine of the City University of New York
New York, NY 10029

INTRODUCTION AND BACKGROUND

In 1968, Puszkin, Berl and coworkers reported the presence of
contractile proteins in whole brain (1). Subsequent work showed
the presence of contractile proteins in nerve-endings (2,3,4). The
properties of various contractile proteins found in brain tissue
resembled, to a large extent, those from muscle tissue (2,4). As
a result, it was proposed that contractile activities are involved
in neurotransmitter release (5). With the advantage of sodium
dodecyl sulfate polyacrylamide gel electrophoresis as a procedure
for distinguishing and comparing the size of various proteins, many
researchers working with other tissues or cells began to report
the presence and involvement of contractile proteins in motility-
related cell functions (6).

Rapid progress was achieved in the mid-seventies toward understanding the biochemistry and physiology of non-muscle contractile
and regulatory proteins. By this time, a more extensive role for
regulatory proteins in brain functions had been advanced, i.e.,
that calcium-regulated proteins modulate the release of putative
neurotransmitters (7). Yet, the exact manner by which cytoskeletal
proteins participate in brain function remains to be determined.
However, several important new areas of work are providing new
insights.

An important aspect of nerve endings is that most of their
contents seem to be stored in subcellular compartments such as
endoplasmic reticulum-like cisternae, coated vesicles and synaptic
vesicles. The latter are believed to contain the neurotransmitter
which is released upon stimulation. In order to maintain

transmission, nerve-endings seem to possess a recycling system which permits recovery of fused vesicle membrane and possibly the reuptake of neurotransmitters as well. The turnover rate of active nerve-endings as well as their lack of a proportional rate of supply of metabolites from axonal transport necessitates an efficient system of recycling.

For recycling to occur, at least two complex components have to be reutilized: the chemical compounds being released from their storage sites, and the vesicle membrane being utilized to hold and release the putative neurotransmitter by exocytosis. These processes require the existence of motile events within this structure, and cytoskeletal proteins coupled with modulatory proteins could be responsible for such motile activity.

Clathrin has been a target of our research because of its involvement in endocytosis, the process by which nerve-endings in particular reutilize membrane and part of receptor-bound materials that previously had been secreted (8).

Clathrin is abundant in brain (9) and has been classified as a mechanochemical protein (10) because it has binding affinity for actin, α-actinin, and tropomyosin -all classical cytoskeletal proteins- and it forms reversible ultrastructural assemblies.

We reported that clathrin assembles ultrastructural entities which are either filamentous or lattice-like, resembling basket or cage shapes. Clathrin also copurifies with tubulin and with tropomyosin, the latter a modulatory protein found in brain nerve-endings (11).

Previous studies indicated that coated vesicles participate in the process of replenishing synaptic vesicles in nerve-endings (8). It has been shown (12,13) after stimulation of neuronal elements, that there was a depression of synaptic vesicles concomitant with transmitter release. Since there was no increase in presynaptic membrane surface area, there was no accounting for the disappearance of synaptic vesicle membrane. It was presumed that synaptic vesicles reformed locally from the plasmalemma, since more transmitters were being released than could be accounted for by the depression in number of synaptic vesicles.

Direct morphological evidence for local synaptic vesicle formation in neuromuscular junctions was obtained by Heuser and Reese (8). Stimulated frog muscle showed synaptic-ending boutons depleted of vesicles accompanied by an increase in the internal cisternae membrane area of the neuronal bouton. Recovery from stimulation involved the plasmalemma and cisternae areas returning to their original states. It was postulated that the coat protein recognized specific vesicle particles in the membrane and retrieved that membrane area for recycling.

The involvement of coated vesicles is widespread and extends to many tissues and cultured cells (14). All seem to possess a similar pattern of surface-membrane increase and recycling of membrane utilizing a coated vesicle mechanism. Biochemical investigation of the proteins forming lattices around vesicles began in 1975 (15). Clathrin was isolated from a fraction enriched with coated vesicles after modification of the subcellular fractionation method of Kaneseki and Kadota (16). Morphologically, the enriched fraction of coated vesicles was sufficiently homogeneous to proceed with meaningful biochemical analysis.

In sodium dodecyl sulfate gels, about half of the total protein present in the coated vesicle fraction migrated into an electro-phoretic zone equivalent to a molecular weight of 180,000. This major protein was named "clathrin," after the Greek word, clathri, meaning latticework (9).

Highly homogeneous clathrin has two polypeptides tightly bound of 30,000 and 28,000 molecular weights (11,17). We have collected evidence that identifies these polypeptides as tropomyosin, a protein that previously had been isolated from brain (18). In this paper we describe studies on the properties of clathrin, its localization and interaction with cytoskeletal proteins.

EXPERIMENTAL PROCEDURES

Coated Vesicles and Clathrin

Fresh bovine brains were cleansed of meninges and the gray matter separated from the white matter by suction. Highly purified coated vesicles were prepared from 400 g of gray matter after homogenization in a Waring blender, essentially as described previously (10,19).

For the preparation of clathrin from vesicles, the gray matter from two or three bovine brains was homogenized in 500 ml of 0.1 M Tris-HCl, pH 7.0, containing 1 mM EGTA, 0.5 mM $MgCl_2$, 1 mM 2-mercapto-ethanol and 0.02% sodium azide. The crude coated vesicle fraction obtained after centrifugation was extracted with 150 ml of 20 mM Tris-HCl, pH 7.5, and 1 mM 2-mercaptoethanol, 1 mM EDTA, 0.02% sodium azide. High purity clathrin was obtained by repeated column chroma-tography using Sepharose 4B equilibrated in a 20 mM Tris buffer, pH 7.5, containing 2 M urea and a final column chromatography run using a 20 mM Tris buffer, pH 7.5, without urea. The peak containing clathrin was pooled, precipitated with 50% ammonium sulfate, dialyzed and stored frozen at -20°C in 10% sucrose.

Clathrin was resolved using linear gradient slab gels prepared from 7.5% to 15-or-20% polyacrylamide, containing 0.1% sodium dodecyl sulfate (10). An electrical current of 8 mA, applied for 16 hours to a Tris-glycine buffer, pH 8.3, containing 0.1% sodium dodecyl sulfate separated and resolved the proteins in the mixture. Before loading, the protein samples were brougth to 1% sodium dodecyl sulfate and reduced with 0.1% 2-mercaptoethanol before application.

Tropomyosin and Actin

A modification of the procedure used to purify brain tropomyosin (18) and platelet tropomyosin (20) was used. Beef brains were cleansed of meninges and the cerebella and brain stems removed. The cortex was homogenized in 20 volumes of 100% ethanol at top speed in a Waring blender. The precipitate obtained by centrifugation at 10,000 x g for 20 minutes was resuspended in 20 volumes of anhydrous ethyl ether. The precipitate was treated with ethanol and ether as described above and the resulting powder air dried. The powder was extracted with 1 M potassium chloride, 2 mM 2-mercaptoethanol, 10 mM Tris, pH 7, in a proportion of 50 ml per g of powder for 16 hours at 4°C. The extract, separated from the debris by centrifugation at 30,000 x g, was placed in a boiling water bath for 10 minutes. Most of the contaminating proteins precipitated, leaving tropomyosin in solution. The extract was cooled, centrifuged at 100,000 x g for 1 hour, and the supernatant acidified with normal HCl to pH 4.1. The pellet obtained by centrifugation at 30,000 x g for 20 minutes was dissolved in 1 volume of 0.2 M potassium chloride containing 30 mM sodium phosphate buffer, pH 7. The solubilized proteins were fractionated by ammonium sulfate precipitation. The 40-to-53% saturated solution was centrifuged and the resulting pellet dissolved in a minimal volume of 50 mM Tris buffer, pH 8, dialyzed for 16 hours, then clarified by high speed centrifugation at 100,000 g for 1 hour. The yield of tropomyosin at this step was 0.5 mg per g of dried powder. From the solution of tropomyosin obtained, paracrystals were formed by dialysis overnight at a concentration of 3 mg per ml, against 50 mM $MgCl_2$ and 50 mM Tris-HCl, pH 8.

Muscle actin was isolated by modification of a procedure published elsewhere (2). A dry acetone powder was prepared and after dryness, the acetone powder was extracted with stirring using a buffer of 20 mM Tris-HCl, pH 8.0, and containing 0.2 mM ATP and 1 mM 2-mercaptoethanol. The extract was centrifuged at 100,000 x g for 2 hours at 4°C, and the proteins in the supernatant concentrated by precipitation with 60% saturation of ammonium sulfate. The pellet was resuspended in 0.2 mM ATP, 10 mM Tris-HCl, pH 7.5, and polymerized by addition of 0.1 M KCl and 1 mM $MgCl_2$. Actin was allowed to polymerize overnight at 4°C, ultracentrifuged at 100,000 x g for 3 hours. Several polymerizations and depolymerizations were carried out as described above until the protein was approximately 80-90% pure.

Alpha-actinin was isolated from a crude coated vesicle fraction from bovine brain, extracted overnight in a low ionic strength buffer containing 14 mM 2-mercaptoethanol (21).

Peptide mapping was carried out using chymotryptic or tryptic digestion followed by peptide dansylation. Fluorescent peptides were visualized and photographed after separation by thin-layer chromatography in polyamide plastic plates (23).

Immunological Techniques

White, New Zealand-strain male rabbits weighing between 2-3 Kg were immunized over a period of 6-to-8 weeks by intradermal and intramuscular injections of highly purified clathrin previously assembled as baskets and emulsified in an equal volume of complete Freund's adjuvant at pH 6.5. Each injection contained 0.5 mg of clathrin in a volume of 1 ml. Rabbits were bled 8-to-10 days after the last injection, and the individual sera tested for antibody reactivity and, when positive, stored at -20°C.

Double gel diffusion was performed in 0.85% agarose gels dissolved in 0.05 M sodium barbital-saline buffer, pH 8.0 as described previously (22). Proteins were allowed to react at room temperature, and the agarose-coated glass slides dried, stained with amido black 10B, and de-stained by diffusion using 7% acetic acid and 5% methanol.

Immunoglobulins were purified from serum by a 20% sodium sulfate precipitation followed by DEAE (DE-52) cellulose chromatography (Whatman, Co., Waltham, Mass.) in 0.02 M sodium phosphate buffer, pH 6.3. Anti-clathrin IgG molecules then were purified by affinity chromatography using CnBr Sepharose 4B coupled with highly purified clathrin, essentially as described for anti-α-actinin (22). Specific anticlathrin antibodies were eluted with a solution containing 0.1 M glycine, 0.5 M sodium chloride, pH 2.8. The pH of the eluted protein peak was neutralized with Tris crystals, dialyzed against 20 mM phosphate saline buffer, pH 7.5, and stored in 0.5% bovine serum albumin at -20°C.

Fibroblasts (American Type Culture Collection Catalogue of Strains ATCC CCL 186), primary endothelial and neuroblastoma cells (ATCC CCL 127) from human, and rat glial (ATCC CCL 107) cells were used for localization of clathrin. These cultures were grown to confluency, trypsinized and replated onto small, circular glass coverslips. Cells grown on coverslips were rinsed in phosphate-buffered saline, pH 7.4, and fixed in 3.8% paraformaldehyde for 15 minutes at room temperature. The coverslips were rinsed in phosphate-buffered saline, followed by a wash in phosphate-buffered saline containing 0.1% Triton X-100 for 15 minutes. The cells then were incubated with the appropriate antibody in humidified chambers for

30 minutes at room temperature. The unbound antibodies were rinsed
from the cells by 3 consecutive washes with phosphate-buffered saline
and visualized by indirect immunofluorescence utilizing goat anti-
rabbit immunoglobins conjugated with either rhodamine or fluorescein.
Slides were examined using phase contrast optics and epifluorescent
illumination with filters for rhodamine or fluorescein on a Zeiss
Universal microscope with a 200 Watt UV Lamp. Photographs were taken
and developed using the same exposure and printing for matched
experimental and control sera.

EXPERIMENTAL OBSERVATIONS AND DISCUSSION

Purification of Clathrin

      Clathrin can be isolated from coated vesicles (10,19). However,
the yield is not very large and the procedure is very time consuming.
The purification procedure for clathrin was simplified initially by
preparing an enriched coated vesicle fraction that retained its
lattice bound on the vesicles surfaces. This required that the pH
of the buffer used for brain homogenization be maintained below 7.0.

      The next task was to maintain clathrin dissociated from the
membrane fragments and from itself. Three possibilities were open:
first, extract clathrin with high salt buffers; second, use low
salt concentration buffers in the alkaline pH region; and third,
use detergents in the medium that could solubilize the lipid membrane
and release clathrin molecules into solution.

      Purification attempts using high salt buffers for extraction
rendered a highly emulsified, opalescent solution containing a
largely complex and extremely heterogeneous population of proteins.
Clathrin comprised a small amount of the total proteins. The use
of a Tris buffer, pH 7.5, containing 1 mM EDTA, rendered a preparation with the highest amount of clathrin in solution. The extracts
produced using detergents were highly lipidic, with clathrin
comprising a small proportion of the total protein content. Thus,
we decided to pursue the second possibility, i.e., the Tris-HCl
buffer pH 7.5, 0.1% 2-mercaptoethanol, 1 mM EDTA in the extraction
medium.

      Subsequent experiments showed that the apparently soluble
preparation of clathrin described above was, in reality, a complex
of membrane fragments and clathrin assemblies that did not sediment
after high speed centrifugation (100,000 x g for 1 hour). We found
that addition of 2 M urea to the Tris buffer and the removal of
EDTA was effective in dissociating these preparations. Therefore,
the extract was purified by chromatography on Sepharose 4B
equilibrated with the 2 M urea buffer. The procedure eluted a first
peak with membrane fragments and little clathrin from a second peak:

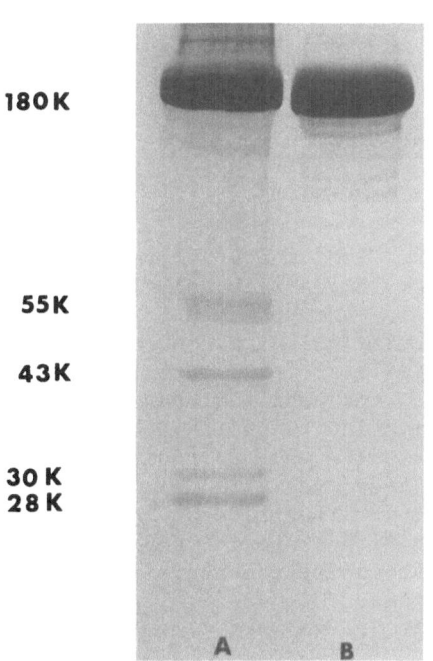

180K

55K

43K

30K
28K

A   B

Fig. 1. Sodium dodecyl sulfate polyacrylamide slab gel electrophoresis--A gradient of polyacrylamide gel of 7.5 to 15% containing 0.1% sodium dodecyl sulfate was loaded with A: purified clathrin (125 µg), and B: highly purified clathrin (135 µg) obtained after brief chymotryptic treatment of the clathrin preparation shown in A. The other proteins copurifying with clathrin in A comigrate with known standards as follows: clathrin (180,000 m/w); tubulin (55,000 m/w); actin (43,000 m/w); and the doublet of polypeptides, identified as tropomyosin (30,000 and 28,000 m/w). Minor amounts of high m/w polypeptides (>250,000) also are detected. After enzyme treatment, most of clathrin accompanying proteins are absent. However, the capacity to form baskets or cages exhibited by clathrin in A is absent in clathrin shown in B.

the clathrin peak. There was always a small third peak containing a largely heterogeneous protein mixture with a low concentration of clathrin. Rechromatography of the second peak yielded a highly purified preparation in which the proteins still accompanying clathrin probably had high affinity for clathrin, possibly forming a complex. Among these proteins, we identified ᴪ-actinin having a molecular weight of 100,000; actin, 43,000; the tubulin dimer, approximately 55,000; and a doublet of approximately 30,000 molecular weight (Fig. 1).

Separation of some of these proteins was accomplished with mixed results. Actin, tubulin and ᴪ-actinin could be separated, but the final purified product still was showing the doublet of 30,000 molecular weight. Since in the preparation of clathrin, the separation of tubulin eventually is accomplished without measurable loss of clathrin properties, it was deemed unimportant at this time to pursue studies to determine the affinity of tubulin for clathrin. Alternatively, we explored the affinity of clathrin for ᴪ-actinin and actin. We found that ᴪ-actinin and actin did exhibit affinity binding for clathrin when clathrin was forming a monolayer on the surfaces of Lytron polystyrene particles (10).

180 K

30K
28K

A   B   C

Fig. 2. Sodium dodecyl sulfate
polyacrylamide slab gel electro-
phoresis--The gradient of polyacrylamide gel electrophoresis
was similar to that described in
Fig. 1. A: purified brain
tropomyosin (10 µg); B: purified
clathrin (60 µg); C: purified
brain tropomyosin (5 µg). In this
figure we illustrate purity and the
comigration of brain tropomyosin
with the doublet observed in
clathrin.

Under these conditions, clathrin-coated Lytron particles adsorbed
α-actinin and actin from solution. These proteins were visualized
subsequently by desorbtion and separation on sodium dodecyl sulfate
polyacrylamide gels. Clathrin showed no binding of muscle tropomyosin
from solution. We attribute this negative result to the fact that
clathrin was saturated with the doublet of 30,000 molecular weight
only recently identified as polypeptides similar to those of brain
tropomyosin (11) (Fig. 2).

The 30,000 molecular weight doublet separated from clathrin
only with difficulty. It remained bound to clathrin after several
chromatography runs. It also remained when clathrin had been
precipitated with ammonium sulfate, with vinblastine (which separated
actin and tubulin and other minor components of clathrin, but not
the doublet) and even after precipitation or aggregation of clathrin
in solution by high concentrations of chlorpromazine (24). In one
instance, however, the task approached success. When clathrin

containing the doublet was chromatographed on DEAE cellulose, it could not be eluted from the column, even after passing a solution of 1 M KCl. When DEAE cellulose chromatography was performed in the presence of 6 M urea, clathrin remained on the fibers of DEAE cellulose. The doublet separated and eluted as a single discrete peak. As a result, this technique was used to obtain sufficient doublet for further analysis.

Since the migration properties of the doublet were strikingly similar to those reported for brain tropomyosin, we proceeded to compare, structurally and functionally, the properties of both brain tropomyosin and clathrin's doublet. Brain tropomyosin (Fig. 2) prepared in highly purified form (18,20) was found to co-migrate on slab gel electrophoresis with the doublet of clathrin. Further similarities were noted when both proteins, dialyzed in the presence of 50 mM $MgCl_2$, pH 8, produced a white opalescence which, when examined by negative staining in the electron microscope, was found to consist of typical tropomyosin paracrystals. They bore the familiar striation pattern produced by staggered side-to-side and/or head-to-tail attachment of tropomyosin molecules (18).

Further similarities between these molecules were established when antibodies elicited in rabbits exhibited similar antigenic properties. The brain tropomyosin antibodies reacted with the clathrin doublet and the antibodies to the clathrin's doublet reacted with the brain tropomyosin preparation. Both produced confluent precipitin lines.

During the course of these studies we compared the molecule of brain clathrin with that extracted from other tissues. Liver clathrin, isolated by a procedure similar to that used for brain, emerged accompanied by a doublet. When the peptide maps of brain tropomyosin and clathrin's doublets from liver and brain were compared they were found to be similar. Highly important to note however, was the fact that the peptide maps obtained with brain and liver clathrin were found to be similar but not identical.

Immunological Studies Utilizing Clathrin

Attempts were made to establish immunologically the localization of clathrin in tissues and cells using highly purified antibodies obtained by affinity chromatography. If clathrin's polypeptides were highly preserved molecules phylogenetically, as previous publications already had reported, and our structural analysis had verified, then the antibodies to one species should have cross reacted with antigens from other species. This proved to be the case. The antibodies prepared to the bovine antigen produced a characteristic staining pattern of fluorescent dots when tested on various cell cultures (Fig. 3).

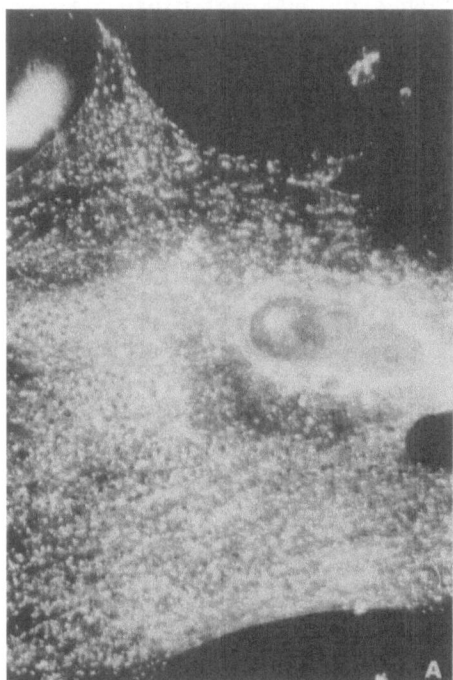 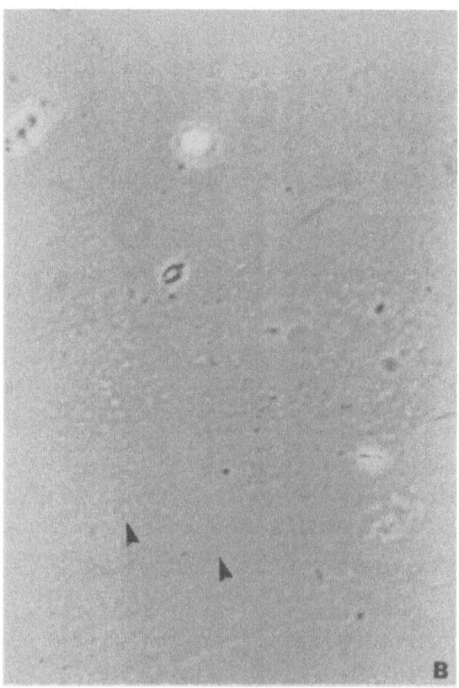

Fig. 3. A: Human fibroblast treated with clathrin antibodies and
stained by indirect immunofluorescent techniques with goat anti-
rabbit antibodies conjugated with rhodamine. Experimental details
are given in text. A large number of fluorescent dots are observed.
They are produced by lattices of clathrin located in the
cytoplasmic side of the membrane. An intense perinuclear staining
may denote the contours of the Golgi apparatus suggesting a high
concentration of clathrin in this organelle. 1,450 X. B: Same cell
as in 3A shown by phase contrast optics. Parallel stress fibers
(arrowheads) are observed. 1,450 X.

        It is apparent to us that the fluorescent dots correlated
closely with the actin stress fibers. Clathrin lattices originally
were attached to the cytoplasmic side of the membrane and the
detergent treatment for membrane permeabilization could have produced
an artifact thus distorting the ultrastructural localization. Also,
such treatment may have diminished by solubilization the visualization of clathrin. Therefore, the possibility cannot be overlooked
that membrane permeation is capable of releasing and preventing
detection of a large amount of clathrin from the cells' interiors.
The lattices of clathrin observed --often parallelly arrayed or
juxtaposed with actin stress fibers-- remained because of their

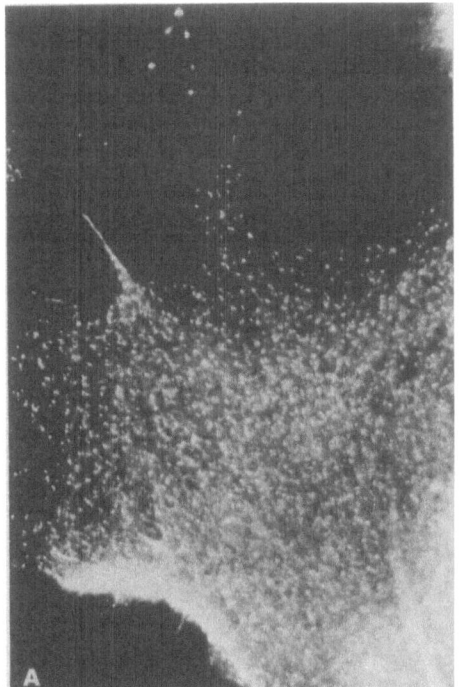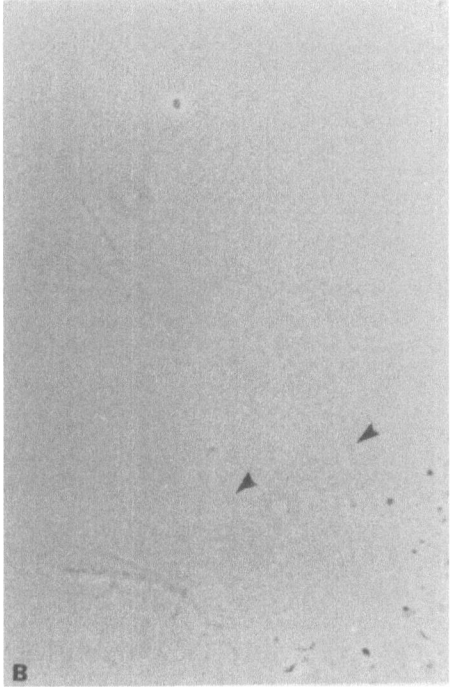

Fig. 4. A: Thinly spread portion of human fibroblast stained as
indicated in 3A. The characteristic dots extend to the end of
the cytoplasm. The linear arrangement of the dots follows the
general direction of the stress fibers. 1,425 X. B: Phase contrast
optics of the same cell shown in A. The stress fibers run along
the long axis of cell. 1,450 X.

binding affinity to actin cables or cross-linkage by fixatives when
in the vicinity of microfilamentous fibers (Fig. 4).

We found that the dot stain pattern given by acti-clathrin in
human fibroblasts was present as well in other cultured cells such
as vascular endothelial cells and glial cells. Neuroblastoma and
granule cerebellar cell cultures showed intense cytoplasmic
fluorescence that did not resolve in dots, probably due to their
sizes and little cytoplasmic contents (Figs. 5-8). More recently,
we detected positive clathrin fluorescent staining in human blood
platelets.

Biophysical Properties of Clathrin

A most startling development was clathrin's capacity to assemble
cages or baskets in vitro. These conformations were produced by a

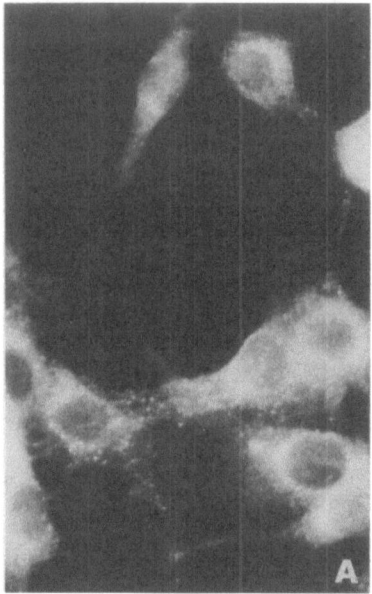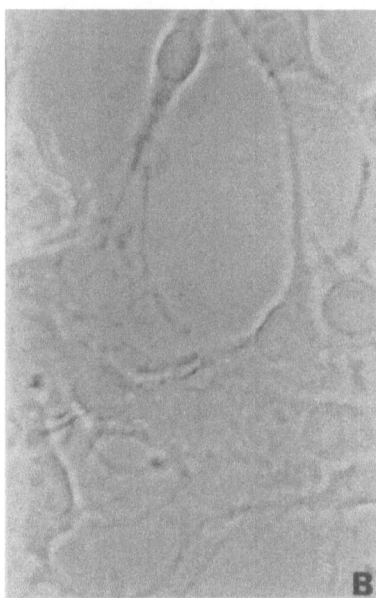

Fig. 5. A: Glial cells in culture stained with clathrin antibodies
as described in Fig. 3A. Fluorescent dots produced by clathrin
lattices appear in the cytoplasm of these cells. Because of the
small size of the cell, actin stress fibers are not easily
visualized or are differently arranged or distributed. 1,150 X. B:
Same area as in 5A shown by phase contrast optics. 1,150 X.

small change in pH. Clathrin is maintained in a monodispersed form
by keeping it in a 20-100 mM Tris-HCl buffer, pH 7.2-7.5, in the
presence of reducing agents like 0.1% 2-mercaptoethanol. When the
pH was lowered to 6.8 or 6.5, by the addition of either MES buffer
pH 6.0 or solid crystals of MES, a white opalescent turbidity
appeared in the solution. Examination of this material in the
electron microscope revealed the presence of numerous cages or
baskets (Fig. 9).

Clathrin assembly into baskets was insensitive to divalent
cations or nucleotides. However, they were inhibited by high salt
concentrations such as 0.5 M KCl. Upon close examination in the
electron microscope, baskets were formed by pentagons and hexagons
giving rise to icosahedral structures (Fig. 10).

Reconstructed models based on electron microscope pictures
of baskets or cages have suggested that at least 54 single clathrin
molecules or 108 molecules as dimers would be needed to assemble a
clathrin cage. The diameter of a given cage was determined to be in
the range of 600-700 Å and was almost identical to that found for

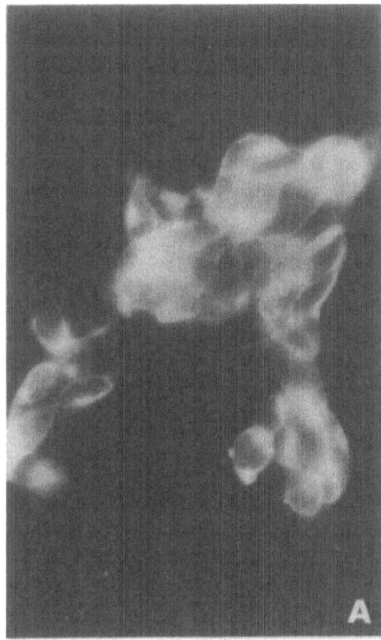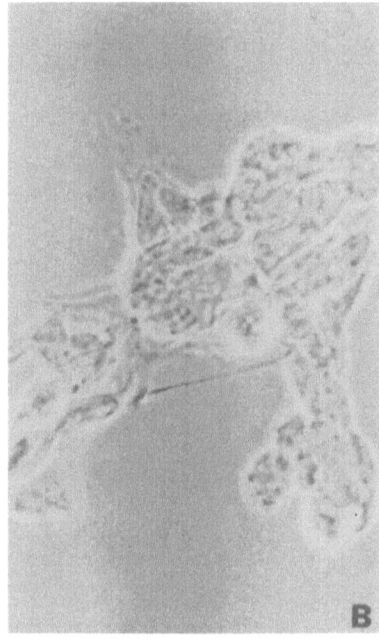

Fig. 6. A: Neuroblastoma cells in culture showing intense cytoplasmic staining with clathrin antibodies. The nuclear contour is
given by the dark area occupying most of the cytoplasm. 1,150 X.
B: Same area as in 6A shown by phase contrast optics. 1,150 X.

many coated vesicles. A model prepared using a simple Minit molecular
building system is included in this text. In all instances, the
formation of the baskets was rapid, within 1-3 minutes after
adjusting the pH (Fig. 11).

Disassembly of Baskets

        Shifting the pH above 7.0 results in the dissociation of clathrin
from the basket conformation. At this point it is not clear if the
disappearance of baskets results from the complete dissociation of
clathrin molecules from each other or from the rearrangement into
different subunits giving rise to polydispersed clathrin. Among the
polydispersed forms that we found, there were dimers, trimers,
tetramers, pentagons, hexagons and a variety of other intermediate
types of formations (Fig. 10). A most striking one was the
appearance of filaments or bundles of filaments. The filamentous
material also was observed by allowing clathrin to stand in the
refrigerator in solution in a pH higher than 7.0, in the absence
of denaturing agents or with high concentrations of salt. Linearly
arranged clathrin bundles seemed to be cross linked. Also they can

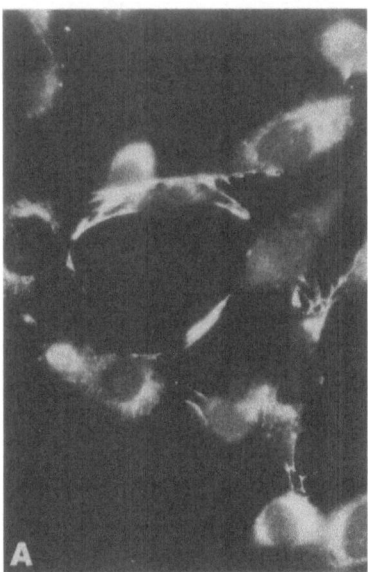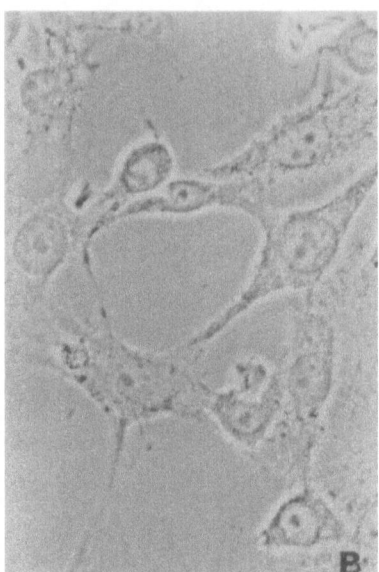

Fig. 7. A: Glial cells showing immunofluorescent staining with an
unrelated antibody, in this case to human fibronectin. The intense
fluorescence found in fibrillar patterns is given by the extracellular matrix of fibronection. The perinuclear fluorescence
corresponds to compartmentalized cytoplasmic fibronectin probably
stored in vesicular organelles. 1,150 X. B: Same area as in Fig. 5A
but shown by phase contrast optics. 1,150 X.

be induced by vinblastine at pH 7.

It is apparent then that the molecules of clathrin are capable
of forming such linear structures probably by head to tail attachment
combined by side to side interaction. If side apposition is inhibited
by altering the proton concentration around molecules, they could
conceivably give rise to geometrical figures by retaining the head-
to-tail or head-to-head binding. Highly cross-linked filamentous
bundles with different thickness could be the result of such a
disorganized type of interaction. Often large aggregates were formed
which by negative staining resembled images of the cytoplasmic ground
substance (Fig. 12).

Sedimentation of Clathrin Structures by Reagents

One way to sediment clathrin from solution by centrifugation
at high speed is by adjusting the buffer pH to 6.0-6.8. Under these
conditions, baskets can be pelleted at 100,000 x g for 30 minutes.
Other conditions to precipitate clathrin are ammonium sulfate 50%

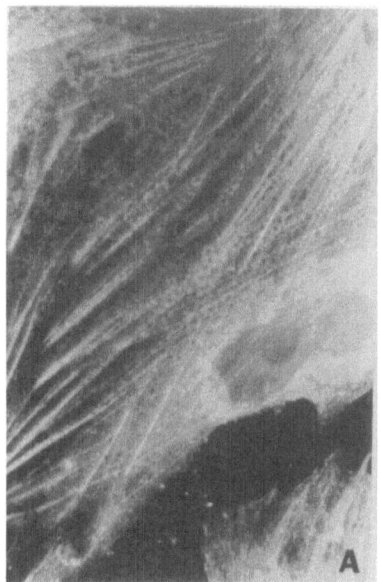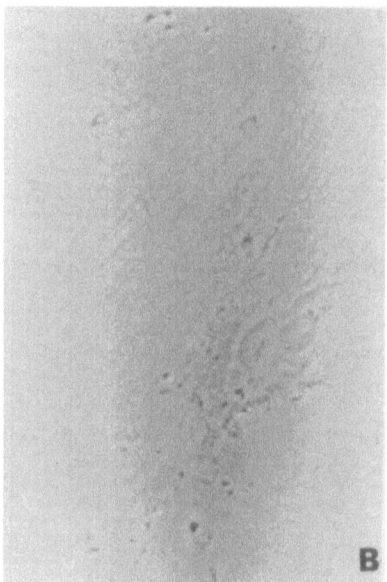

Fig. 8. A: Human fibroblast stained with anti-tropomyosin
illustrating the affinity of tropomyosin for the actin stress
fibers. The nucleus is negative. In this cell the actin stress
fibers crisscross each other. 1,450 X. B: Same as in 8A but shown
by phase contrast optics. 1,450 X.

saturation, the addition of up to 1 mM vinblastine at pH 7 or up to
1 mM of chlorpromazine or imipramine. Clathrin sedimented after
centrifugation can be dissolved in concentrated buffers (0.5 M Tris)
at pH greater than 7.0 or in diluted Tris-HCl buffers containing 2
M urea. After dialysis to remove urea, clathrin still retains its
biological activity, i.e., the formation of baskets at pH 6.5 to 6.8,
and binding of cytoskeletal proteins.

Effect of Proteolytic Enzymes on the Clathrin Complex

In our search for means to separate the doublet from clathrin,
we subjected clathrin preparations to brief proteolytic digestion
with papain, trypsin or chymotrypsin. The enzyme that showed best
results was chymotrypsin at pH 6.5. When clathrin, as baskets, was
put in the brief presence (15-60 minutes) of low concentrations of
the enzyme, it was found that the accompanying proteins were being
hydrolyzed at a much higher rate than clathrin. Clathrin polypeptides
were not cleaved during short periods of exposure. However, the
capacity of clathrin to form baskets apparently was lost. This
would imply that the separation or absence of accompanying proteins
resulted in the inability of clathrin to form baskets. However,
at this point, the possibility cannot be excluded that small

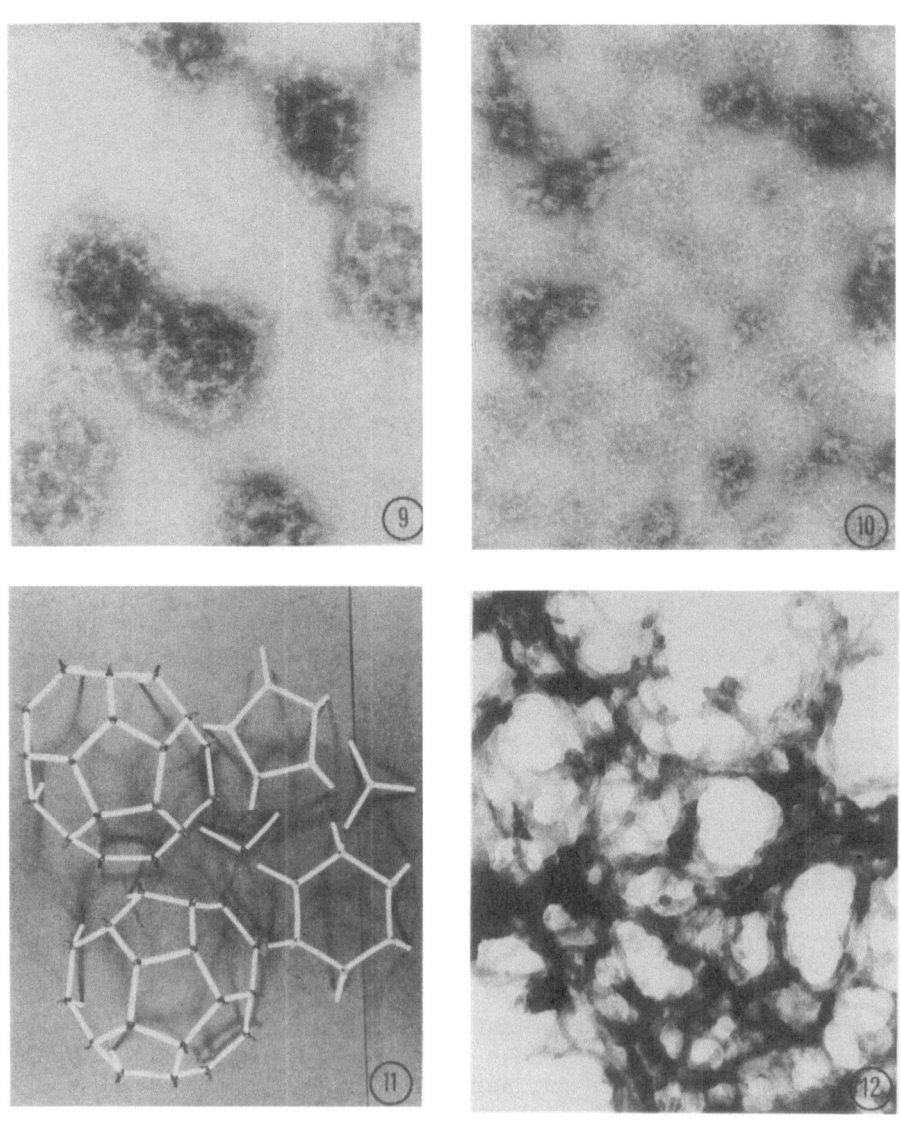

undetected polypeptide fragments were cleaved from clathrin rendering it incapable of forming baskets (Fig. 1).

Viscometric Properties of Clathrin

Clathrin in solution at pH 7.5 exhibits relatively high specific viscosity. The specific viscosity followed a normal pattern when plotted versus concentration. When the pH of a given clathrin preparation was adjusted to 6.5 the specific viscosity decreased considerably. Raising the pH above 7 the specific viscosity returned to its original value. Correlated with electron microscopic observations, the drop in viscosity was attributed to the formation of baskets. The increase in viscosity was attributed to the depolymerization or disassembly of baskets concomitant with the formation of lattices, filaments or bundles of them. Concurrent with these changes in viscosity the turbidity of clathrin in solution was also affected. At pH's above 7, turbidity was low; at pH's below 7, turbidity increased.

At present we believe that clathrin, assembled as baskets, offers reduced shear resistance to passage through a capillary tube resulting in lower specific viscosity values. The increased turbidity may be the result of light scattering by the cage- or basket-like structures. When these determinations are standardized, they may become useful in establishing the state of assembly of clathrin obviating the use of electron microscopic observations.

Fig. 9. Highly purified brain clathrin at pH 6.5 forms baskets or cage-like structures that are visualized by electron microscopy. Negative staining with 1% uranyl acetate. 350,000 X.

Fig. 10. Negative staining of clathrin assemblies at different states. Partial assemblies of clathrin molecules are found as hexagons, pentagons, and trimers. Some completely assembled baskets or cages also are found. 120,000 X.

Fig. 11. Schematic representation of the structures observed in Fig. 10.

Fig. 12. Aggregated clathrin obtained after several passages through the capillary tube of a viscometer. This type of amorphous configuration is strikingly similar in morphology to that of the microtrabecular lattice.

Fig. 13. Idealized representation of an intermediate stage in the endocytosis mechanism. Compounds on the outside of the cell membrane are bound by receptor anchored to the membrane producing an invaginated coated pit with clathrin assembled into its characteristic lattice structure below the membrane, binding to actin stress fibers in the cytoplasm.

CONCLUDING REMARKS

Our latest studies show that under certain conditions clathrin forms in vitro clusters below the membrane of synaptosomal preparations. At present, we have not found the conditions under which clathrin in vitro forms lattices around plain synaptic vesicles. Moreover, clathrin's affinity for membrane may be mediated by actin, tropomyosin, α-actinin, a combination of these, or by still unidentified components.

The physiological significance of clathrin in brain, particularly in nerve-endings, is under intense scrutiny, since clathrin may be a crucial element in the economy of cells and tissues (Fig. 13). Future studies will establish whether or not clathrin response to pH determined in vitro results in the assembly or disassembly of clathrin structures in vivo. Also, the significance of cofactors of clathrin activity, like tropomyosin, classically a regulatory protein of contractility, awaits elucidation. Most important for the regulation of clathrin in cell functions will be to determine the manner by which clathrin becomes activated in the presence of calcium to initiate assembly of its lattice. These goals are under intense study in our laboratories.

Acknowledgements.   This work was supported by N.I.H. grants #NS
12467, #HL 20718, #AA 03671 and #GM 26829.

REFERENCES

1. Puszkin, S., Berl, S., Puszkin, E. and Clarke, D.D., Actomyosin-
    like proteins isolated from mammalian brain, Science, 161 (1968)
    170-171.
2. Berl, S. and Puszkin, S., Mg^{2+}-Ca^{2+}-activated adenosine tri-
    phosphatase system isolated from mammalian brain, Biochemistry,
    9 (1970) 2085-2096.
3. Puszkin, S., Nicklas, W.J. and Berl, S. Actomyosin-like proteins
    in brain: subcellular distribution, J. Neurochem., 19 (1972)
    1319-1330.
4. Puszkin, S. and Berl, S., Actomyosin-like proteins from brain:
    separation and characterization of the actin-like component,
    Biochim. Biophys. Acta, 256 (1972) 695-707.
5. Berl, S., Puszkin, S. and Nicklas, W.J., Actomyosin-like proteins
    in brain, Science, 179 (1973) 441-448.
6. Goldman, R. and Knipe, D., Functions of cytoplasmic fibers in
    non-muscle cell motility, Cold Spring Harbor Symp. Quant. Biol.,
    37 (1973) 523-534.
7. Puszkin, S. and Kochwa, S., Regulation of neurotransmitter
    release by a complex of actin with relaxing protein isolated
    from rat brain synaptosomes, J. Biol. Chem., 249 (1974) 7711-7715.
8. Heuser, J.E. and Reese, T.S., Evidence for recycling of synaptic
    vesicle membrane during transmitter release at the frog neuro-
    muscular junction, J. Cell. Biol., 57 (1973) 315-344.
9. Pearse, B.M.F., Clathrin: a unique protein associated with
    intracellular transfer of membrane by coated vesicles, Proc.
    Natl. Acad. Sci USA, 73 (1976) 1255-1259.
10. Schook, W. Bloom, W.S., Ores, C., Kochwa, S. and Puszkin S.,
    Mechanochemical properties of brain clathrin: interaction with
    actin and ⍺-actinin and polymerization into basket-like structures
    or filaments, Proc. Natl. Acad. Sci. USA, 76 (1979) 116-120.
11. Puszkin, S., Maimon, J. and Schook, W., Clathrin association
    with low molecular-weight proteins acting as cofactors for the
    assembly/disassembly of baskets, J. Cell Biol., 83 (1979) 293a.
12. Heuser, J.E. and Reese, T.S., Redistribution of intramembraneous
    particles from synaptic vesicles: direct evidence for vesicle
    recycling, Anat. Rec., 181 (1975) 374-384.
13. Heuser, J.E., Reese, T.S., Dennis, M.J., Jan, Y. and Evans, L.,
    Synaptic vesicle exocytosis, J. Cell. Biol., 57 (1973) 499-524.
14. Roth, T.F. and Porter, K.R., Yolk protein uptake in the oocyte of
    the mosquito Aedes aegypti L., J. Cell Biol., 20 (1964) 313-332.
15. Pearse, B.M.F., Coated vesicles from pig brain, purification
    and biochemical characterization, J. Mol Biol., 97 (1975), 93-98.
16. Kanaseki, T. and Kadota, L., The vesicle in a basket: a mor-
    phological study of the coated vesicle isolated from the nerve-
    endings of the guinea pig brain with special reference to the

mechanism of membrane movements, J. Cell Biol., 42 (1969) 202–220.

17. Woodward, M. and Roth, T.F., Influence of buffer ions and divalent cations on coated vesicle disassembly and reassembly, J. Supramolec. Struct., 11 (1979) 237–250.

18. Fine, R.E. and Blitz, A.L., Tropomyosin in brain and growing neurons, Nature, New Biol., 245 (1973) 182–185.

19. Bloom, W.S., Schook, W., Ores, C., Feageson, E. and Puszkin, S., Brain clathrin: viscometric and turbidimetric properties of its ultrastructural assemblies, Biochem. Biophys. Acta, in press.

20. Cohen, I. and Cohen, C., A tropomyosin-like protein from human platelets, J. Mol. Biol., 68 (1972) 383–387.

21. Schook, W., Ores, C. and Puszkin, S., Isolation and properties of brain α-actinin, Biochem. J., 175 (1978) 63–72.

22. Puszkin, S., Puszkin, E., Maimon, J., Rouault, C., Schook, W., Ores, C., Kochwa, S. and Rosenfield, R., α-actinin and tropomyosin interactions with a hybrid complex of erythrocyteactin and muscle myosin, J. Biol. Chem., 252 (1977) 5529–5537.

23. Gerday, C., Robyns, E., Gosselin-Rey, C., High resolution techniques of peptide mapping: separation of bovine carotid actin peptides on cellulose thin layers and of the corresponding dansyl-peptides on polyamide thin layers, J. Chromatog., 38 (1968) 408–411.

24. Lisanti, M., Schook, W., Beckenstein, K, and Puszkin, S., Brain clathrin ultrastructural assemblies: effects of neuroactive compounds and cytoskeletal disruptors, J. Neurochem., (submitted).

EFFECT OF NERVE GROWTH FACTOR ON MICROTUBULES: EVIDENCE FOR A
PROTECTIVE ACTION AGAINST DEPOLYMERIZATION INDUCED BY CALCIUM
IONS OR COLD TREATMENT

P. Calissano, D. Mercanti, G. Monaco and L. Castellani

Laboratorio di Biologia Cellulare, C.N.R., Via Romagnosi
18A, and Instituto di Anatomia Comparata
Universita di Roma, Via Borelli 50, Roma

INTRODUCTION

Although the powerful and specific promoting activity of nerve
growth factor (NGF) was first reported more than 25 years ago (1),
still little is known about its molecular mechanisms and physiological
action. This is frequently the case in the field of hormones and
growth factors, whose phenomenological aspects (spectrum of actions,
number and type of target cells, etc.) become known even decades
before the understanding of their mechanism of action. This gap is
mainly due to lack of a precise knowledge of how several eukariotic
cells respond to a given external stimulus. The discovery of second
messengers is a remarkable exception to this rule, but it would be
unwise, in our opinion, to consider them as transducers or
amplifiers for all hormones and growth factors.

The case of NGF is, in this respect, most instructive. As first
suggested by Levi Montalcini (2) this molecule is a protein differing
in some respects from a typical hormone. In contrast to a typical
hormone, NGF is synthesized in minute quantities in many different
type of cells, and in large amounts in exocrine (snake venom glands,
mouse salivary glands, guinea pig prostate) rather than endocrine
glands. Moreover, several attempts to detect changes in second
messengers or in intracellular substances which could mimick their
action, e.g. Ca ions, following NGF binding to target cells, have
given negative or controversial results. For instance, the negligible
changes in cAMP and Ca^{++} fluxes previously described (3) in a clonal,
NGF-receptive cell line (PC12) derived from a rat pheochromocytoma
tumor (4) have not been confirmed (5).

These considerations, as well as other results obtained in our

261

laboratory which will be briefly presented in this report led us to postulate an alternative, second messenger independent, action of this protein on target cells via an interplay with the contractile elements which constitute the cytoskeleton of a nerve cell, i.e. microtubules (MTs) and microfilaments (MFs) (6,7,8). The crucial role played by these filamentous structures in several intracellular events culminating in axonal growth and elongation is now well established. For instance, colchicine or cytochalasin B completely prevent growth of neurites induced by NGF, while 95% inhibition of protein synthesis by cycloheximide does not result in any substantial block of this event (9,10). These findings, as well as other experimental evidences, indicate that while assembly and organization of MTs and MFs are crucial for neurite growth, the beginning of this process does not require synthesis of new proteins. NGF target cells, such as sympathetic or sensory cells explanted in vitro, are equipped with the proteins necessary for early steps of axonal growth, but axonal growth takes place only when sparked by a triggering message. The aim of this presentation is to submit a molecular model which prospect the mechanism by which NGF could exert such triggering action.

Previous studies have shown that addition of NGF to a 105,000 g supernatant of a mouse brain homogenate, followed by centrifugation and SDS-acrylamide gel electrophoresis of the precipitate reveals the existence of two major proteins plus several minor components (6). These two proteins have been subsequently identified as tubulin and actin, precursors of MTs and MFs, respectively. Subsequent studies (unpublished data) have shown that if NGF is incubated with an analogous 105,000 x g supernatant of 11 day chick embryo brain, the precipitate obtained is much more selectively constituted by the two proteins mentioned above. Other co-precipitating proteins are barely detectable, suggesting that in adult mouse brain there may be some other proteins which either are non specifically precipitated by NGF, or that they are bound to tubulin and/or actin, and sediment with them. Whatever the correct explanation, these and many other experiments which will be mentioned in this paper indicate that NGF interacts in a quite specific fashion with the major constituents of MTs and MFs.

These findings raised the question of the possible biological meaning of such interactions. As a first approach to this problem we analyzed in detail the properties of NGF-tubulin and NGF-actin interactions (stoichiometry, possible substances interfering with binding, dissociation constants of the complexes, etc.). We shall briefly summarize NGF-tubulin interactions and then present some new properties of the complexes that they form.

SUMMARY OF NGF-TUBULIN INTERACTIONS

NGF binds to tubulin, purified from calf brain with 3 cycles of assembly-disassembly as described by Shelanski et al. (11), with an apparent dissociation constant of $1.5 \times 10^{-8}M$ (12). At saturation there are 2 moles of NGF bound (m.w. 26,500) per tubulin dimer (m.w. 110,000). Colchicine, vinblastine, GTP or divalent cations such as Mg^{++} or Ca^{++}, do not significantly alter NGF binding, nor does this protein, at saturating concentrations, affect the binding of any of the substances mentioned above (6). These experiments indicate that NGF interacts with tubulin at a different site(s).

NGF markedly accelerates the rate of MTs assembly and increases, in a concentration dependent fashion, the total amount of MTs formed (7).

Finally, when the vinca alkaloid vinblastine is added to pre-formed MTs, within 10-30 minutes they undergo a progressive disassembly which, at the electron microscope, is characterized by the appearance of spirals and rings. Such disassembly has been interpreted (13) as the consequence of a shift in the equilibrium between MTs and dimers in favour of the latter. An analogous situation would occur with other MTs disrupting drugs such as colchicine and related substances (13). According to this hypothesis, tubulin-vinblastine complexes would add to the growing tubule and block any further addition of free tubulin dimers. Since MTs are in a dynamic state with their precursor dimers, the alkaloid would, even at largely substoichiometric concentration, inhibit MT formation and shift preformed MTs toward their unpolymerized state. It has been calculated that colchicine at an approximate ratio of 1 per 500 tubulin dimers inhibit 50% MTs formation.

We have found that if NGF is added to preformed MTs, vinblastine is no more effective in inducing disassembly (14). This antagonistic action is detectable not only in vitro, but also in vivo in newborn mice injected with the alkaloid plus NGF (15). Thus, while the administration of vinblastine alone causes, within a few days, a profound reduction in the size and number of cells forming the superior cervical ganglia, simultaneous administration of NGF completely prevents the destructive action exerted by the vinca alkaloid. While such in vivo antagonistic action between NGF and vinblastine is still difficult to interpret at the molecular level, the analogous in vitro competition can be visualized as follows. We may exclude a direct competition between NGF and vinblastine for the same site, or even for different sites via an induced conformational change, since the two substances do not interfere with their reciprocal binding to tubulin (14,15). On the other hand, it was previously shown and confirmed in our laboratory that vinblastine cannot bind to intact tubules because its binding site is unaccessible, probably hidden in the tubule. Thus, the vinca alkaloid

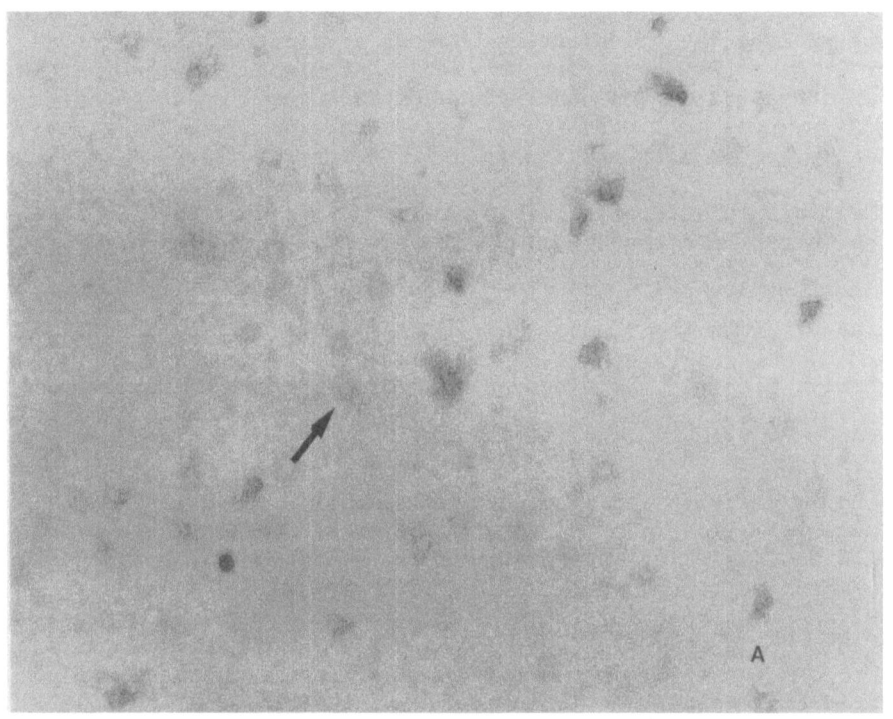

Fig. 1. Effect of cold treatment on MTs in the presence or absence
of NGF. Tubulin, 1.0 mg/ml in reassembly buffer (MES 0.1M pH 6.6,
Mg^{++} 0.5 mM, EGTA 0.5 mM, GTP 1.0 mM) after 3 cycles of assembly-
disassembly, was incubated in the absence (A) or in the presence (B,C)
of 2.5 S NGF (75 µg/ml) for 30 min. Samples of these solutions were
then transferred to ice, allowed to stand for 10 minutes and
negatively stained with 1.0% uranyl acetate. Notice the absence of
intact microtubules in A and the presence of several rings (arrow).
When NGF is present (B,C) several microtubules/field are detectable
together with many rings in the background. At higher magnification
(C) several MTs preincubated with NGF have the appearance of duplex
microtubules. A, X 110,000; B, X 27,000; C, X 110,000. NGF, in
the form of 2.5 S, was purified by the method of Bocchini and
Angeletti (21). Tubulin was purified as described by Shelanski et al.
(11).

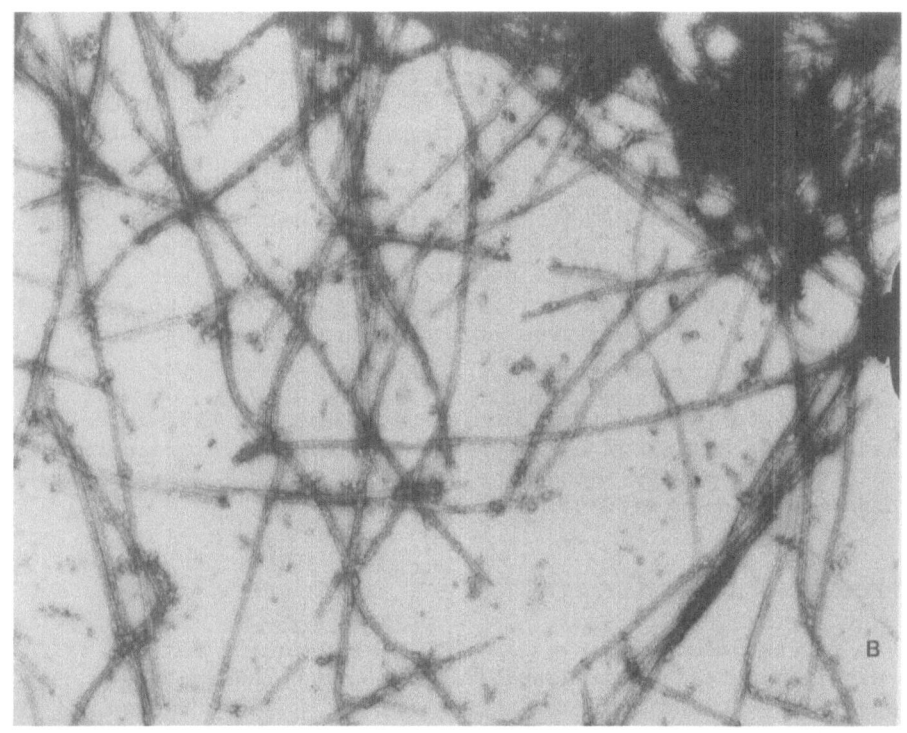

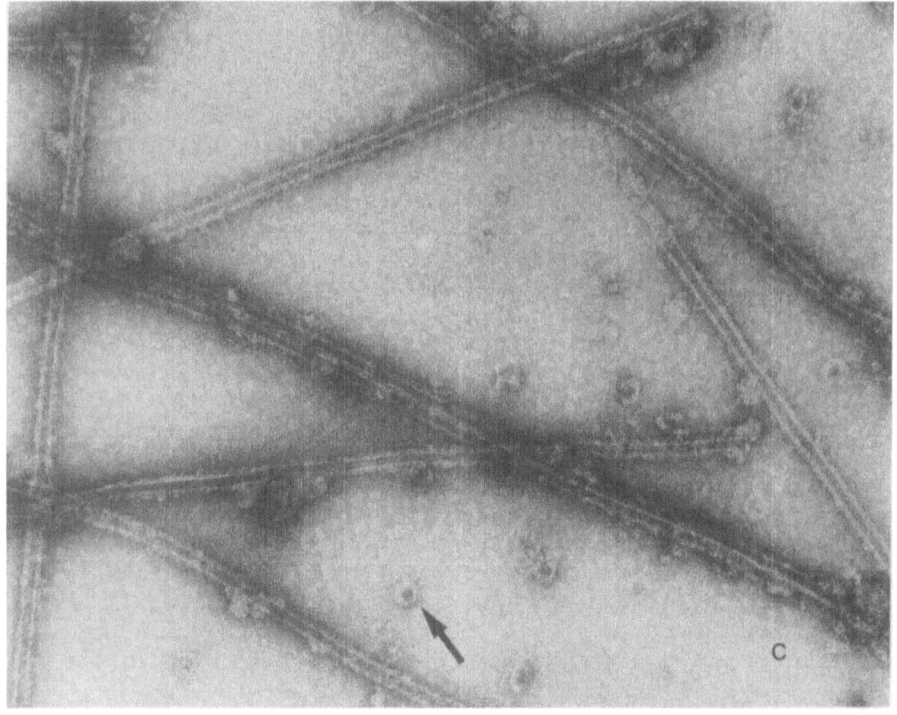

may only bind to tubulin dimers and, as mentioned before, block their addition to the growing tubule, but is uneffective whenever tubulin is in its assembled state. Since NGF favours, by a still unknown mechanism, MTs assembly, its antagonistic action on vinblastine effect may be visualized as an indirect consequence of a stabilization of the intact tubules. According to this hypothesis, NGF would bind at the terminal end of MTs where it would prevent or markedly slow down the detachment of dimers occurring in NGF-free solutions. This in turn would prevent vinblastine interactions with dimers. The net result would be the presence of intact MTs even in the presence of exceedingly high concentrations of a depolymerizing agent such as the vinca alkaloid.

In order to further assess this hypothesis, we have investigated the stabilizing action of NGF by testing its possible protective effect against two other disassembling treatments: low temperatures and high Ca^{++} concentrations.

NGF EFFECTS ON MTs DEPOLYMERIZATION BY CA^{++} OR COLD

As it can be seen in Fig. 1, when a solution of MTs is cooled at 4°C, within minutes MTs disassemble and no more tubules are detectable after 10 min of cold treatment (Fig. 1A). However, if MTs are preincubated with NGF, depolymerization takes place at a much slower rate and several MTs/field are detectable by electron microscopy (Fig. 1,B-C).

An analogous situation occurs when Ca^{++} is added to preformed MTs. Following addition of this cation at 1.0-2.0 mM concentrations MTs very rapidly disassemble (Fig. 2) and turbidity of the solution drops by more than 50% within 1-2 min. Electronmicroscopic analysis shows that Ca^{++}-depolymerized MTs form a layer of rings with no visible intact tubules (Fig. 3A). In the presence of NGF, on the contrary, two major events are detectable. Before addition of Ca^{++}, the rate of MTs formation and their total amount is markedly higher than in controls, as indicated by the rapid increase in turbidity and its much higher extent when light scattering reaches a plateu (Fig. 2). This effect, already reported by our laboratory (7), has been interpreted as stimulation of MTs assembly, probably through the formation of nucleation sites during the phase of preincubation of tubulin dimers with NGF. As it can be seen in Figs. 2 and 3 moreover, MTs preformed in the presence of NGF exhibit a remarkable resistance to depolymerization induced by Ca^{++}. Thus, when evaluated turbidimetrically, the fall induced by this cation does not even lead to the values of light scattering of control MTs before treatments with Ca^{++}. In other words, if light scattering is taken as a quantitative measure of the amount of MTs in solution, even in the presence of 2.0 mM Ca^{++} MT-NGF complexes are as abundant as MTs

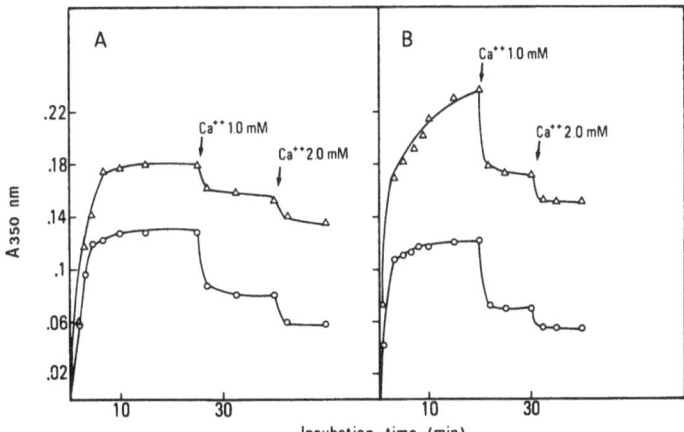

Fig. 2. Effect of Ca^{++} on MTs preincubated with or without NGF.
Tubulin 1.4 mg/ml in reassembly buffer was incubated in the absence
(o) or in the presence (Δ) of NGF at a concentration of 70 μg/ml
(A) or 140 μg/ml (B) for 10 minutes at 0°C. Solution plus or minus
NGF were transferred to cuvettes thermostated at 37°C and the
increase in light scattering monitored at 350 nm. When turbidity
reached a plateau, Ca^{++} was added (arrows) to reach the indicated
concentrations. Before the addition of the second aliquot of Ca^{++},
samples from each cuvette were withdrawn and negatively stained
(see Fig. 3).

before addition of the divalent cation. This conclusion is supported
by a direct electron microscopic analysis of MTs-NGF complexes after
addition of Ca^{++} (Fig. 3B-C). It can be seen that, in contrast to
controls, MTs are abundant and have a normal appearance. These
findings clearly indicate that MTs in the presence of NGF not only
assemble at a much faster rate and in larger amount, but that they
exhibit a marked resistance to depolymerization induced by Ca^{++}.

An intriguing aspect of these findings is whether the increased
rate of assembly and resistance to depolymerization induced by
vinblastine (14), Ca^{++}, or cold treatments, reflects two distinct
sites of action of NGF, e.g. at the level of nucleation of dimers
and of detachment from the terminal ends, or whether NGF operates
at one single site in the polymer but its action is somewhat
"translated" over the entire microtubule.

DISCUSSION AND CONCLUSIONS

The interaction of NGF with tubulin reported in this and previous

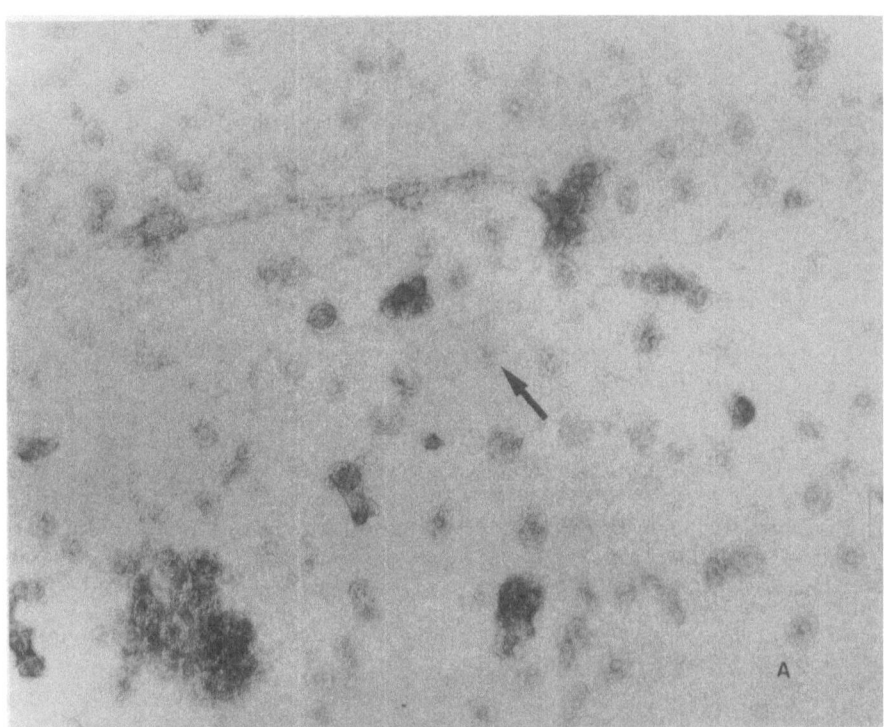

Fig. 3. Microtubules after treatment with 1.0 mM Ca^{++}. Notice the
almost complete absence of MTs and the presence of a dense layer of
rings in control samples treated with Ca^{++} (A). B and C show MTs
preincubated with NGF before addition of 1.0 mM Ca^{++}. Notice the
presence of many short microtubules (B) which, at higher magnification, have an almost normal appearance (C). A, X 108,000; B, X
11,000; C, 67,500.

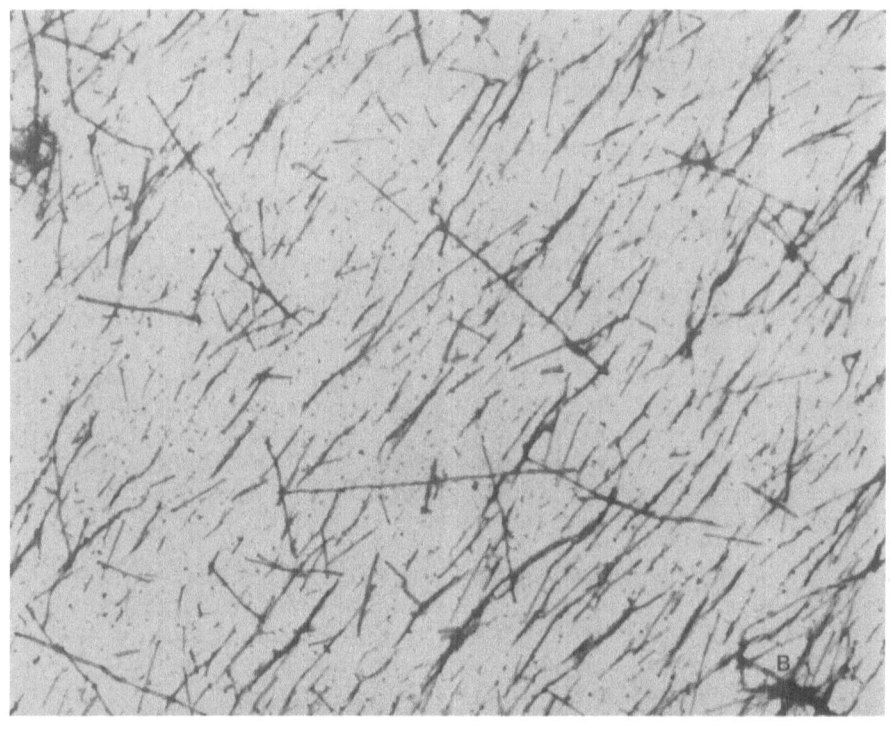

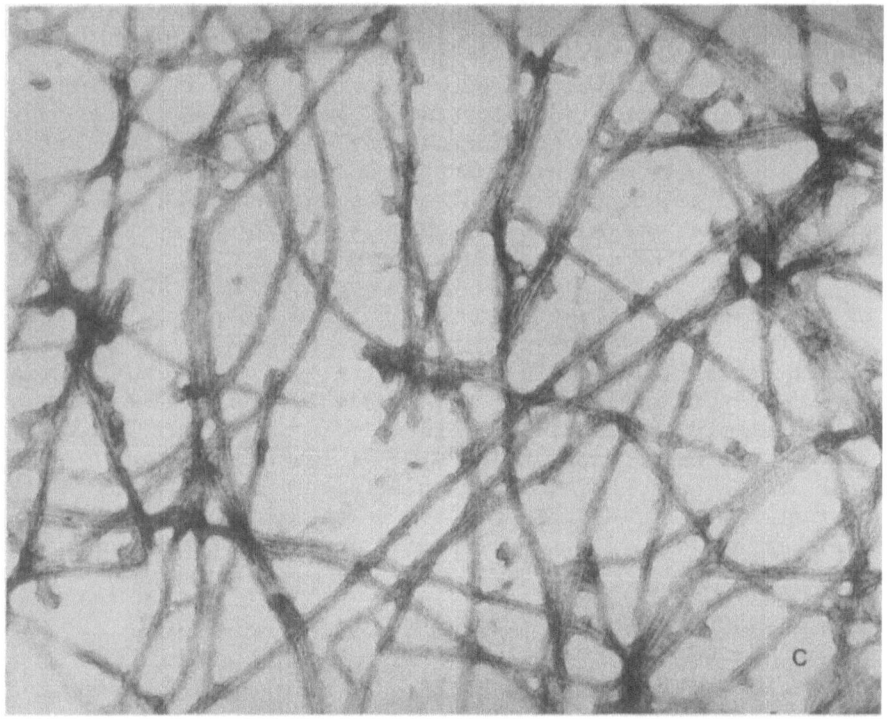

papers (6,7,12,14,15), as well as that of NGF-actin (8), raises
several questions about their relevance to the process of neurite
growth induced by this protein: why only very few types of cells
respond to NGF since tubulin and actin are present in all eukariotic
cells?; how can nanogram amounts of NGF affect the properties of
micrograms of tubulin and actin, and how could these interactions
result in neurite growth? The presence of NGF receptors in target
cells is now a well established fact (16), and a mechanism to
internalize NGF-receptor complexes has been recently described (17,
18,19). It is even possible that recognition and internalization of
NGF are essential but not sufficient conditions for NGF action,
which occurs only when an intracellular mechanism brings about the
actual contact between NGF and a pool of actin, of tubulin, or both,
in strategic points of the cell. Thus, the specificity of NGF action
is visualized more as a capacity of its target cells to recognize,
internalize, and allow the proper contact with the contractile
elements than of the growth factor itself. Other proteins may play
an analogous role to NGF in other types of cells.

If we assume that indeed NGF comes in contact with these cytoskeletal elements, we may attempt to answer the second crucial
question: how the interaction of nanogram concentrations of NGF with
these proteins becomes a signal which, within hours, results in
neurite growth and elongation?

As mentioned in the Introduction, this event is directly
dependent not only upon the assembly of tubulin or of actin but from
both processes, since it is sufficient to block MTs or MFs formation
with colchicine or cytochalasin B, respectively, to arrest the whole
process of neurite growth (9,10). Studies performed on the mechanism
of MTs or MFs poisoning by these drugs have shown that their action
is exerted at largely substoichiometric concentrations (13,20). Both
drugs would bind to the growing polymer and prevent further
elongation, so that one single molecule is sufficient to arrest the
whole assembly of a MT or of a MF.

We may conceive the existence of proteins having the property
of stimulating, rather than inhibiting, assembly by forming nucleation
sites, or by stabilizing preformed polymers at their terminal ends.
NGF may act in such substoichiometric concentrations and either
stimulate assembly or stabilize preformed MTs or MFs, or it may exert
both effects depending upon the particular functional need or point
of attack in the cell. If a process similar to that which takes place
in a test tube between NGF and tubulin or actin occurs also in a
nerve cell it could well set in motion a series of events culminating
in neurite growth.

The precise molecular mechanism by which assembly and
organization of MTs and MFs are related to axonal growth and
elongation has still to be elucidated, not only for NGF target cells

but also for almost all eukariotic cells employing these contractile elements as force generating structures. Only when such mechanisms will be understood shall we be able to depict in more detail the possible role played by NGF in this process. At the beginning of our investigation, NGF-tubulin or NGF-actin complexes were only odd associations between apparently unrelated molecules. Now these complexes provide a basis for understanding the mechanism of action of NGF. This model is now ready to be tested in the living nerve cell to assess the still missing step in this hypothesized chain of events: the demonstration that NGF comes in contact with tubulin and actin.

Acknowledgement. We wish to thank Dr. R. Levi Montalcini for helpful and stimulating suggestions during the preparation of the manuscript.

REFERENCES

1. Levi Montalcini, R., Effects of mouse tumor transplantation on the nervous system, Ann. N.Y. Acad. Sci., 55 (1952) 330-343.
2. Levi Montalcini, R. The nerve growth factor: its mode of action on sensory and sympathetic nerve cells, Harvey Lectures, 60 (1966) 217-250.
3. Shubert, D. La Corbiere, M., Whitlock, C. and Stallup, W., Alterations in the surface properties of cell responsive to nerve growth factor, Nature, 273 (1978) 718-723.
4. Greene, L.A. and Tichler, A.S., Establishment of a noradrenergic clonal line of rat adrenal pheochromocytoma cells which respond to nerve growth factor, Proc. Natl. Acad. Sci. USA, 73 (1976) 2424-2427.
5. Landreth, G., Cohen, P. and Shooter, E.M., Ca^{++} transmembrane fluxes and nerve growth factor action on a clonal cell line of rat pheochromocytoma, Nature, 283 (1980) 202-204.
6. Calissano, P. and Cozzari, C., Interaction of nerve growth factor with mouse neurotubule protein(s), Proc. Natl. Acad. Sci. USA, 71 (1974) 2131-2135.
7. Levi, A., Cimino, M., Mercanti, D., Chen, J.S. and Calissano. P., Interaction of nerve growth factor with tubulin. Studies on binding and induced polymerization, Biochim. Biophys. Acta 399 (1975) 50-60.
8. Calissano, P., Monaco, G., Castellani, L., Mercanti, D. and Levi, A., Nerve growth factor potentiates actomyosin adenosinetri-phosphatase, Proc. Natl. Acad. Sci. USA, 75 (1978) 2210-2214.
9. Yamada, K.M., Spooner, B.S. and Wessell, K.N., Ultrastructure and function of growth cones and axons of cultured nerve cells, J. Cell Biol., 49 (1971) 614-635.
10. Bray, D., Thomas, C. and Shaw, G., Growth cone formation in cultures of sensory neurons. Proc. Natl. Acad. Sci. USA, 75 (1978) 5226-5229.
11. Shelanski, M.L., Gaskin, F. and Cantor, C.R., Microtubule assembly

in the absence of added nucleotides, Proc. Natl. Acad. Sci. USA, 70 (1973) 765-768.

12. Calissano, P., Monaco, G., Levi, A., Menesini-Chen, G.M., Chen, J. S. and Levi Montalcini, R. In Perry, S.V., Margreth, A. and Adelstein, R. S. (Eds.), Contractile Systems in Non Muscle Tissues, North-Holland, 1976, pp. 201-214.

13. Margolis, R.L. and Wilson, L., Addition of colchicine-tubulin complex to microtubule ends: the mechanism of substoichiometric colchicine poisoning, Proc. Natl. Acad. Sci. USA, 74 (1977) 3466-3470.

14. Monaco, G., Calissano, P. and Mercanti, D., Effect of NGF on in vitro preformed microtubules. Evidence for a protective action against vinblastine, Brain Res., 129 (1977) 265-274.

15. Menesini-Chen, G.M., Chen, J.S., Calissano, P. and Levi Montalcini, R., Nerve growth factor prevents destructive effects on sympathetic ganglia in newborn mice, Proc. Natl. Acad. Sci. USA, 74 (1977) 5559-5563.

16. Herrup, K. and Shooter, E.M., Properties of nerve growth factor receptor of avian dorsal root ganglia, Proc. Natl. Acad. Sci. USA, 70 (1973) 3884-3888.

17. Stöeckel, K., Paravicini, V. and Thoenen, H., Specificity of the retrograde axonal transport of nerve growth factor, Brain Res. 76 (1974) 413-421.

18. Calissano, P. and Shelanski, M.L., Interaction of nerve growth factor with pheochromocytoma cells. Evidence for tight binding and sequestration. Neuroscience (1980) in press.

19. Yankner, B.A. and Shooter, E.M., Nerve growth factor in the nucleus: interaction with receptors on the nuclear membrane, Proc. Natl. Acad. Sci. USA, 76 (1979) 1269-1273.

20. Chang Lin, D., Dugan-Tobin, K., Grumet, M. and Lin, S., Cytochalasin inhibits nuclei-induced actin polymerization by blocking filament elongation, J. Cell Biol., 84 (1980) 455-460.

21. Bocchini, V. and Angeletti, P.U., The nerve growth factor: purification as a 30,000 molecular weight protein, Proc. Natl. Acad. Sci. USA, 64 (1969) 787-794.

SELECTIVE RE-INNERVATION OF TWITCH AND SLOW MUSCLE FIBERS OF THE FROG

Enrique Stefani

Department of Physiology and Biophysics, Centro de
Investigación y Estudios Avanzados, Instituto
Politécnico Nacional, México 14, D.F.

INTRODUCTION

Re-innervation in adult mammalian skeletal muscle fibers is non specific. A motor nerve supplying several different muscles regenerates forming inappropriate connections as readily as appropriate ones (35,2,24). Furthermore, in hyperinnervated skeletal muscle there is no indication that the original nerve can displace or repress foreign innervation (11). Moreover, the establishment of foreign synapses can lead to suppression of original inputs (4). In lower vertebrate, however, there is considerable evidence for specificity of muscle innervation, and some evidence for the ability of native nerve fibers to displace foreign ones (20,29,7). In this article the re-innervation of skeletal muscle fibers of the frog will be analyzed.

Frog skeletal muscles possess two distinct group of skeletal muscle fibers: twitch fibers innervated by large motor axons, and slow fibers innervated by small motor axons (34,15,16). These two types of muscle fibers are found intermingled in different skeletal muscles (i.e. iliofibularis and pyriformis), although slow fibers are generally packed in bundles close to the nerve entrance into the muscle. Slow fibers are unable to generate action potentials (5) and respond with a sustained contracture to prolonged depolarization (17). Twitch fibers when stimulated propagate action potentials and give a transient contracture when they are depolarized. Both the electrical and mechanical properties of slow fibers depend upon the innervation. Slow fibers after denervation develop a sodium action potential mechanism (25), which appears on the membrane in spots or patches (28), and they lose their ability to maintain tension after cross-innervation with a fast conducting nerve (sartorius nerve)

273

(23). A similar finding was obtained in the early stages of re-
innervation after crushing or cutting the sciatic nerve. It was
speculated that after sectioning a large mixed nerve, containing
large and small motor axons, the muscle fibers would have a greater
chance of being re-innervated by large motor axons because they are
more numerous than small motor axons. In addition, since large axons
degenerate faster, it was conceivable that they may also regenerate
more rapidly (10). The main purpose of the experiments reported here
was to identify the types of motor axons innervating fast and slow
motor fibers after cutting or crushing the sciatic nerve, or the
nerve near its entrance into the muscle. The experiments were done
in collaboration with H. Schmidt and A. Elizalde (31,32,29,30,9).

EXPERIMENTAL PROCEDURES

The experimental procedure has been described in detail by
Stefani & Schmidt (31) and Schmidt & Stefani (29,30). Experiments were
performed on the iliofibularis and pyriformis muscles of frogs (Rana
temporaria and Rana pipiens).

Operations

Surgical procedures were performed on the right side of the
animals, either under ether or tricaine (1:1000) anesthesia.

1) Sciatic nerve: The right sciatic nerve was crushed inside
the pelvis with a forceps whose tip was about 1.5 mm wide;
subsequently, the nerve was ground over the edge of another forceps
until only a thin bridge of connective tissue was left. The crushed
region was about 4 mm central to the point where the pyriformis nerve
branched off the sciatic nerve, and 6-8 mm from the centre of the
pyriformis muscle. In a small group of frogs the sciatic nerve was
cut at the same level.

2) Pyriformis nerve: The pyriformis nerve was crushed near its
entry into the muscle. The nerve was cut until a small bridge of
connective tissue was left which avoided the retraction of both nerve
stumps. The skin was closed with one or two stitches. The frogs were
kept at room temperature and were fed once a week with a mixture of
beef liver and cod liver oil.

Intracellular Recordings

The muscles were removed together with spinal nerves VIII, IX
and X and mounted in a way to reduce mechanical artifacts. In brief,
the muscle was rolled and simultaneously stretched along a polyethylene rod which could be rotated around its longitudinal axis.
This enabled one to always place the end-plate region below the
micropipettes. Detectable twitching of muscle fibers ceased when

the muscles had been stretched to 140-150% of their in situ length.

Conventional intracellular techniques for potential recording
and current application were used. Micropipettes were filled with
3 M KCl (for potential recording) or 2 M K-citrate (for passing
current). These micropipettes had high resistances and a shank of
about 25 mm; they were, therefore, very flexible, and this helped
them to remain inside the muscle fibers even if twitching had not
been suppressed completely by the method of mounting. For measuring
the membrane time constant and effective resistance between inside
and outside of the fibers the current passing microelectrode was
placed intracellularly at a distance of 50-200 μm from the recording
electrode. Rectangular current pulses were applied which hyper-
polarized the membrane of muscle fibers by 20-40 mV; the duration of
the pulses was 100 msec for twitch, and 1 sec for slow fibers. The
voltage deflection (V_o) at the end of the current pulse (I_o) was
used to calculate the effective resistance of the membrane R_{eff} =
V_o/I_o . The resting potentials of the fibers could be modified by
passing d.c. current; this was particularly necessary for slow muscle
fibers because of the voltage drop generally produced by the insertion of the current micropipette.

Normal slow muscle fibers generally had a membrane time constant
well above 200 msec; twitch fiber membrane time constants were below
30 msec (33). This difference persisted throughout the experimental
period up to 446 days after denervation, and identification of slow
fibers was, therefore, not difficult.

Nerve Stimulation

The nerves were carefully sealed into a perspex chamber which
contained two pairs of platinum electrodes. Anode and cathode of each
pair of electrodes were 1.5 mm apart. The proximal pair was placed at
5-10 mm from the central end of the spinal nerves. The distal pair
was positioned 10-15 mm further down on the sciatic nerve; this
position corresponded to a distance of 4-10 mm proximal to the
crushed or cut region. Square pulses of 0-1 msec duration were
applied; the threshold of the most excitable motor axons was usually
0.5-1.5 V (depending mainly on the diameter of the nerves). When the
stimulus strength was raised above threshold there was generally a
small reduction of the latency between stimulus and beginning of end-
plate potentials, e.p.p.s (22). Latencies were, however, constant
at stimulus strengths above 1.2 times threshold; stimuli of 1.5 to
twice threshold value were, therefore, used throughout. According
to the site of stimulation e.p.p.s. appeared with different latencies;
from their time interval and the distance between the stimulating
cathodes the conduction velocities of single motor axons could be
calculated.

The temperature of the nerve-muscle bath in most experiments

Table 1. Membrane properties of twitch and slow muscle fibers and properties of their motor axons. Results obtained in twelve pyriformis muscles of Rana pipiens. The membrane time constant was measured as the time of voltage deflection to reach 0.7 (slow fibers) or 0.85 (twitch fibers) of its final value. In slow fibers values are underestimated since the current pulse (ca 1 sec) was too short to reach plateau. In both fibers the membrane potential was set to about −90 mV.

Fiber Type	Effective Resistance (MΩ)	Membrane time constant (msec)	Conduction velocity of motor axons (m/sec)	Stimulation threshold (V)	Latency of e.p.p. (msec)
Twitch	1.7±0.12 (11)	10±0.7 (11)	16.5±0.5 (35)	0.41±.01 (35)	5.5±0.2 (35)
Slow	9.4±0.7 (16)	305±22 (16)	4.0±0.12 (36)	2.3±0.1 (36)	16.2±0.8 (36)

was held between 7 and 9°C by a Peltier element attached to the bottom of the chamber. Ringer solution had the following composition (mM): NaCl, 110.4; KCl, 2.5; NaHCO$_3$, 2.4; CaCl$_2$, 7.2. The high Ca concentration helped to identify the slow fibers because their effective resistance increases more than that of the twitch fibers in calcium rich medium (33). Values are given as mean ± standard error (SE) of the mean. The number of observations is in brackets.

PROPERTIES OF NORMAL MUSCLES

Electrical Properties

Slow muscle fibers can be easily distinguished from twitch fibers for their much higher R_{eff} and membrane time constant (Table 1) (33). Fig. 1 shows voltage responses (upper traces) to square pulses of current (lower traces) injected intracellularly in a slow muscle (A and B) and in a twitch muscle fiber (E and F). The long time constant and large R_{eff} of the slow fiber can be appreciated. Furthermore, normal slow fibers, opposite to twitch fibers (E), are unable to generate action potentials. In Fig. 1A, depolarizing pulses of current of increasing amplitude failed to produce an action potential. The small hump observed in the larger pulse is due to the delayed rectifier (1). After insertion of the second microelectrode the resting potential of slow fibers was usually between -50 to -70 mV. In these experiments the fibers were polarized by a small inward current to about -90 mV. With this procedure, the condition of the fibers improved considerably and the action potential mechanism could be tested.

Innervation

Slow muscle fibers generally showed composite synaptic potentials upon nerve stimulation (Fig. 1D). In most cases (63 out of 80) synaptic potentials of slow fibers had three components with different stimulating thresholds indicating that the fibers are polyneurally innervated. Fig. 1C shows the difference in latency when the nerves are stimulated at two points 10 mm apart. The ingoing notch observed between the stimulus artifact and the synaptic potential is due to extracellularly recorded action potentials from neighboring twitch muscle fibers. This fiber was innervated by two motor axons with a conduction velocity of 4 m/sec and a threshold of 2.6 and 3.6 V.

Fig. 1G shows the synaptic potential and action potential recorded in a twitch fiber upon nerve stimulation. End-plate potentials recorded in iliofibularis and pyriformis muscle fibers located near the nerve entry have relatively low quantal content and occasionally failed to trigger action potentials. These two populations of muscle fibers, with a distinct innervation and electrical properties, were clearly defined. Fig. 2 shows the amplitude histogram of conduction velocities and latencies, and in Fig. 3 the threshold is plotted against the conduction velocity. No overlap between the two populations can be seen. Fibers with larger conduction velocity have smaller thresholds (large motor axons), and fibers with smaller conduction velocity have larger threshold (small motor axons). Table 1 summarizes the electrical properties and the pattern of innervation in slow and twitch muscle fibers.

Fig. 1. (next page) Electrical properties and innervation of slow (left row) and twitch (right row) fibers of normal pyriformis muscle. In A, B, E and F square pulses of current (lower traces) were delivered intracellularly and voltage responses (upper traces) were simultaneously recorded. Note the large R_{eff} (6.6 MΩ) and the long time constant (τ_m) (200 msec) of the slow fiber (B). The twitch fiber had a R_{eff} of 1.2 MΩ and a τ_m of 10 msec. Depolarizing pulses elicited action potential (a.p.) in the twitch fiber (E). No a.p. could be evoked in the slow fiber (A). In C and G spinal nerves were stimulated at two points separated 10 mm. The calculated conduction velocities were 4 m/sec and 18 m/sec respectively. D shows composite synaptic potentials in the slow fiber with stimulus intensities of 2.6 V and 3.8 V.

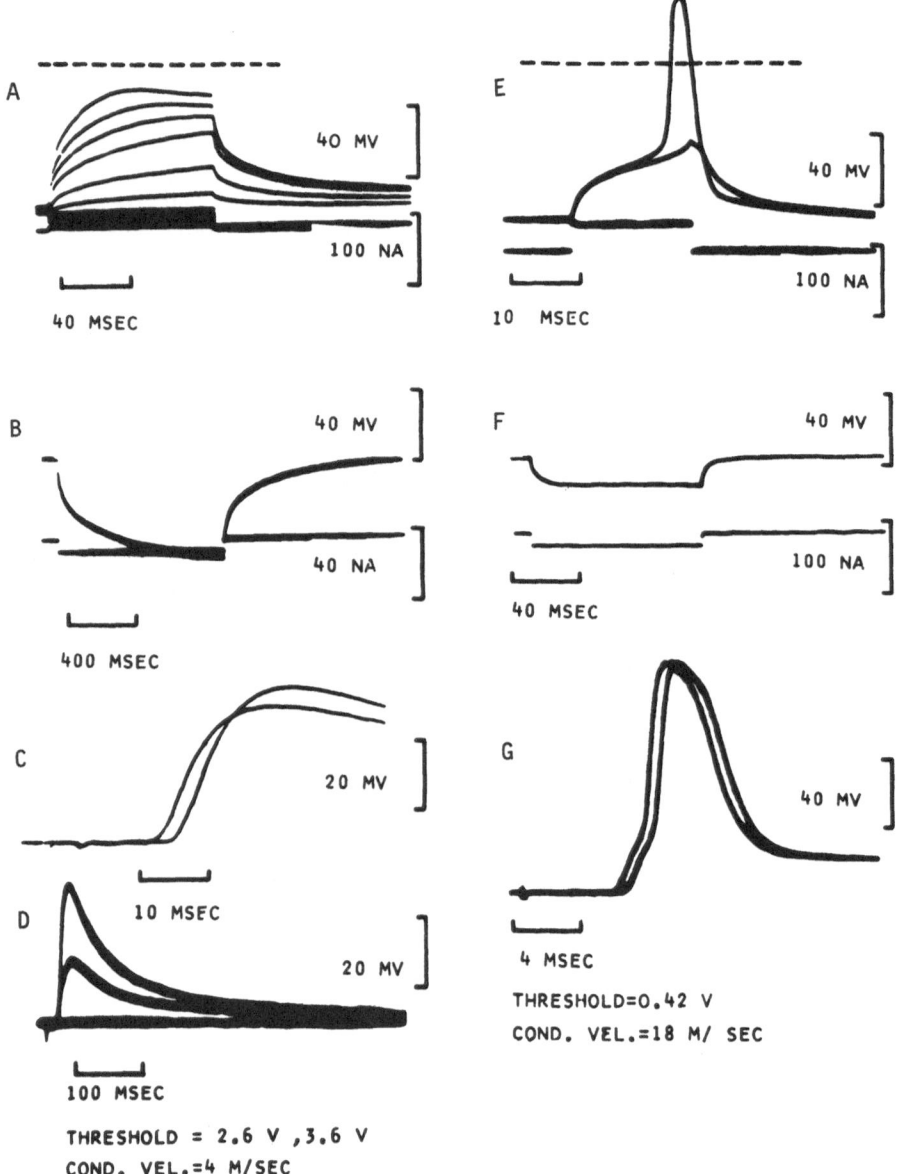

Fig. 1. The legend for this figure is on the preceding page.

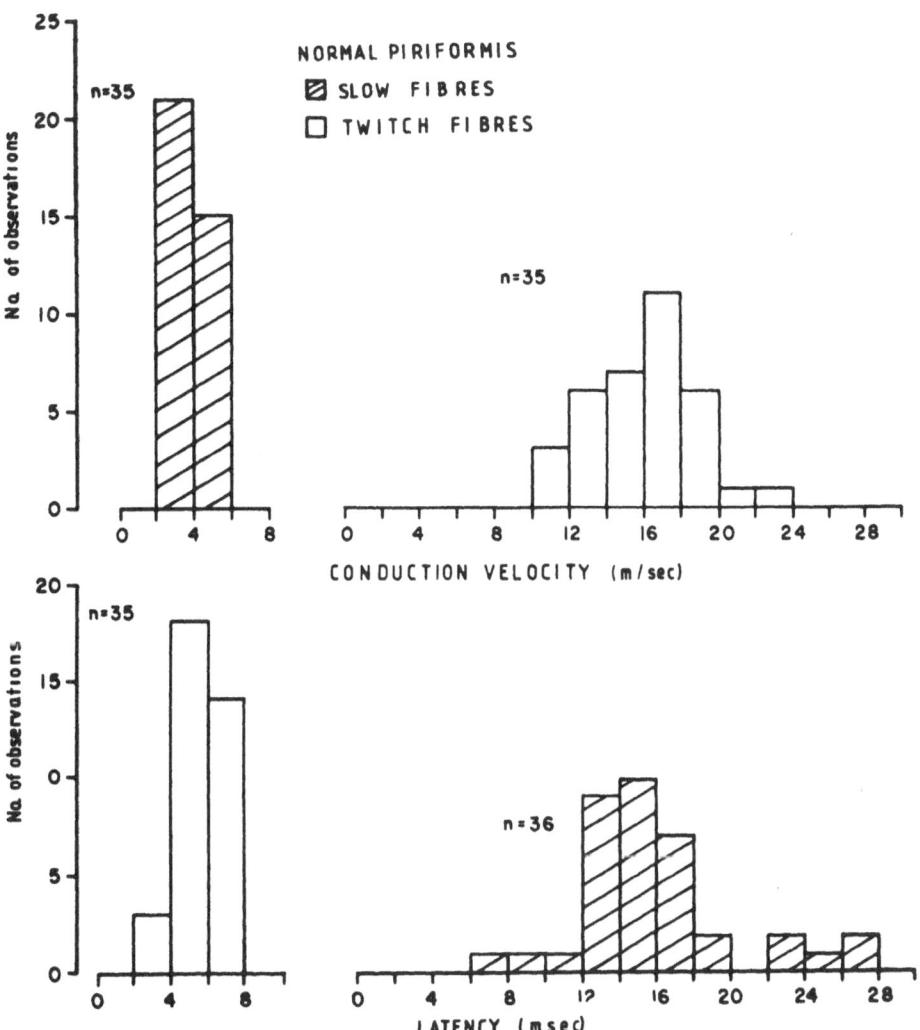

Fig. 2. Conduction velocity (upper) and latency of synaptic potentials (lower) histograms of motor axons that innervate slow (dashed columns) and twitch (empty columns) muscle fibers. Note the clear cut separation of conduction velocities of the two populations. The latency histograms show a small degree of overlap.

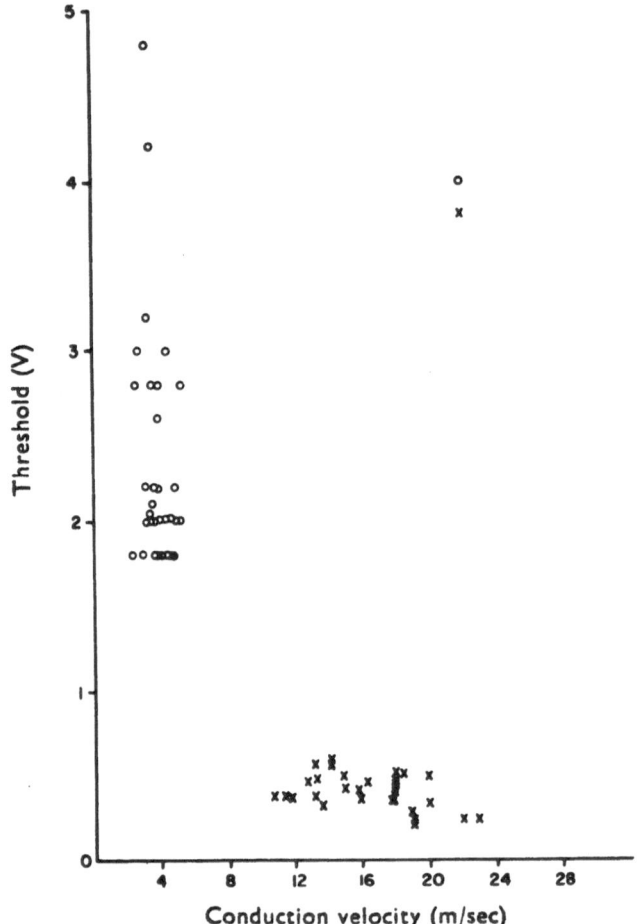

Fig. 3. Relation between conduction velocity (abscissa) and stimulation threshold (ordinate) for motor axons innervating slow (circles) or twitch (crosses) muscle fibers of normal pyriformis muscles. Twitch and slow fibers were identified according to their passive electrical properties.

MUSCLES UNDERGOING RE-INNERVATION

Sciatic Crushed

In several muscles analyzed during the first 24 days following denervation, all tested muscle fibers were not functionally re-innervated. In a muscle investigated on the 28th day, seven twitch fibers were re-innervated by low threshold axons indicating that

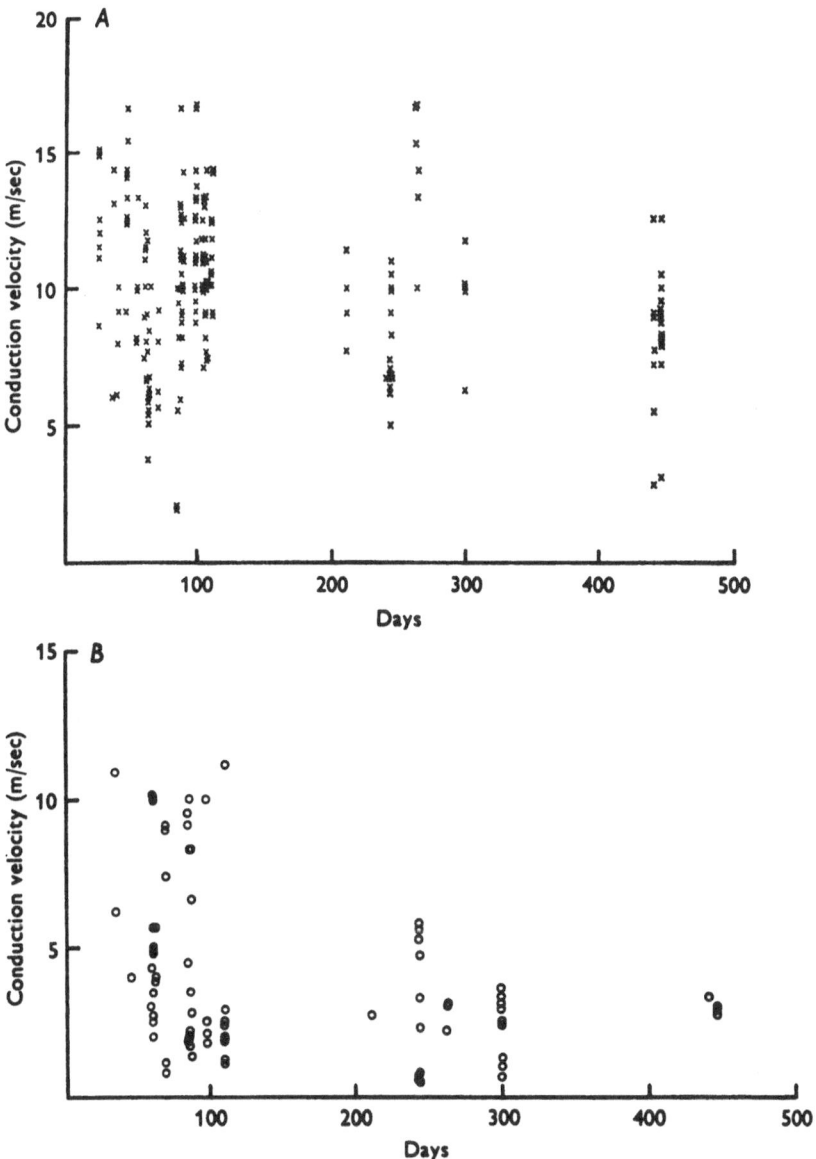

Fig. 4. Conduction velocities of motor axons re-innervating twitch
(A) and slow muscle fibers (B) of pyriformis muscle after crushing
the sciatic nerve. Note that slow fibers were initially re-innervated
by axons faster than 5 m/sec. In late stages they were only re-
innervated by slower axons. Printed with permission of the
Physiological Society from (29).

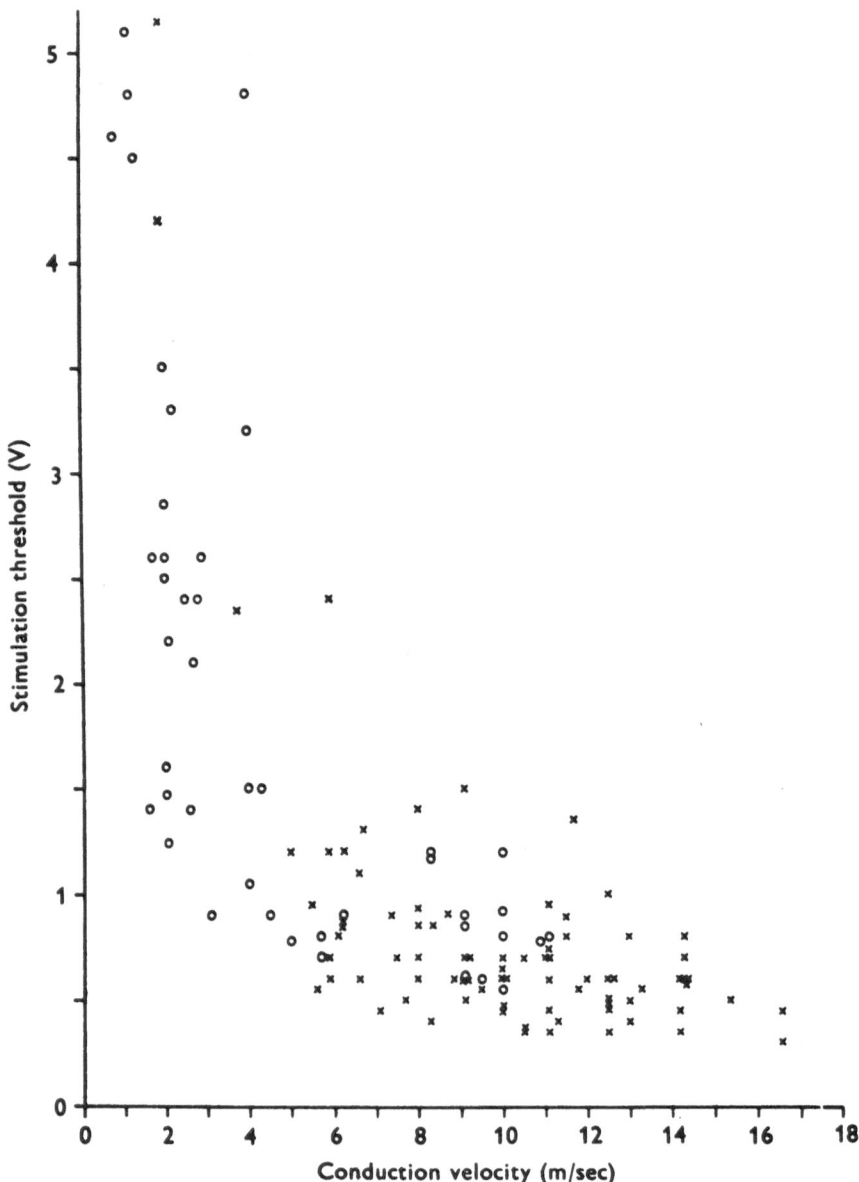

Fig. 5. Relation between conduction velocity (abscissa) and
stimulation threshold (ordinate) of motor axons innervating slow
(circles) and twitch (crosses) muscle fibers of pyriformis muscle
35 to 110 days after crushing the sciatic nerve. Note that a
proportion of slow fibers (circles) was innervated by low threshold
axons conducting faster than 5 m/sec. Printed with permission of
the Physiological Society from (29).

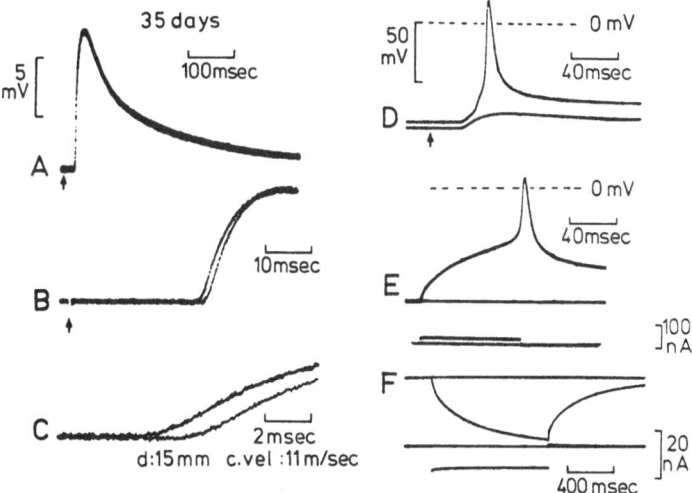

Fig. 6. Slow muscle fiber re-innervated by a <u>large</u> <u>motor</u> <u>axon</u> (11 m/
sec), examined 35 days after crushing the sciatic nerve. <u>A</u>, <u>B</u> and
<u>C</u> end plate potential; the nerve was stimulated (arrow) at two
points 15 mm apart. Note different sweep speeds. In <u>D</u>, the end
plate elicited an action potential; the lower trace shows a
subthreshold end plate potential after hyperpolarization. <u>E</u>: action
potential to direct stimulation. <u>C</u>: hyperpolarizing electrotonic
potential, $\underline{R_{eff}}$ >5.9 MΩ and τ_m >310 msec.

large <u>motor</u> <u>axons</u> had reached the muscle. The results obtained were
deliberately divided into three groups according to the time elapsed
after the operation: 1) period I, 0-33 days, 2) period II, 35-110
days and 3) period III, more than 4 months after the operation. In
period I, most fibers were not innervated. Only 16% of twitch fibers
tested were functionally re-innervated. All slow fibers found were
not innervated. In period II about 50% of all fibers tested were
re-innervated and finally in period III most fibers were re-
innervated. Fig. 4 shows conduction velocities of motor axons which
innervated twitch (A) and slow (B) muscle fibers. Twitch muscle
fibers were from the beginning re-innervated by motor axons with a
conduction velocity faster than 5 m/sec. A substantial number of
slow fibers were initially re-innervated by axons faster than 5 m/sec.
As re-innervation progresses, slow muscle fibers lose the fast
conducting axons and are re-innervated by axons conducting below 5
m/sec. The clear cut separation between <u>large</u> and <u>small</u> <u>axons</u> was
lost. The majority of axons had intermediate conduction velocities
not normally observed (5-10 m/sec). In re-innervating axons the
relation between conduction velocities and threshold was maintained.
Axons with low threshold have higher conduction velocity and vice-
versa (Fig. 5). This suggests that peripheral damage of motor axons

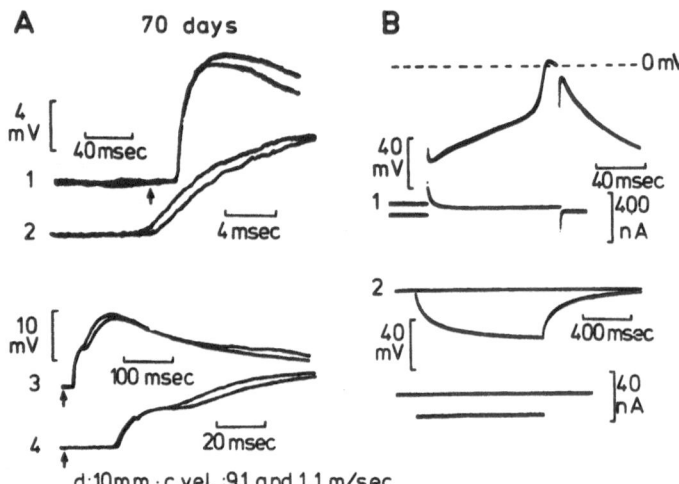

Fig. 7. Slow muscle fiber simultaneously re-innervated by <u>large</u> and <u>small</u> <u>motor</u> <u>axons</u>, examined 70 days after crushing the sciatic nerve. <u>A1</u> and <u>2</u>, low threshold end plate potential by stimulation of nerve. <u>A3</u> and <u>4</u>, end plate with two components stimulating the nerve with 1.6 V; the early component corresponds to that shown in record <u>A1</u> and <u>2</u>. Note different sweep speeds. Latency differences of the two end plate potential components correspond to conduction velocities of 9.1 and 1.1 m/sec. In <u>B1</u> the fiber fires an action potential to intracellular depolarization. In <u>B2</u> an hyperpolarizing electrotonic potential is shown; $\underline{R_{eff}}$ >2.4 MΩ and τ_m >240 msec.

is followed by retrograde changes which reduce conduction velocities and raise the threshold. Similar observations have been reported after nerve section in the recovery of conduction velocities in mammalian nerves (13,3). We therefore can tentatively conclude that conduction velocities above 5 m/sec were <u>large motor axons</u>, and below, <u>small motor axons.</u>

Slow fibers re-innervated by large motor axons. Fig. 6 shows a slow fiber re-innervated by a <u>large motor axon</u> 35 days after the operation. The fiber was characterized as slow according to its passive electrical properties (Fig. 6F). The motor axon innervating this fiber had: a 0.8 V threshold and a conduction velocity of 11 m/sec (Fig. 6<u>A</u>, <u>B</u>, <u>C</u>). Slow fibers in this period gained the ability of producing action potentials to nerve (D) and direct stimulation (E). In a subsequent section the time course of action potential development and disappearance following re-innervation will be discussed in detail. As many as twenty seven slow fibers were found to be incorrectly re-innervated by <u>large motor axons</u>. Thus, we can conclude that the re-innervation of slow fiber is <u>non selective.</u>

Slow fibers re-innervated by large and small motor axons. Eleven
slow fibers in seven muscles, 61-100 days post operatively were found
to be simultaneously re-innervated by both axons. An example is given
in Fig. 7. The fiber retained the ability to produce action potential
and was characterized as slow by the electrical properties (B, 1 and
2). In A, 1 and 2, the low threshold e.p.p. (0.6 V) is shown. The
conduction velocity was 9.1 m/sec. In A, 3 and 4 (slower time base)
when the stimulus strength was increased above 1.3 V a second
component appeared. The measured conduction velocity was 1.1 m/sec.
As expected, the component with lower conduction velocity had a
longer delay.

Slow fibers re-innervated by slow motor axons. During period II
about 50% of slow fibers were re-innervated by small motor axons,
but in period III, practically all slow fibers were re-innervated by
small motor axons. Fig. 8 shows a slow fiber 61 days after the operation. The fiber had lost the action potential mechanism (B, 1); only
a very small anode break response remained after a large hyperpolarizing pulse (B, 2). The fiber was multiple innervated (A, 1) by
small motor axons (A, 2) with high threshold (2.3 V) and slow
conduction velocity (1.6 m/sec). Thus, after a period of non
selective re-innervation, the original innervation pattern is being
restored and most slow fibers were again multineuronally innervated.

Action potentials in slow fibers. Approximately 2 weeks after
cutting or crushing the sciatic nerve, the slow fibers in the
operated side were able to produce propagated action potentials (25).
Fig. 9A shows action potentials in a slow fiber from the pyriformis
muscle 13 days after crushing the sciatic nerve. The action potential
has a clear threshold and an overshoot. For comparison in B, an
action potential from a twitch fiber from the pyriformis muscle 30
days after denervation, is shown. The figure also illustrates the
voltage changes evoked by square pulses of current used to identify
the fiber type. Action potentials in slow fibers are sodium dependent
and are blocked by tetrodotoxin. They are propagated with a conduction
velocity of 0.2-0.6 m/sec. They have an amplitude of 70-100 mV and
a rate of rise of about 25 V/sec. Schalow and Schmidt (28) found in
an early period after denervation action potentials of variable size
and shape. These action potentials were often composed of 2-4
components and the size of individual components depended on the
position of the recording microelectrode. Their results suggested
that following denervation Na channels were built into distinct
areas or patches and that this process depends on the amount of
denervation of individual fibers.

In the early phase of re-innervation slow fibers (32,29) were
initially re-innervated by large motor axons. In this period, slow
fibers produced full-sized action potentials (Fig. 6D, E). Action
potential of similar shape were recorded from slow fibers already
re-innervated by small motor axons. As re-innervation progresses,

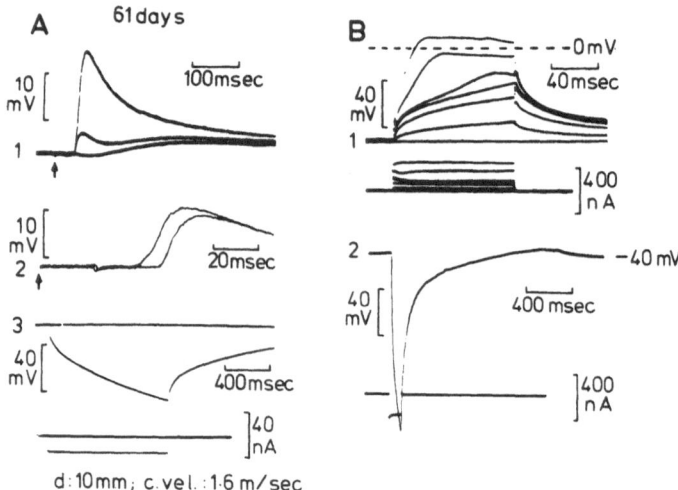

d: 10 mm; c. vel. : 1·6 m/ sec

Fig. 8.   Slow muscle fiber re-innervated by two <u>small</u> <u>motor</u> <u>axons</u>
61 days after crushing the sciatic nerve. <u>A1</u>: Composite synaptic
potential. <u>A2</u>: Synaptic potential elicited upon stimulating nerve
at two different points. The calculated conduction velocity was
1.6 m/sec. <u>A3</u> and <u>B2</u> hyperpolarizing electrotonic potential. The
fiber lost the ability to fire action potentials (<u>B1</u>).

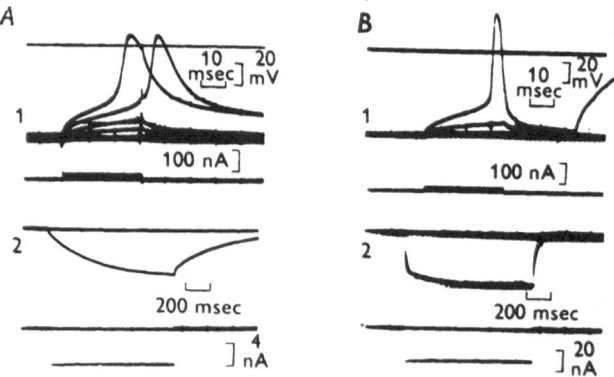

Fig. 9. Action potentials in slow fibers from pyriformis muscle 13
days after crushing the sciatic nerve (<u>A1</u>). <u>B1</u>: action potential in a
fast fiber in a muscle denervated for 30 days. <u>A2</u> and <u>C2</u>: hyperpolarizing electrotonic potential to identify fiber types. Printed
with permission of the Physiological Society, modified from Fig. 1
of (25).

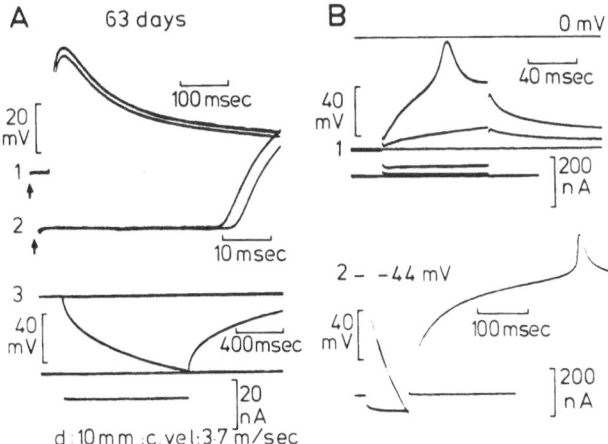

Fig. 10. Action potential in a slow muscle fiber re-innervated by
a small motor axon 63 days after crushing the sciatic nerve. A1 and
2: End plate potential after nerve stimulation at two different
points. The conduction velocity of the motor axon was 3.7 m/sec.
A3: hyperpolarizing electrotonic potential to identify fiber type.
Action potential to direct stimulation (B1) or after a strong
hyperpolarizing pulse (B2).

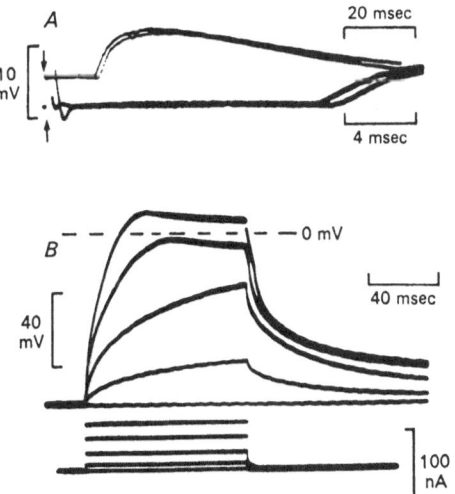

Fig. 11. Slow fiber re-innervated
by a large motor axon 85 days after
crushing the sciatic nerve. A: end
plate potential after nerve
stimulation at two different points;
records at two sweep speeds. The
conduction velocity of the motor
axon was 9 m/sec. B: depolarizing
current pulses (lower traces) do
not initiate regenerative response.
Printed with permission of the
Physiological Society from (30).

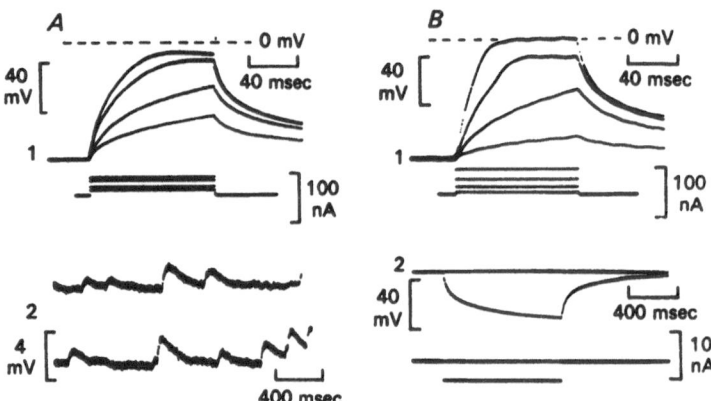

Fig. 12. Two slow fibers examined 88 (A) and 87 (B) days after
crushing the sciatic nerve. The fibers did not show synaptic
potentials upon nerve stimulation. Both fibers did not have action
potentials upon intracellular injection of depolarizing current
pulses (A1 and B2). A2: miniature end plate potentials. B2 hyperpolarizing current pulses to identify the fiber type. Printed with
permission of the Physiological Society from (30).

slow fibers became again re-innervated by small motor axons. The
action potential decreased in size (Fig. 10B) and thereafter
completely disappeared (Fig. 8B). Functional re-innervation by
small motor axons in the late stages of re-innervation is not a
necessary condition for the suppression of the action potential.
Fig. 11A shows a slow fiber of a pyriformis muscle 85 days after
the operation, which is re-innervated by a large motor axon and
which is unable to produce action potentials. Denervated slow fibers
(functionally denervated) in this late period of re-innervation are
also unable to produce action potentials (Fig. 12A, B).

The experiments show clearly that the action potential mechanism
is triggered by the denervation and is suppressed in close temporal
relationship with the regeneration of motor axons. The restoration
of original membrane properties is therefore probably related to the
appearance of small motor axons in the muscle.

The most likely explanation of these results seems to be the
assumption that there is a "trophic" influence of regenerating slow
motor axons, which is independent of their synaptic activity, and
which could be due to the secretion of a trophic substance. This
substance could be released from the regenerating motor nerves
already before they make contact with the surface of the muscle
fibers. Alternatively, small motor axons may form non-functioning
synapses from which the "trophic" substance is released prior to

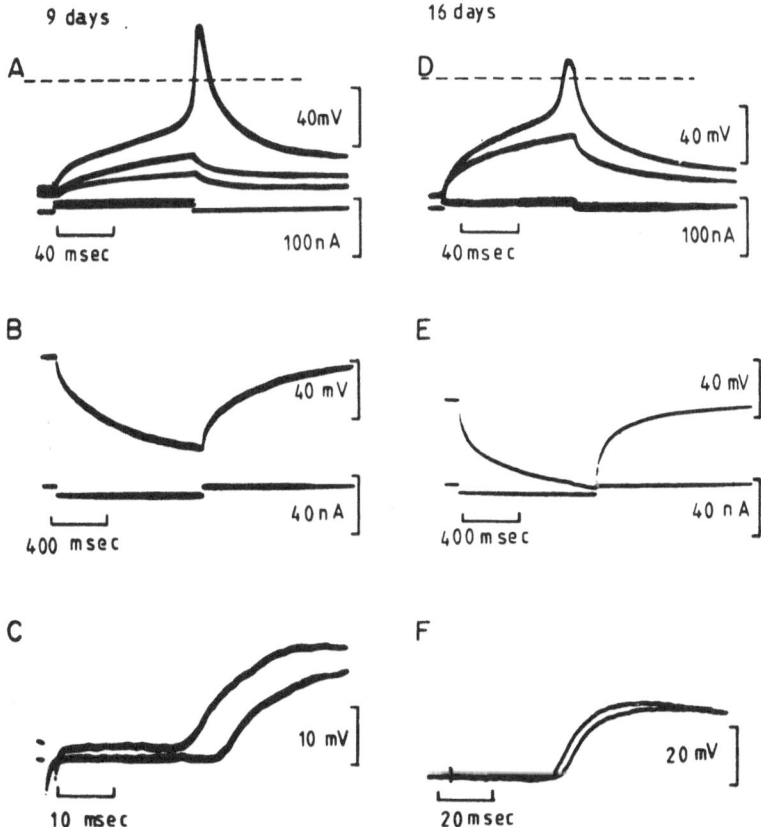

Fig. 13. Electrical properties and innervation of slow muscle fibers
of pyriformis muscle 9 (A, B and C) and 16 (D, E and F) days after
crushing the pyriformis nerve. A and D: action potentials elicited
upon intracellular stimulation. B and E: electrotonic hyperpolarizing
potential to identify fiber type. C and F end plate potentials after
stimulating the nerves at two different points. The calculated
conduction velocities were 1.8 and 2.4 m/sec respectively.

neuromuscular transmission (21).

Pyriformis Nerve Crushed

The initial non-selective re-innervation of slow muscle fibers
could be explained by the fact that after the sciatic crush, large
motor axons reach the muscle earlier than small motor axons. In

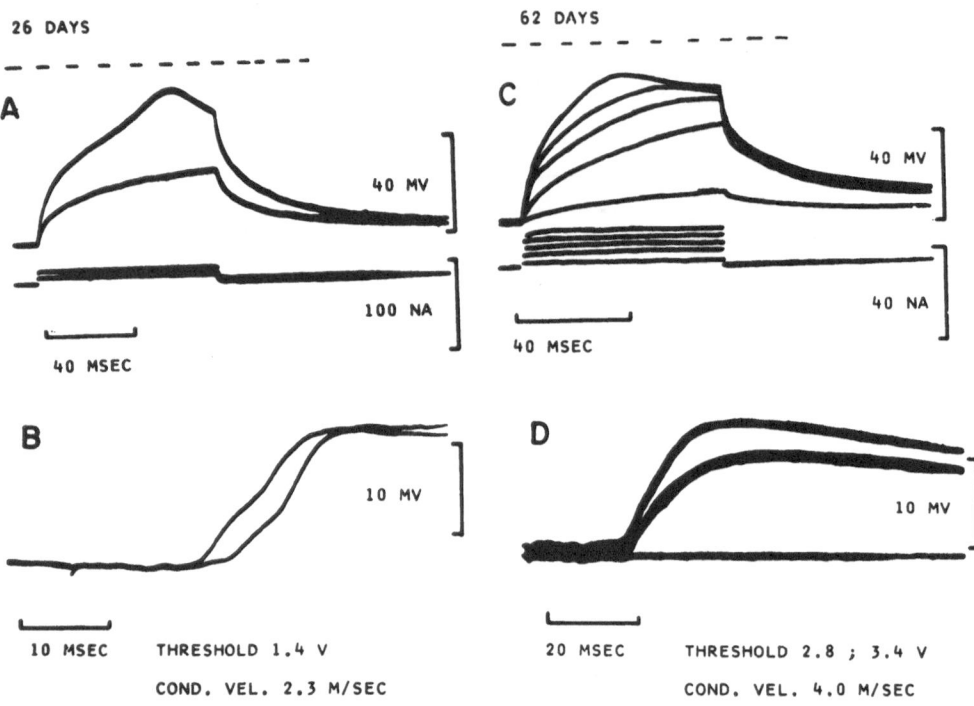

Fig. 14. Electrical properties and innervation of slow muscle fibers
of pyriformis muscle 26 (A and B) and 62 (C and D) days after
crushing the pyriformis nerve. Both fibers were re-innervated by
small motor axons. Note the progressive disappearance of the action
potential. In A a small hump remains. In C no traces of regenerative
activity is detected.

these series of experiments we have reduced the regenerating distance
by crushing the nerve before its entrance into the muscle; thus,
one may expect that large and small motor axons will reach the muscle
almost simultaneously. Functional re-innervation started 9 days
after the operation. From the initial stages of re-innervation,
slow and twitch muscle fibers were re-innervated by the corresponding
axons. Fig. 13 shows two slow fibers selectively re-innervated 9
and 16 days after the operation. The conduction velocity and the
threshold of the axons correspond clearly to the small motor type.
The fibers developed action potentials during the denervation period
and as re-innervation progresses, they become smaller and finally
disappear. During the initial periods most slow fibers are singly
innervated. In later stages they become again multi-innervated.
Fig. 14 shows two slow fibers 26 and 62 days after the pyriformis
nerve was crushed. Note the components of the synaptic potentials
(B and D) and the progressive disappearance of the action potential.

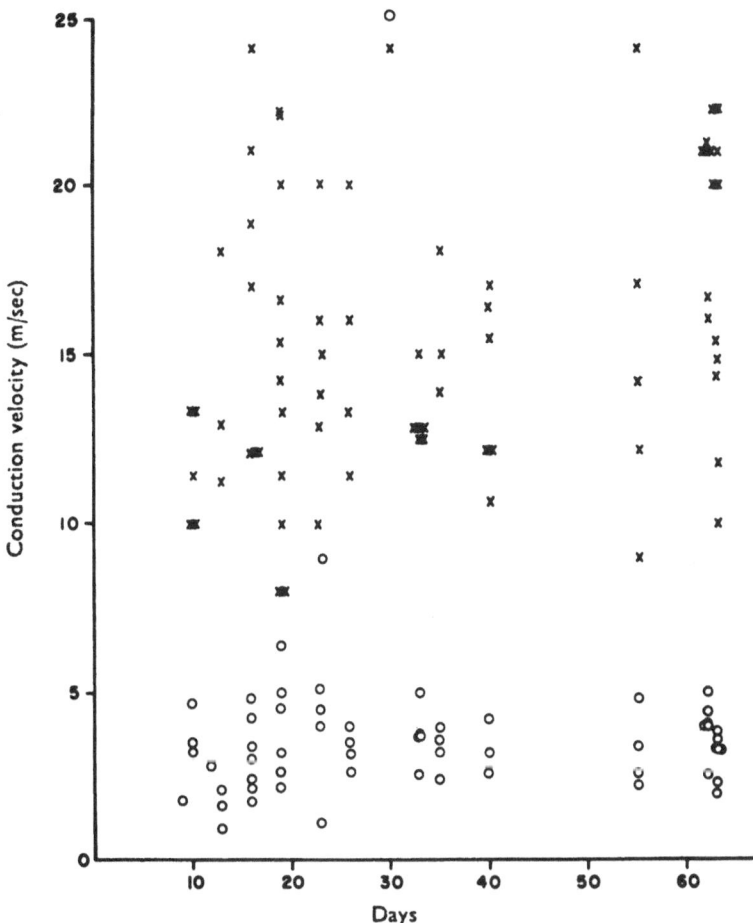

Fig. 15. Conduction velocity of motor axons innervating twitch (crosses) and slow (circles) muscle fibers of pyriformis muscle at different times after crushing the pyriformis nerve. Slow muscle fibers were from early stages selectively re-innervated by small motor axons.

Fig. 15 shows the conduction velocities of motor axons innervating twitch and slow muscle fibers at different times after the operation. Slow muscle fibers were re-innervated by axons with conduction velocities slower than 5 m/sec and twitch fibers by axons faster than 8 m/sec. Only two slow fibers out of 54 were simultaneously innervated by small motor axons (2.6 and 1 m/sec) and large motor axons (6.4 and 9 m/sec).

A similar selective re-innervation was observed after cutting

the pyriformis nerve (Huerta, M., personal communication). These
results indicate that re-innervation is selective for fiber type and
that regenerating axons can recognize the corresponding muscle fibers.

DISCUSSION AND CONCLUSIONS

Normal Muscles

The results obtained in the present investigation have confirmed
and extended previous observations on the innervation of muscle fiber
types of skeletal muscle of the frog (14). Twitch muscle fibers in
pyriformis were found to be innervated exclusively by fast conducting
axons having low threshold. On the other hand, slow muscle fibers
were innervated by slowly conducting axons with high threshold. In
general, slow fibers were polyneurally innervated. The ranges of
conduction velocity and threshold were clearly separated. The fastest
small motor axon conducted at 5 m/sec while the slowest large motor
axon had a conduction velocity of 10 m/sec; the higher threshold
for large motor axon was 0.7 V, and the smallest one for small motor
axons was 1.7 V. Similar results were obtained from the motor axons
innervating the iliofibularis muscle.

In conclusion, the fiber types were clearly defined according
to their electrical properties and innervation. We could not find a
single fiber with mixed properties of an intermediate type. However,
it is important to stress that the analyzed muscle fibers were mainly
from superficial layers since we were impaling them with two separate
microelectrodes. Further work will be carried out analyzing fibers
located in different layers using double barreled microelectrodes
to achieve a more careful search of an intermediate muscle fiber.
An intermediate fiber was recently reported according to mechanical
properties in single muscle fiber studies (19).

Re-Innervated Muscles After Crushing the Sciatic Nerve

In the early period of re-innervation, slow muscle fibers were
re-innervated by axons with conduction velocities higher than normal
(5 m/sec). It is important to assess the nature of these motor axons.
Two possibilities have to be considered: 1) the axons belong to
neurons which normally innervate twitch muscle fibers, i.e. large
motor axons; 2) they belong to neurons which normally innervate slow
muscle fibers and that their conduction velocity was speeded up
after the operation. For the following reasons it is more likely that
axons conducting faster than 5 m/sec are large motor axons. It is
well established that injury of mammalian nerve fibers is followed
by retrograde changes which result in a marked reduction in their
conduction velocity (12,8,6,18). An increased conduction velocity
would have to be the result of an increase in axon diameter in the
proximal nerve stump. Nerve fiber diameters were reported to decrease

after axotomy, thus ruling out this possibility (12). We conclude
therefore that in the early stages of re-innervation, slow muscle
fibers are non selectively re-innervated by large motor axons. As
re-innervation progresses, slow motor fibers were re-innervated by
large and small motor axons. In several slow muscle fibers,
simultaneous innervation of large and small motor axons could be
demonstrated. In this period, cholinesterase staining has shown the
presence of two end plate types on the same slow muscle fibers, end
plate "en grappe", typical of slow fibers, and end plate "en plaque",
typical of twitch fibers (Huerta, M., personal communication). This
observation further supports the above conclusion. Finally, slow
muscle fibers became exclusively polyneurally re-innervated by small
motor axons, and large motor axons synapses were withdrawn from them.

These observations suggest that small motor axons have a
preference for slow muscle fibers. The initial non selective-re-
innervation of slow fibers could be explained by the fastest
degeneration rate of large motor axons following by a possible
fastest regeneration speed (10). There is no indication on the
mechanism responsible for the connections of small motor axons to
their normal target cells. It can be even argued that this mechanism
is non existing. For example, the first regenerating large motor axons
innervate all muscle fibers they find in their way. As slow muscle
fibers have multiple synaptic sites, they have remaining free sites
which became re-innervated by the more slowly regenerating small
motor axons. Thereafter, large motor axons, withdraw from slow
muscle fibers possibly due to a progressive polyneural innervation
of small motor axons. A similar mechanism was found on adult
salamander muscle where foreign synapses were eliminated (7).

Re-Innervated Muscles After Crushing or Cutting the Pyriformis Nerve

Muscle fibers after this operation are selectively re-innervated
from early states. It was speculated that by reducing the regenerating
distance, large and small motor axons will reach the muscle almost
simultaneously. In this way the selectivity of re-innervation could
be more accurately tested. The results thus obtained indicate that
there is some mechanism responsible for forming synapses with normal
target cells. The mechanism by which selective synaptogenesis between
distinct classes of regenerating nerves and target cells occur remains
unexplained. Different possibilities can be considered (26). 1) The
formation of random connections, with later elimination of incorrect
contacts, and functioning only correct synapses. 2) The initial
formation of correct connections by virtue of some extraneural
directing mechanisms or some specific interaction between pre and
post-synaptic elements. These possibilities are not mutually
excluding. Synapse elimination appears to occur after crushing the
sciatic nerve, and may occur as well during initial re-innervation
as well as after cutting or crushing the pyriformis nerve. However,
the fact that the formation of new synaptic contacts does not occur

randomly favours the second hypothesis. There is evidence that regenerating axons seem to recognize the site of the neuromuscular junction by virtue of a specialized region of the extracellular basal lamina (27). It can be speculated that the basal lamina of slow and twitch muscle fibers are of different nature giving some basis for the regenerating axons to recognize their normal target cells.

Acknowledgements. This work was partially supported by Conacyt of Mexico, grant PNCB 0059. The secretarial work of Emma Bensimón is acknowledged.

REFERENCES

1. Adrian, R.H., Chandler, W.K. and Hodgkin, A.L., Voltage clamp experiments in striated muscle fibres, J. Physiol., 208 (1970) 607–644.
2. Bernstein, J.J. and Guth, L., Nonselectivity in establishment of neuromuscular connections following nerve regeneration in the rat, Exp. Neurol., 4 (1961) 262–275.
3. Berry, C.M. and Hinsey, J.C., The recovery of diameter and impulse conduction in regenerating nerve fibres, Ann. N.Y. Acad. Sci., 47 (1946) 559–574.
4. Bixby, J.L. and Van Essen, D.C., Competition between foreign and original nerves in adult mammalian skeletal muscle, Nature (Lond.), 282 (1979) 726–728.
5. Burke, W. and Ginsborg, B.L., The electrical properties of the slow muscle fibre membrane, J. Physiol., 132 (1956) 586–598.
6. Cragg, B.G. and Thomas, P.K., Changes in conduction velocity and fibre size proximal to peripheral nerve lesion, J. Physiol., 157 (1961) 315–327.
7. Dennis, M.J. and Yip, J.W., Formation and elimination of foreign synapses on adult salamander muscle, J. Physiol., 274 (1978) 299–310.
8. Eccles, J.C., Krnjević, K. and Miledi, R., Delayed effects of peripheral severance of afferent nerve fibres on their efficacy of their central synapses. J. Physiol., 145 (1959) 204–220.
9. Elizalde, A., Huerta, M. and Stefani, E., Selective re-innervation of twitch and slow muscle of the frog after severing the nerve near its entrance into the muscle, (in preparation).
10. Elul, E., Miledi, R., Stefani, E., Neural control of contracture in slow muscle fibres of the frog, Acta physiol. Latinoam., 20 (1970) 194–226.
11. Frank, E., Jansen, J.K.S., Lomo, T. and Westgard, R.H., The interaction between foreign and original motor nerves innervating the soleus muscle of rats, J. Physiol., 247 (1975) 725–743.
12. Gutmann, E. and Holubar, J., The degeneration of peripheral nerve fibres, J. Neurol. Neurosurg. Psychiat., 13 (1950) 89–105.
13. Gutmann, E. and Sanders, F.K., Recovery of fibre numbers and diameters in the regeneration of peripheral nerves. J. Physiol.,

101 (1943) 489-518.

14. Hess, A., The structure of extrafusal muscle fibres of the frog and their innervation studied by the cholinesterase technique, Am. J. Anat., 107 (1960) 129-135.

15. Kuffler, S.W. and Gerard, R.W., The small nerve motor system to skeletal muscle, J. Neurophysiol., 10 (1947) 383-394.

16. Kuffler, S.W. and Vaughan Williams, E.M., Small nerve junctional potentials. The distribution of small motor nerves to frog skeletal muscle and the membrane characteristics of the fibres they innervate, J. Physiol., 121 (1953a) 289-317.

17. Kuffler, S.W. and Vaughan Williams, E.M., Properties of the slow skeletal muscle fibres of the frog, J. Physiol., 121 (1953b) 318-340.

18. Kuno, M., Miyata, Y. and Muñoz-Martínez, E.J., Differential reaction of fast and slow α-motoneurones to axotomy, J. Physiol., 240 (1974) 725-739.

19. Lännergren, J., An intermediate type of muscle fibre in Xenopus laevis. Nature, (Lond.), 279 (1979) 254-256.

20. Mark, R.F. and Marotte, L.R., The mechanism of selective re-innervation of fish eye muscle. III. Functional, electro-physiological and anatomical analysis of recovery from section of the IIIrd and IVth nerves, Brain Res., 46 (1972) 131-148.

21. Mark, R.F., Marotte, L.R. and Johnstone, J.R., Re-innervated eye do not respond to impulses to foreign nerves, Science (N.Y.), 170 (1970) 193-194.

22. Miledi, R., The strength latency relaxation of axons, Acta physiol. Latinoam., 7 (1957) 155-185.

23. Miledi, R. and Orkand, P., Effect of a fast nerve on slow muscle fibres of the frog, Nature (Lond.), 209 (1966) 717-718.

24. Miledi, R. and Stefani, E., Non-selective re-innervation of slow and fast muscle fibres in the rat, Nature (Lond.), 222 (1969) 569-571.

25. Miledi, R., Stefani, E. and Steinbach, A.B., Induction of the action potential mechanism in slow muscle fibres of the frog, J. Physiol., 217 (1971) 737-754.

26. Njå, A. and Purves, D., Specificity of initial synaptic contacts made on guinea-pig superior cervical ganglion cells during regeneration of the cervical sympathetic trunk, J. Physiol., 281 (1978) 45-62.

27. Sanes, J.R., Marshall, L.M. and McMahan, U.J., Reinnervation of muscle fiber basal lamina after removal of myofibers. Differentiation of regenerating axons at original synaptic sites, J. Cell Biol., 78 (1978) 176-198.

28. Schalow, G. and Schmidt, H., Local development of action potentials in slow muscle fibres after complete or partial denervation, Proc. R. Soc. Lond. B., 203 (1979) 445-457.

29. Schmidt, H. and Stefani, E., Re-innervation of twitch and slow muscle fibres of the frog after crushing the motor nerves, J. Physiol., 258 (1976) 99-123.

30. Schmidt, H. and Stefani, E., Action potentials in slow muscle

fibres of the frog during regeneration of motor nerves, J. Physiol., 270 (1977) 507–517.

31. Stefani, E. and Schmidt, H., A convenient method for repeated intracellular recording from the same muscle fibre without membrane damage, Pflugers Arch. ges. Physiol., 334 (1972a) 276–278.

32. Stefani, E. and Schmidt, H., Early stage of re-innervation of frog slow muscle fibres, Pflugers Arch. ges. Physiol., 336 (1972b) 271–275.

33. Stefani, E. and Steinbach, A.B., Resting potential and electrical properties of frog slow muscle fibres. Effects of different external solutions, J. Physiol., 203 (1969) 383–401.

34. Tasaki, L. and Mizutani, K., Comparative studies on the activities of the muscle evoked by two kinds of motor nerve fibres. Part I. Myographic studies, Jap. J. Med. Sci. Biol., 9 (1944) 237–244.

35. Weiss, P. and Hoag, A., Competitive re-innervation of rat muscles by their own and foreign nerves, J. Neurophysiol., 9 (1946) 413–418.

PRESYNAPTIC CONTROL OF INFORMATION TRANSMISSION IN THE VERTEBRATE

SPINAL CORD

Pablo Rudomín

Department of Physiology and Biophysics,
Centro de Investigación y de Estudios Avanzados
Instituto Politécnico Nacional, México 14, D.F.

INTRODUCTION

Despite their different ways of looking at the nervous system, neurophysiologists, neuroanatomists and neurochemists seem to have a similar concern with the regulation of synaptic transmission, in its broadest sense. The problem is to define what do we mean by regulation, and this is not an easy task. In this context we may distinguish two types of mechanisms controlling synaptic transmission. One, involving actions tending to maintain a constant operating level in the system, for example, those mechanisms involved in keeping a constant pool of available transmitter, and second, those mechanisms involving a change in the operating level of the system, such as those involved in learning. Perhaps the first, but not the second mechanism implies the existence of a state detector, a command action and an effector (100). Figure 1 shows some of the steps involved in the generation of impulses in motoneurons by activity in muscle spindles. In principle any procedure interfering with any of these steps will affect synaptic transmission (90). This can be achieved by external influences or can be a consequence of the functioning of the nervous system itself. From the point of view of transmitted information, what seems to be relevant is the change in the state. If there is no change, there will be no information transfer. The change, however, may or may not be permanent. If we take the membrane potential as an indicator of the state of the system, there is a more or less constant state – the "resting potential" – which may be transiently (few msec) and repetitively (up to 100/sec) disturbed. Transitority results from actions tending to restore the original potential level. This is achieved by several mechanisms: passive ionic diffusion, active ionic transport requiring energy consumption (16), etc. When the impulse arrives

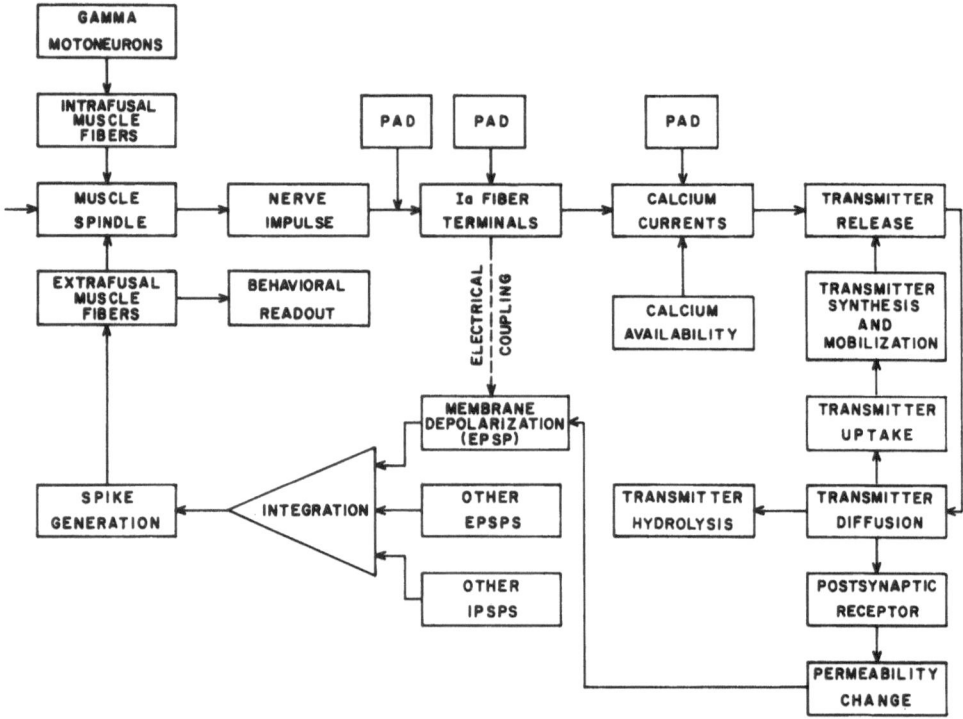

Fig. 1. Diagram of the steps involved in impulse transmission from muscle spindles to alpha motoneurons. Explanations in text. From ref. 90.

to the terminals another process is set up which disturbes, also transiently, a stage of equilibrium. Calcium enters the terminals and transmitters are expelled to the outside, probably by activation of a "contractile like" mechanism (77,107).It takes time for the presynaptic terminal to regain its original conditions: the released transmitter must be restored, either by replenishment from another compartment or by synthesis (107). The speed with which the original conditions are regained will certainly limit the frequency behaviour of the system which may be of relevance for information transmission, for motor control, or for learning processes.

In this presentation I shall review briefly some of the problems involved with the presynaptic control of synaptic transmission which the central nervous system can exert via specific neuronal pathways. The problem will be centered on two issues. The possible <u>specificity</u> of such mechanism and its relevance for information transmission. In addition, I will discuss the problem of homosynaptic depression as a part of a more general presynaptic mechanism setting the frequency

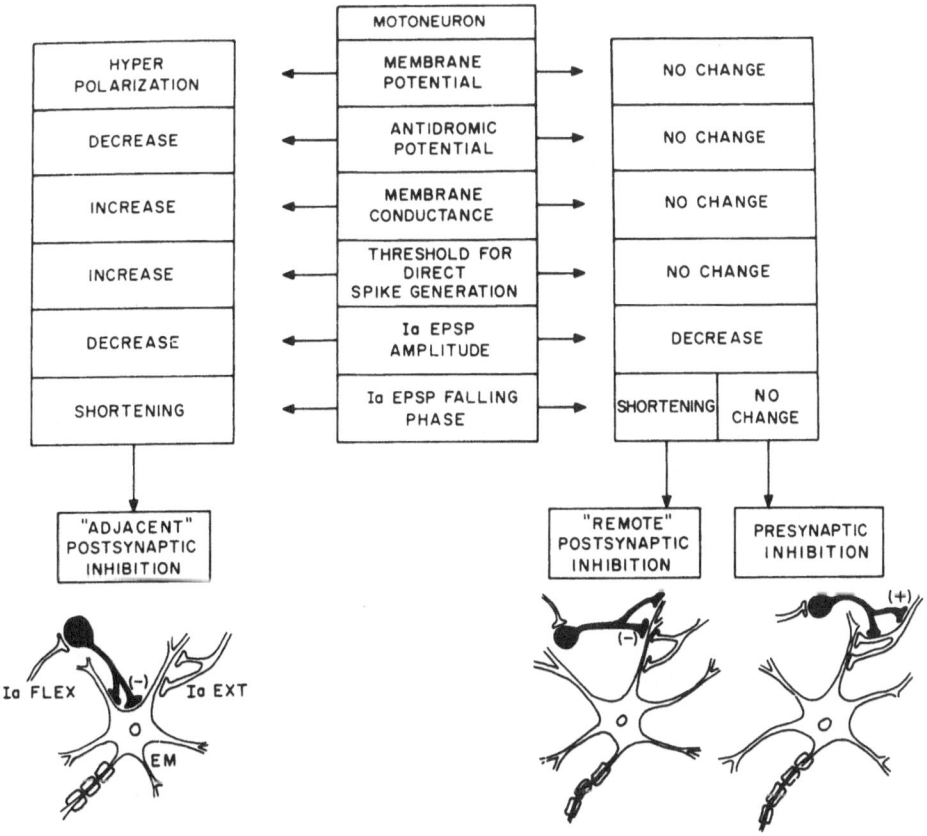

Fig. 2. Characteristics of adjacent and remote postsynaptic inhibition and presynaptic inhibition in the Ia fiber-alpha motoneuron pathway. The diagram summarizes the alterations of gastrocnemius motoneuron properties and of Ia monosynaptic EPSPs produced by conditioning volleys to the biceps-semitendinosus nerve. Further explanation in text.

behaviour of the system.

PRESYNAPTIC INHIBITION IN THE Ia-MOTONEURON PATHWAY

    The possibility that impulse transmission in afferent fibers
can be modified by stimulation of other sensory nerves was established
by Howland et al (58). Frank and Fourtes showed in 1957 (41) that
monosynaptic Ia EPSPs recorded from extensor motoneurons were reduced
in size by conditioning volleys in group I afferents from flexor
nerves. The conditioning volleys did not appeared to alter the
membrane potential or the excitability of the postsynaptic neurons.
It was later considered that the interaction occurred between the
conditioning volley and the test EPSP at loci remote from the
motoneuron cell body (40). Such a "remote inhibition" could be due
either to interaction between EPSPs and postsynaptic IPSPs in
electrotonically distant dendritic sites, or to a reduction in
synaptic effectiveness caused by a mechanism operating directly on
the presynaptic terminals, i.e. presynaptic inhibition (Fig. 2).

    Considerable work has been done subsequently to decide between
these two possibilities, which are not mutually exclusive. One of
the main approaches has been to devise more sensitive tests to
detect changes in the electrical properties of the postsynaptic
membrane, either by measuring conductance changes (12,37) membrane
potential changes (18,33), repetitive firing by injected currents
(18,48), or changes in the peak and falling phase of the Ia EPSP
itself (37,97). In all cases it has been possible to find examples
of Ia EPSPs which are depressed without any evidence of changes in
the postsynaptic membrane properties. Therefore, by exclusion, the
EPSP depression has been attributed to a presynaptic inhibitory
action (37). In this regard a more direct demonstration of the
existence of presynaptic inhibition is the increase in the number of
failures of single fiber Ia EPSPs produced by conditioning volleys
to sensory nerves (69). Increased failures could result from either
conduction failure or reduction in the quantum content of the Ia
EPSP.

    Presynaptic inhibition in the Ia fiber-motoneuron monosynaptic
pathway has been attributed to depolarization of the terminal
arborizations of the Ia terminals (33,34). This assumption is based
on the observation that both the depression of the Ia EPSP and the
depolarization of the Ia fibers change in a parallel fashion in all
tests so far performed: they both have a similar time course, and
are produced by the activation of the same type of afferent fibers.
Also, the two appear to have the same pharmacological behaviour (36).
This is a necessary, but not sufficient condition since it is
possible that both the primary afferent depolarization (PAD) and the
Ia EPSP depression have a common cause and are not causally related.

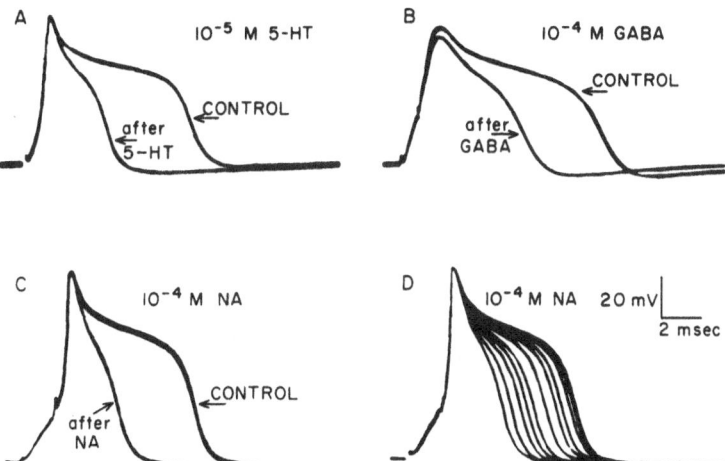

Fig. 3. Selective block of Ca^{++} inward current in primary afferent neurons by transmitter amino acids. Action potentials of dorsal root ganglion cells normally show a plateau which is due to Ca^{++} inward current. In each record there is a control trace of spike (with plateau) and a briefer spike produced after application of A) 10^{-5}M 5-HT; B) 10^{-4}M GABA and C, D) 10^{-4}M NA; progressive recovery of Ca^{++} component is shown in D, where successive traces were recorded at 10 sec intervals. From ref. 30.

In fact, it has been suggested that the presynaptic depolarization of the optic tract fiber terminals may not be associated with depression of synaptic transmission (103). This raises a question which will be considered below, on whether there are different kinds of presynaptic depolarizations and on the extent to which all presynaptic depolarizations lead to depression of synaptic transmission.

The knowledge of how presynaptic depolarization leads to depression of synaptic transmission is far from being established because the nature of the mechanisms producing the presynaptic depolarization is also unknown. The most commonly accepted view is that the presynaptic depolarization reduces the amplitude of the action potential in the presynaptic terminals (32,33). Since transmitter release appears to be proportional to the latter, synaptic transmission would be depressed. Another possibility is that pre-synaptic depolarization produces conduction blockade in the fine terminal arborizations (9,52,53,58,105). At first sight it is not immediately obvious why depolarization should block conduction. If conduction were critical in branching points one would expect a mild depolarization to relieve the block. At least this is what happens

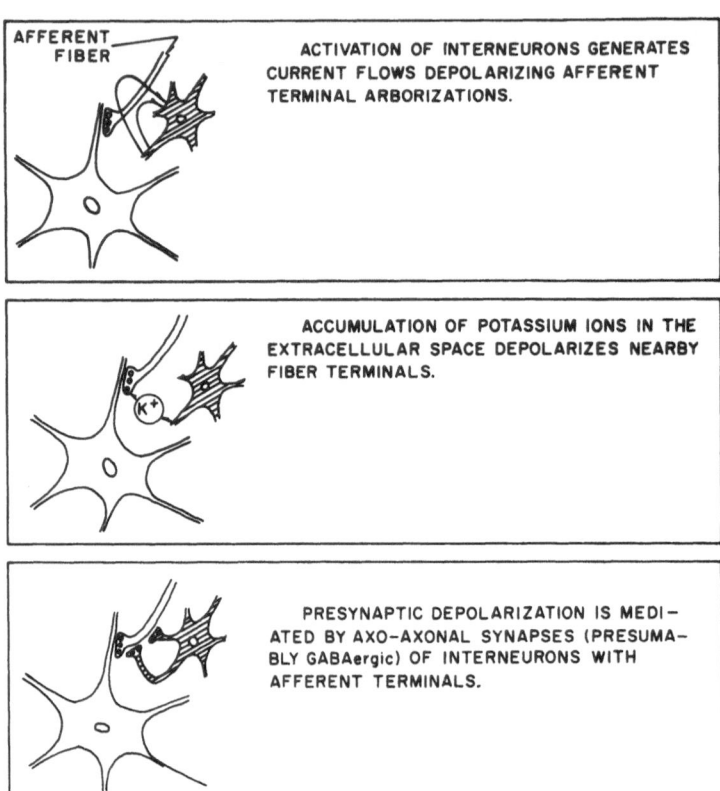

Fig. 4. Diagram of the mechanisms involved in PAD generation.
Explanation in text.

with the antidromic invasion of motoneurons (32). A strong depolarization, however, would prevent invasion by inactivation of sodium
currents. In other words, if invasion by the action potential in the
fiber terminal arborizations were critical in resting conditions, a
mild depolarization would be expected to produce presynaptic facilitation and a strong depolarization presynaptic inhibition. So far
there is no experimental support for presynaptic facilitation
produced by mild depolarizations, since depression of Ia EPSPs
results even with the weakest conditioning volleys which produce a
small PAD (33,34). It is of course possible that a conditioning
stimulus restricted to very few or even one afferent fiber produces
sufficient presynaptic depolarization as to block conduction in
critical nodal points. In this context Shapovalov and Shiriaev (101)
have shown recently that activation of a single motoneuron in the
frog spinal cord produces a substantial PAD which can be recorded in
single fibers, and it is not altogether impossible that direct

activation of one or few interneurons in the i.termediate nucleus of
the cat spinal cord may produce detectable PAD (92,93).

A third possibility is that the calcium currents involved in
transmitter release are voltage dependent and are reduced by the
presynaptic depolarization, or by a direct action of the transmitter
generating the PAD. In fact, there is recent evidence obtained in
cultured ganglion cells (30) that GABA, the presumed transmitter
for presynaptic depolarization (see below), may indeed reduce the
Ca^{++} currents associated with the action potential without changing
the membrane potential of the cells (Fig. 3).

MECHANISMS INVOLVED IN THE GENERATION OF PAD

There have been three hypotheses to explain primary afferent
depolarization in the spinal cord (see Fig. 4). Already in 1939
Barron and Matthews (8) assumed that current flows in interneurons
could depolarize the terminal arborizations of afferent fibers. They
also suggested an alternative possibility, namely, that activity
in interneurons would result in accumulation of potassium ions with
consequent depolarization of nearby neurons and afferent fibers.
Finally, in the early sixties, Eccles and collaborators (36)
suggested that PAD might be due to chemical synapses apposed to
presynaptic terminals (49), releasing a specific depolarizing
transmitter substance, presumably GABA (70).

The hypothesis that PAD may result from interaction by current
flows has not received experimental support by recent experiments
performed in the frog spinal cord (46). Implicit in this hypothesis
is the idea that activation of the responsible interneurons would
produce the required extracellular fields irrespective of the
presence of afferent arborizations since the latter would act merely
as passive detectors. This point has been tested in the frog spinal
cord, where long lasting PAD in afferent terminals can be produced by
antidromic activation of motoneurons (15). In the frog, slow
potential fields corresponding to afferent terminal PAD produced by
ventral root stimulation disappear after the degeneration of the
dorsal root fibers (46) which suggests, provided there is no
transneuronal degeneration (47), that the intraspinal dorsal root
afferents generate the currents associated with PAD instead of acting
simply as passive detectors of such currents. It should be pointed
out, however, that electrical interactions by current flow between
motoneurons and afferent fibers do exist, both in the frog (42,51)
and in the cat (22,26,54) spinal cord, but these are of rather brief
duration because they are generated by the current flows associated
with the antidromic activation of motoneurons.

It is now well established that repetitive stimulation of
sensory nerves may increase the concentration of potassium ions in

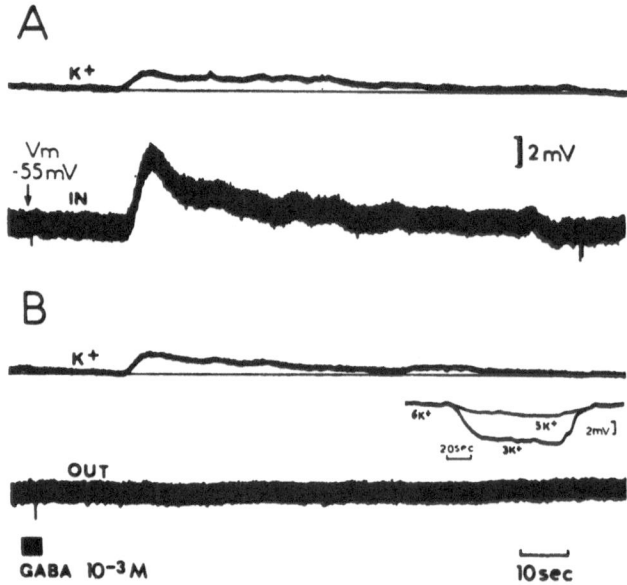

Fig. 5. Rise of extracellular potassium concentration (K+) is associated with the GABA-induced depolarization of dorsal root ganglion neurons. A: recording obtained while the intracellular micropipette was located within a ganglion neuron. Extracellular potassium potential (K+) recorded simultaneously with the change of membrane potential (VM). GABA was applied at time shown in B (black bar). B: the same, but with the recording micropipette withdrawn to a extracellular position. Insert, in situ responses of the K+ sensitive microelectrode to superfusion of the ganglion with Ringer's solution containing 5 mM and 3 mM KCl; potassium concentration of normal Krebs solution, 6 mM. From ref. 27.

the dorsal horn and intermediate nucleus, and that intraspinal afferents are sensitive to the observed changes in extracellular potassium (65-67,74,111). In addition, there is anatomical evidence suggesting that PAD could be produced by a specific transmitter substance released by interneurons: axoaxonic synapses have been found in the dorsal horn, motor nuclei, medial ventral horn and Clarke's column (17,47,86,104,115). Pharmacological evidence further suggests that GABA may be the transmitter involved in the generation of PAD: GABA depolarizes afferent fibers and increases their excitability to direct stimulation in the cat and frog spinal cord (4,5,23,24,70,108,109); GABA antagonists such as picrotoxin and bicuculline reduce the PAD (5,19,20,23,24,87); semicarbazide-induced depletion of GABA reduces the dorsal root potentials and the long lasting inhibition of monosynaptic reflexes attributed to presynaptic inhibition (2,6). In the frog spinal cord, inhibition of GABA

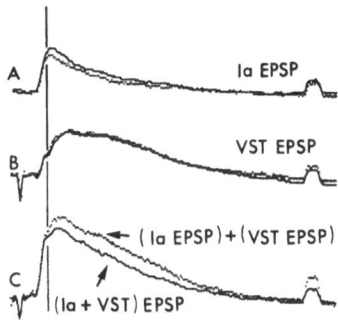

Fig. 6. Selective depression of monosynaptic Ia-EPSPs but not of VS-EPSPs produced by conditioning volleys to sensory nerves. A: monosynaptic EPSPs produced by supramaximal stimulation of group 1 fibers in the lateral gastrocnemius nerve. B: EPSPs produced by stimulation of the ipsilateral ventromedial funiculus (VST) at the midthoracic level (2xT strength of the most excitable fibres in the tract). Continuous traces in A and B are the control responses. Dotted traces show responses obtained 30 msec after BST conditioning stimulation (three shocks at 300/sec, 1.3 times threshold of the most excitable fibers in the nerve). C: continuous trace is the post-synaptic potential obtained during simultaneous stimulation of Ia and descending fibers with the same parameters as in A and B. The dotted trace shows the synaptic potential obtained by algebraic addition of the two unconditioned EPSPs illustrated in A and B. The vertical line was traced at the peak of the monosynaptic component of the VST EPSP. All are averaged records of 200 responses elicited once every 6 sec. Calibrations, 1mV, 1 msec. From ref. 97.

transaminase, a major degradative enzyme for GABA, increases the dorsal root potentials and the excitability of afferent terminals and also increases the concentration of GABA within the cord (25). The concentration of GABA is greatest in the dorsal horn-intermediate nucleus region (45,71,79), the location of presumed synapses generating PAD (45,46; Jankowska, McCrea, Rudomín and Sykova, unpublished observations).

Considering the available evidence, it is not possible to decide between the potassium and the GABA hypothesis for the generation of PAD, particularly because of recent evidence (27) of a GABA-induced increase in extracellular potassium in dorsal root ganglia (Fig. 5). Nevertheless the implications of the two hypothesis are sufficiently important form the point of view of information encoding and specificity of transmission as to warrant further investigation, because generation of PAD by a transmitter substance acting on presynaptic fibers implies specific synaptic connections to the afferent terminals, whereas potassium accumulation may result, at

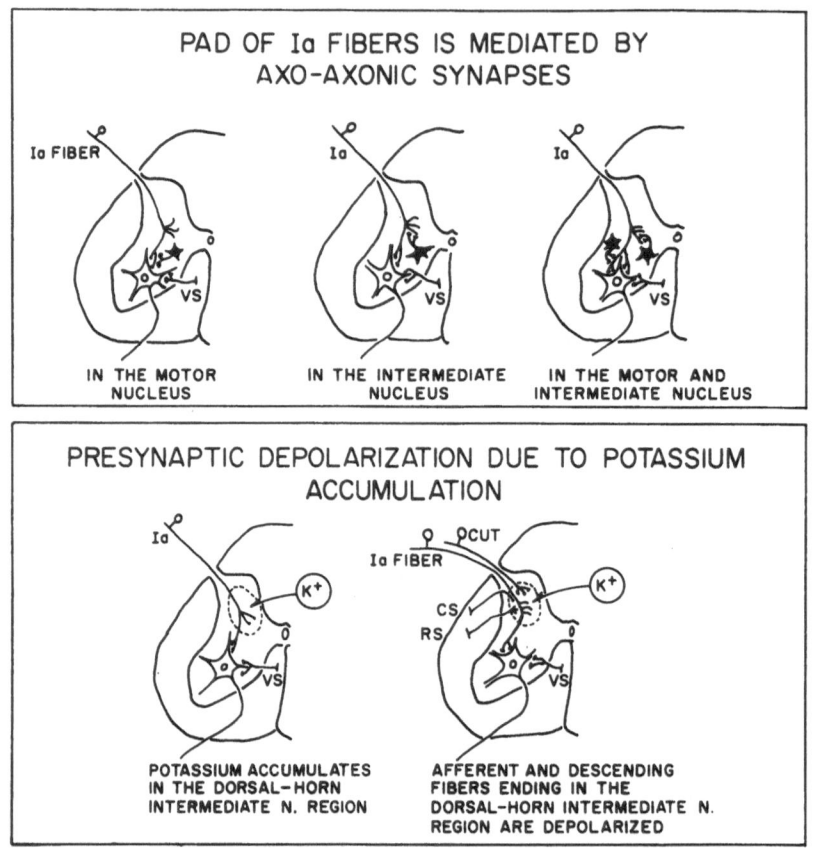

Fig. 7. Diagram of the possible pathways and mechanisms involved in presynaptic depolarization of afferent and descending fibers. Explanation in text.

least in principle, in less- specific actions affecting any pre-synaptic terminal in the vicinity (90).

PRESYNAPTIC DEPOLARIZATION OF AFFERENT AND DESCENDING FIBER TERMINALS

The problem of the specificity in the pathways and mechanisms producing presynaptic depolarization can be studied in various ways. The usual approach has been to determine the relative amount of PAD produced by stimulation of various kinds of sensory fibers on a given afferent fiber. This has shown that in the cutaneous PAD-generating system there is a remarkable spatial and sensory modality specificity with respect to mechanoreceptor afferents from the skin, perhaps to provide an automatic gain control mechanism. However, the patterns of PAD in muscle afferents do not appear to fit a simple negative feedback model (9; see 99 for review).

The existence of specific input-output patterns in the PAD system is compatible with both the potassium and the transmitter hypothesis because specificity could arise from the interneuronal sets that ultimately serve as a source of extracellular potassium or GABA (9). A more precise test for specificity in the PAD system would be the analysis of the depolarization patterns of the terminal arborizations of fibers ending on adjacent areas of the same cell. Demonstration of specificity of PAD or presynaptic inhibition would support the transmitter hypothesis, since accumulation of extracellular potassium would not be expected to generate specificity in such a microenvironment.

Ia and vestibulo-spinal (VS) fibers converge onto the same motoneurons in the lumbosacral segments (37,50,97,114). We have shown several years ago (97) that monosynaptic Ia EPSPs were depressed (without changing their falling phase) by conditioning volleys in flexor group I afferents, whereas monosynaptic EPSPs produced by stimulation of descending tracts (presumably vestibulo-spinal) were unaffected. The two species of EPSPs interacted non-linearly, implying electrotonic proximity between the two sets of active synapses (Fig. 6). This suggested that the observed changes in amplitude of monosynaptic Ia EPSPs were not due to remote post-synaptic conductance changes but to a mechanism of presynaptic inhibition specifically acting on the Ia fiber terminals, but not on the VS fiber terminals. This finding admits several explanations as illustrated in Fig. 7. If PAD of Ia fibers were mediated by axo-axonic synapses releasing a specific transmitter substance (presumably GABA), axo-axonic synapses could be present either in the motor nucleus, in the intermediate nucleus, or in both regions. The effects of PAD on synaptic transmission would then depend on whether the afferent fiber terminals were actively depolarized by the transmitter substance or passively by electrotonic spread of a depolarization generated elsewhere. It is clear that with any of these arrangements the VS fibers would not be depolarized because they would not receive axo-axonic synapses.

If presynaptic depolarization results from potassium accumulation, VS fibers would not be depolarized because they don't give collaterals, or pass through the regions where potassium accumulates after repetitive activation of sensory nerves (mostly in the intermediate nucleus and dorsal-horn) (65,66,74). With this model (see Fig. 7) the Ia fiber collaterals in the intermediate nucleus would be depolarized and this depolarization would spread electrotonically to the branches in the motor nucleus. This arrangement may have interesting functional implications, but it is certainly not the

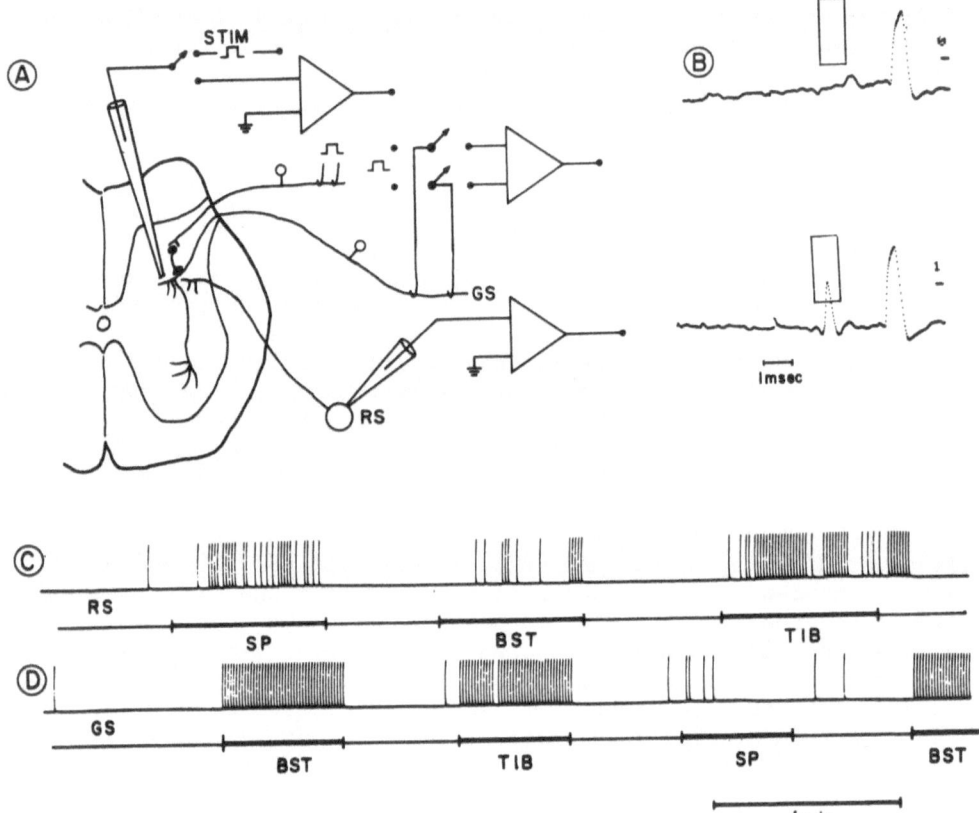

experimental paradigm needed for the demonstration of <u>specificity</u> in the PAD mechanisms and pathways. The tests for specificity would be more critical if performed with afferent and descending fibers terminating in the dorsal horn and intermediate nucleus, <u>precisely in those areas where potassium accumulates</u> after repetitive stimulation of sensory nerves. There are two systems of fibers where this requirement is met. Group I (Ia and Ib) and rubro-spinal (RS) fibers synapse with the same interneurons in the intermediate nucleus (56, 57) and cutaneous and cortico-spinal (CS) fibers converge in the dorsal horn (57,94), as indicated in Fig. 7.

As a first step in the study of presynaptic modulation of synaptic effectiveness of RS and CS fibers we asked the question on whether or not these fibers were subjected to a mechanism of presynaptic depolarization of the type of PAD already documented for group I and cutaneous fibers. To this end, we first studied the excitability changes of RS and group I fibers produced by conditioning volleys to sensory nerves. Figure 8 shows data obtained from one experiment where the excitability testing micropipette was placed in the intermediate nucleus, in such a position where antidromic responses of a <u>single</u> RS and of a <u>single</u> group I gastrocnemius (GS) fiber (presumably Ia) were generated by the same stimulus. It can be seen that the antidromic firing of the RS cell was increased by conditioning stimulation of sensory nerves and that volleys to the

---

Fig. 8. Changes in the antidromic firing index of a single rubro-spinal fiber and of a single group I afferent fiber, both ending in the intermediate nucleus. A: diagram of the experimental method. The excitability testing micropipette was placed in the intermediate nucleus, in such a position where antidromic responses of a single RS and of a single group I GS fiber were produced by the same stimulus. B: samples of antidromic unitary discharges recorded in the red nucleus by constant current stimulation in the intermediate nucleus (at 2.938 mm from the cord surface). Recordings are of computer display and show the floating window discriminator which generated a pulse whenever there was an event falling within the window. In C the window discriminator was triggered by the RS unit and in D by the GS unit. Upper traces in C and D are the window discriminator output; lower traces, the time of conditioning stimulus application. Stimulus to SP and TIB was a single shock at two times threshold. BST stimulus was three shocks at 300/sec, strength maximal for group I fibers. Conditioning stimuli were applied 30 msec before the test stimulus, which was applied once per second. Test stimulus strength was adjusted to produce few responses of the RS and GS fibers (17 and 10 µA, respectively). Note that for RS terminals SP was more effective than BST, whereas for the GS fiber the pattern was the opposite. TIB conditioning increased the antidromic firing of both the RS and GS fibers probably because it contains cutaneous and muscle afferents. From ref. 93.

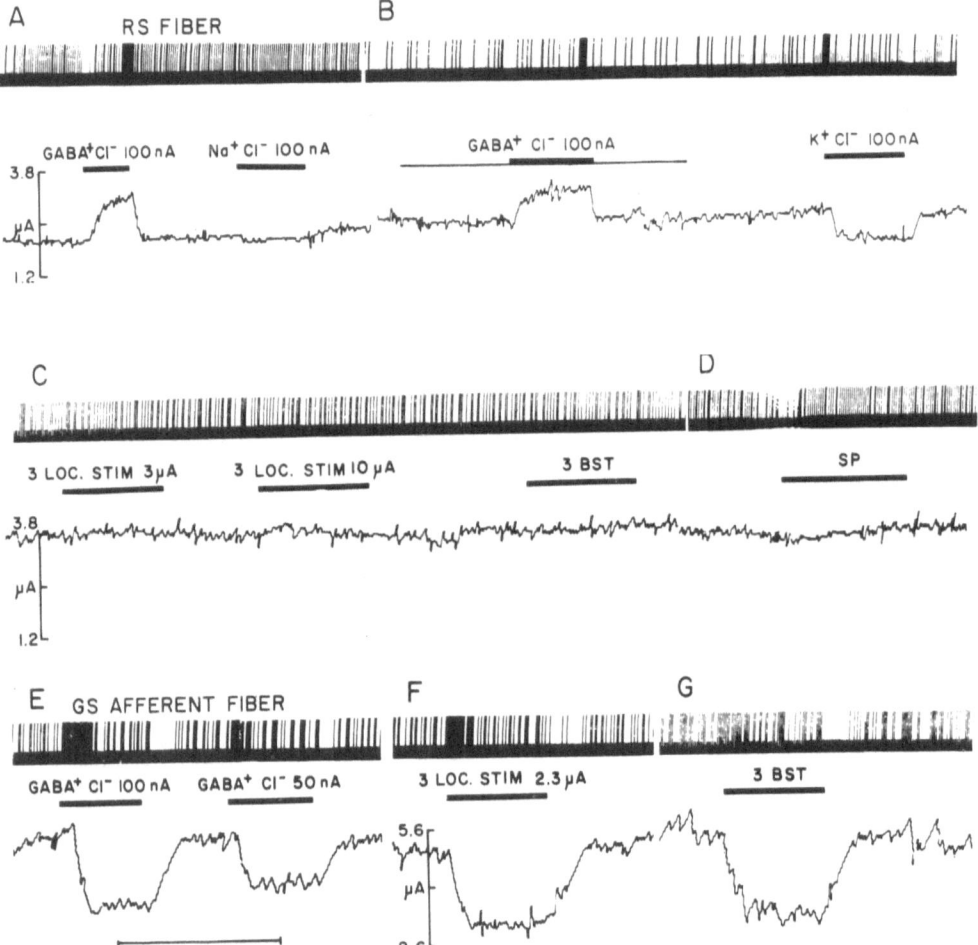

superficial peroneus (SP) and tibialis (TIB) nerves were more
effective than volleys to biceps-semitendinosus (BST; Fig. 8C).
However, for the GS afferent fiber stimulation of BST was more
effective than stimulation of the SP nerve (Fig. 8D). This finding
shows that presynaptic depolarization (PD) of RS terminals results
from activation of different pathways (or mechanisms) that the PAD
of Ia fibers. Similar results have been found with CS fibers
projecting in the dorsal horn, since their antidromic activation
was also more effectively increased by cutaneous volleys than by
volleys to group I afferent in flexor muscles (90,94).

Figure 8 further shows that there is a difference in the
building up of the effects produced by BST conditioning stimulation
on the antidromic firing of RS and group I GS fibers. The
excitability of the group I GS fibers increased very effectively with
minimal delay after the application of the conditioning stimulus,
while the increase in excitability of the RS fibers build up slowly
with successive trials. This suggests also that the mechanisms
involved in the depolarization of these two sets of fibers may be
different, although it is possible that these differences resulted
from different settings of the test stimulus intensity relative to
threshold level of the fibers.

A more detailed analysis on the possible mechanisms involved
in the PD of fiber terminals requires methods which allow a

Fig. 9. Effects of GABA, sensory nerve stimulation and intraspinal
stimulation on the threshold of RS and group I GS fiber terminals
in the intermediate nucleus. All tests performed with the multibarrel
array placed in a fixed position (2.350 mm from the cord surface).
Stimulating pulses (0.5 msec duration) were generated by the computer
at a constant rate. Their amplitude was automatically adjusted every
cycle to keep a constant firing index of the unit (set to 0.5 in
this experiment). Upper traces, window discriminator output (see
Fig. 8B). Lower traces, integral of the test stimulus current
which was maintained until the next stimulating cycle. A-D, threshold
changes of a single RS fiber, and E to G of a single GS afferent
fiber. GABA was applied iontophoretically by passing current between
the GABA and the NaCl electrode (indicated by the GABA+Cl⁻ label).
Duration of current flow is indicated by the black bar above the
current traces. In B, the line bar indicates additional injection of
a small retaining balanced current (5nA) passed through the GABA and
NaCl electrode. Intraspinal stimulation (LOC-STIM) was applied
through the same electrode as used for threshold determination and
was of 3 shocks at 300/sec, 0.5 msec duration preceeding the test
stimulus by 25 msec. Stimulus strength is indicated in the figure.
BST stimulus was also 3 shocks at 300/sec, maximal for group I, and
SP one shock, two times threshold for the most excitable fibers in
the nerve. Current calibration is the same for A-D and for E-G.(R.93).

quantitative assessment of the excitability changes. Studies based
on the changes in the firing index may not allow comparison of
actions which depolarize the terminal arborizations above the level
required to fire the unit in all trials. To overcome these limitations
we implemented a method which allows a <u>continuous measurement</u> of the
test current required to maintain a constant firing index (76). If
the excitability of the tested fiber increases because of presynaptic
depolarization, the test current will be reduced. The experiment of
Fig. 9 shows the threshold changes of a RS and a group I fiber
tested with this method. The threshold of the RS fiber did not appear
to be modified by BST conditioning stimulation (Fig. 9C) and was
only slightly reduced by SP codntioning volleys (Fig. 9D). However,
the threshold of the group I fiber tested close in time at the same
site was considerably reduced by BST volleys (Fig. 9G). These
results agree qualitatively with those obtained by examining the
changes in the firing index (Fig. 8) in the sense that cutaneous
volleys appear to be more effective than BST volleys in
depolarizing RS terminals. The <u>magnitude of the threshold changes</u>
<u>was, however, rather small</u> compared with the threshold changes of
the group I GS fibers.

We also analyzed the changes in threshold of VS and Ia fibers
ending in the GS motor nucleus. BST conditioning volleys, which were
very effective in reducing the threshold of the Ia fibers, produced
no threshold changes of the VS fibers tested at the same site in
the motor nucleus. This finding is in keeping with the depression
of monosynaptic Ia EPSPs in motoneurons without changes in VS-EPSPs
produced by these same conditioning volleys (Fig. 6) and provides
further evidence on the specificity of the PAD pathways (see below).

The reduced effects of sensory nerve conditioning on the
excitability of RS and VS fibers can be explained by assuming that
sensory pathways are mostly effective in depolarizing the terminal
arborizations of afferent fibers but not so of descending fibers,
which may be instead depolarized more effectively by activation of
descending or propriospinal pathways (13,14,56). Alternatively, one
could also assume that descending fiber systems are not at all
subjected to a mechanism of presynaptic depolarization of the type
of PAD (46,97).

This question was approached assuming that the last order
interneurons in the pathways producing presynaptic depolarization
of fiber terminals can be directly activated by intraspinal stimulation (89,92,93,96; Jankowska, McCrea, Rudomín and Sykova, unpublished
observations). As shown in Fig. 9F, intraspinal stimulation applied
through the same micropipette as used for excitability testing, with
strengths below the resting threshold of the fiber whose excitability
was being measured, reduced the threshold of group I fibers ending
in the intermediate nucleus. This, together with the very short

latency of the excitability increase (0.7-0.8 msec) is in agreement
with the assumption that this effect results from the activation
of the last order interneuron in the PAD pathway (Jankowska, McCrea,
Rudomín and Sykova, unpublished observations). If any interneuronal
pathways depolarize RS fibers terminals, one could expect intraspinal stimulation to reduce the threshold of these fibers as well.
We have not found any example of intra-spinal conditioning stimulation that lowered the threshold of the RS fiber terminals to any
extent near that observed for afferent fiber terminals (93) (Fig. 9C,
local stimulation of 3 μA). Only when the conditioning stimulus
fired the fibers tested, was there an effect which consisted, however,
of a slight increase in threshold (Fig. 9C, local stimulus 10 μA),
presumably due to fiber after-hyperpolarization (35). Intra-spinal
stimulation in the motor nucleus was also unable to reduce the
threshold of VS fiber terminals although it was quite effective
in reducing the threshold of nearby Ia GS terminals (to about the
same extent as with BST conditioning).

The selectivity of intraspinal stimulation in depolarizing
afferent terminals, both in the intermediate and motor nuclei without
affecting nearby RS and VS fibers, suggests that the activated
pathways do not involve a massive release of potassium. Otherwise,
the RS and VS fibers ending in the same neighborhood would also be
depolarized since they were sensitive to iontophoretic injections of
this ion (Fig. 9B).

A possible explanation for the ineffectiveness of intraspinal
stimulation to change the excitability of RS terminals is that this
procedure was unable to excite existing elements that can evoke
presynaptic depolarization. If such hypothetical elements acted via
mechanisms similar to those proposed for PAD in afferent terminals
(axo-axonic synapses where GABA is supposed to be the transmitter),
one may expect GABA to hyperpolarize the RS terminals as it does
with the cell bodies of these neurons (1,83). This would imply that
the excitability increase of the RS terminals produced by sensory
nerve stimulation would not be due to activation of GABAergic pathways. Alternatively, it is possible that GABA depolarizes the RS
fiber nerve terminals. If so, this would be a very interesting
situation because of its implications on the mechanisms controlling
the location in the cell membrane of the systems of ionic transport.
The hyperpolarizing action of GABA on the CS and RS cell bodies
appears to result from movement of chloride ions towards the cell
interior, according to their electrochemical gradient (1, 31, 68, 83,
110). A depolarization produced by GABA would imply movement of
chloride in the opposite direction. This may be possible provided that
there are active transport mechanisms pumping chloride inside the
cell as in the spinal ganglion cells (27,39,44). In other words, if
GABA were to depolarize the RS fiber terminals this would imply the
existence of two chloride pumps, working in opposite directions,
depending on their spatial location either on the cell body or in

the axon terminals of the RS neurons (see also 81, p. 248).

To test the above possibility we have analyzed the effects of GABA iontophoretically applied to group I afferent fibers and to descending (VS and RS) fibers in the region of their terminal arborizations (93). In no case did we record any lowering of threshold for RS and VS fiber terminals by GABA applied with currents from 10 nA up to 400nA. By contrast, the threshold of all GS afferent fibers tested was considerably reduced by GABA with currents in the same range (Fig. 9E). This difference in the action of GABA was probably not at all related to the proximity of the iontophoretic barrels to the fibers tested, since the resting thresholds for antidromic activation of the RS and VS fibers were similarly low (1-2.4 μA). Moreover, in several cases we have found clear evidence of increase in threshold of RS and VS fibers by iontophoretic injections of GABA (Fig. 9A and B). The validity of these observations was systematically assessed by current controls (passing Cl⁻ from the GABA barrel and Na⁺ barrel as in Fig. 9A), and also by showing that potassium ejections from a neighboring barrel reduced the threshold of these fibers (Fig. 9B).

The reduction in excitability produced by GABA on VS and RS terminals appears to result from hyperpolarization of the fiber terminals, and not to inactivation of spike generation by excessive depolarization. Otherwise we should have seen an excitability increase with the smaller doses of GABA. The finding of hyperpolarization of the RS and VS fiber terminals by GABA is therefore consistent with the action of this amino acid on the cell somata of these cells where it is also hyperpolarizing (1,83). This, together with the finding of a depolarizing action of GABA on axon terminals and the cell body of afferent neurons (21,27,39,44,93), suggests that the membrane characteristics are similar in the terminals and cell bodies of these neurons. This may be important in relation to the recent generalization on the depolarizing action of GABA on unmyelinated axons in the vagus nerve (7), since this action would be expected to be different, at least in principle, in motor and sensory axons.

The analysis of the excitability changes of afferent and descending fiber terminals ending in overlapping regions in the spinal cord has provided a unique opportunity to test the specificity of the pathways and mechanisms producing presynaptic depolarization. Our observations indicate that there is a substantial difference in the mechanisms of presynaptic depolarization of fiber terminals in the spinal cord. Group I afferent terminal arborizations both in the intermediate and motor nuclei (but not RS and VS fiber terminals) appear to be the target of specific presynaptic depolarizing (presumably GABAergic) pathways. RS (but not VS) fiber terminals may also be depolarized by stimulation of sensory nerves, but such PD does not appear to result from the activation of GABAergic pathways.

Activation of pathways involving other transmitter substances (i.e.
glutamate, c.f. 78) seems unlikely because of the ineffectiveness of
intraspinal stimulation to reduce the threshold of the RS fiber
terminals. Rather, extracellular accumulation of potassium (65,66,74,
111), due to massive activation of neuronal elements may account for
the PD of RS terminals.

The possibility that RS terminals may be depolarized by increased
extracellular potassium implies that group I fibers ending in a
close vicinity would be equally depolarized. However, judging from
the threshold reduction of RS fiber terminals produced by sensory
nerve conditioning, we would expect the depolarization of the group
I afferent terminals due to the potassium component to be rather
small, except, perhaps, during convulsive activity where there is
a massive activation of interneurons (Engberg, Jiménez and Rudomín,
unpublished observations).

Depolarization of fiber terminals by GABA is probably associated
with a significant increase in ionic conductances (44) and perhaps
also with a reduction of the calcium currents involved in
transmitter release (30). This implies that the presynaptic
depolarization resulting from transmitter action may have a far more
significant influence on synaptic transmission than PD resulting
from potassium accumulation. This brings a word of caution to the
usual extrapolation that presynaptic depolarization leads to presynaptic inhibition, particularly in those cases where the presynaptic depolarization has been inferred only from changes in the
antidromic firing index fiber terminals, and there is no information
on the ultimate mechanisms producing such a PD, as it is the case
for the following fibers: the cortico-spinal fibers ending in the
dorsal horn (94), the pyramidal tract fiber terminals in the
trigeminal nucleus (28,29) the vagal and carotid sinus fibers ending
in the solitary tract nucleus (3,112), and the C fibers ending in the
dorsal horn (10,55).

Finally, some speculations on the physiological role of PAD are
probably relevant in the context of the present findings. We have
shown that the membrane potential of Ia fibers, but not that of VS
fibers synapsing with motoneurons, appears to be modulated by the
activity of specific sets of interneurons. Apparently these
interneurons affect many Ia fibers in a more or less synchronous
(correlated) manner (98), and this will be reflected in the activation
of motoneurons via Ia fibers, which is also correlated (91), as well
as in the information content transferred by the motoneuron ensemble
(95). The absence of interneuronal sets acting onto the VS fiber
terminals implies that the information content transferred by the
motoneuron ensemble will be different (i.e. uncorrelated) if
activation occurs via VS fibers. From the information point of view
the correlating system operating on the Ia fiber terminals introduces

redundancy into the line at the expense of changing the original
information, which may be useful for pattern recognition processes.
But there are other systems where the original information must be
preserved for survival functions. For example, afferent fibers in the
vagus, depressor and carotid sinus nerve do not appear to be
subjected to presynaptic control mechanisms (43,60,88,89; however,
see 3,112). These afferents convey information from arterial and
pulmonary baro- and chemorreceptors (84) and glucoreceptors (82)
among others, which is utilized for the control of the internal
environment. A mechanism of presynaptic modulation operating on
these sets of afferents would probably interfere seriously with
homeostasis by introducing undesired noise into the system. Perhaps
something similar occurs with the cortico-spinal, rubro-spinal and
vestibulospinal fiber systems where commands must be preserved
unaffected for correct motor performances which are centrally
programmed and depend to a very small extent on peripheral input
(102,113).

LONG LASTING CHANGES IN SYNAPTIC TRANSMISSION

Synaptic transmission in monosynaptic pathways may not be
modulated only by the specific mechanisms of presynaptic modulation
discussed above. Also, the activation patterns of the presynaptic
fibers may induce significant changes in their synaptic effectiveness,
some of which can be traced to a presynaptic mechanism. Perhaps the
most interesting aspect of these changes is their long duration.
They may last minutes or hours, instead of milliseconds, and have
been therefore taken as a model of short term memory processes. Among
these changes, two have received particular attention: those related
with the increased synaptic effectiveness following a short period
of high frequency activation, i.e. post-tetanic potentiation (73,106),
and those related with depression of the synaptic responses during
iterative activation (usually at low frequency) of that same pathway.
The depression of synaptic responses has several characteristics
resembling those of behavioural habituation. This has prompted a
fair amount of work addressed to the elucidation of the mechanisms
involved in this phenomenon, which is considered one of the
elementary modes of learning (38,61,62).

Farel et al (38) have studied in the isolated frog spinal cord
the behaviour of the monosynaptic responses of motoneurons produced
by stimulation of the lateral columns. These responses decrease
during repetitive stimulation. In agreement with what has been found
in other systems (see 61,62), the habituation is not related with
changes in the properties of the motoneurons, and has therefore been
ascribed to a presynaptic mechanism. A building up of presynaptic
inhibition seems to be excluded because descending fibers in the frog
spinal cord do not appear to be the target of specific depolarizing

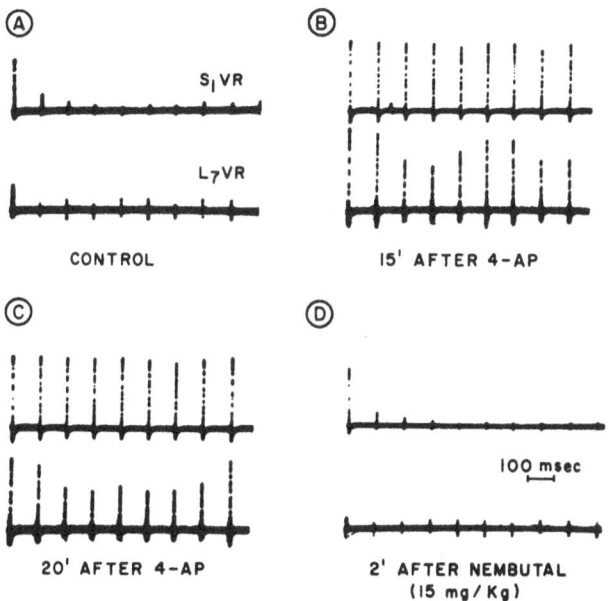

Fig. 10. Effects of 4-AP and pentobarbital on homosynaptic depression
of monosynaptic reflexes in the cat spinal cord. A: control monosynaptic reflexes produced by a train of pulses at 100 msec interval
applied to the GS afferent nerve at a strength maximal for group I
fibers. Monosynaptic reflexes were simultaneously recorded from the
S1 and L7 ventral root central ends, as indicated. B and C: 15 and
20 minutes after 0.5 mg/kg IV 4-AP, respectively. D: 2 minutes after
15 mg/kg of pentobarbital. Note restoration of homosynaptic depression.
From Rudomín, Madrid and González, unpublished observations.

pathways (46). Depression by repetitive stimulation is also seen in
the Ia fiber-motoneuron pathway in the spinal cord of the cat (11),
but this occurs at higher stimulation rates than in the frog spinal
cord (Figs. 10 and 11). It thus appears that in these monosynaptic
pathways there is a frequency limiting factor which results from the
unability of the system to regain its original conditions after
prior activation. In some cases, as in Aplysia, this frequency
behaviour can be controlled externally by activation of serotoninergic
pathways (62,64). This is not so clear in the frog and cat spinal
cord (see below).

4-Aminopyridine (4-AP) increases excitatory and inhibitory
synaptic transmission in the cat spinal cord (59). Monosynaptic Ia
EPSPs are increased after 4-AP without associated changes in the
motoneuron excitability, membrane potential, or antidromic action
potential. This has been taken as indication that 4-AP exerts its
facilitatory action presynaptically. However, the mechanism of this

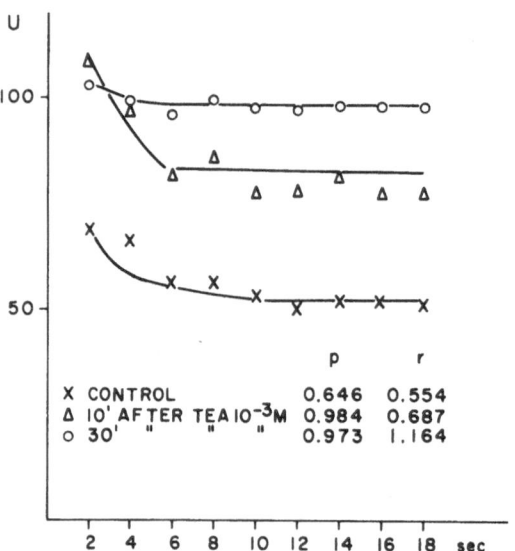

Fig. 11. Effects of 10^{-3}M TEA on lateral column-motoneuron monosynaptic responses in the isolated spinal cord of the frog. A train
of stimuli at 2 sec intervals was given to the lateral column at
the brachial plexus level and monosynaptic responses were recorded
from the Xth ipsilateral ventral root. Stimulus strength was set two
times higher than the intensity giving maximal responses in control
conditions. Note the decrement of the responses produced by successive
stimuli. After TEA 10^{-3}M the first monosynaptic response in the train
was increased and the decrement with successive pulses greatly
reduced. Curves are the best fittings of the data calculated using
the discontinuous model of Capek and Esplin (11; see text). The
calculated values of p and r are shown in the figure. From Galindo,
González and Rudomín, unpublished observations.

action is not completely clear. It is known that 4-AP decreases the
potassium permeability increase during the action potential (72,80,
85). This would prolong the action potential and increase the
transmitter release. There is also the suggestion that 4-AP may
increase directly the calcium currents in presynaptic terminals (75).
It was therefore interesting to investigate the effects of 4-AP on
the frequency behaviour of monosynaptic responses, both in the cat
and in the frog spinal cord. Fig. 10A shows the monosynaptic
reflexes produced by a train of stimuli applied at 100 msec intervals
to the GS afferent fibers in the cat. The size of the successive
reflex responses was not constant but declined with an exponential
time course. Fifteen minutes after 1mg/kg of 4-AP the reflex
responses were increased in size, as expected from our previous
findings (59), but in addition the <u>response decrement</u> practically

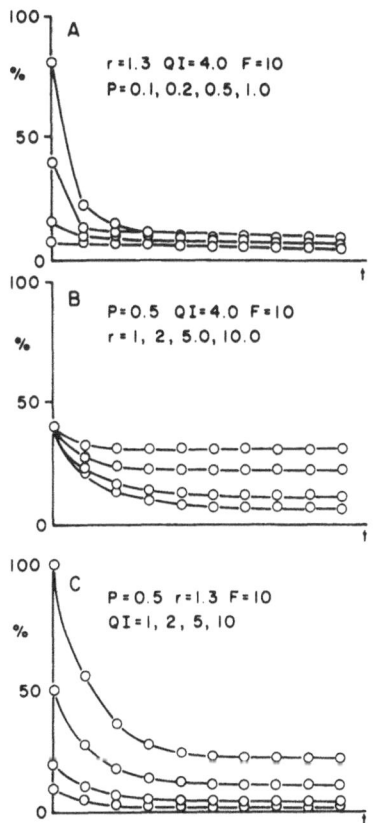

Fig. 12. Effects of p, r and Q on habituation calculated from the discontinuous release model of Capek and Esplin (11). A: effects of changing p (0.1, 0.2, 0.5 and 1.0) while r and Q_1 are kept constant (1.3 and 4.0, respectively). B: effects of changing r (1, 2, 5 and 10) while p and Q_1 are kept constant (0.5 and 4 respectively). C: effects of changing Q_1 (1, 2, 5 and 10) while p and r are kept constant (0.5 and 1.3, respectively). In all cases f was set to 10. Values obtained for each setting of p, r and Q are joined with a continuous line. See text for further explanation.

disappeared in one of the ventral roots and was atenuated in the other (Fig. 10B and C). This effect of 4-AP usually lasted for several hours and could be reversed by relatively small amounts of pento-barbital (Fig. 10D). 4-AP and TEA had a similar action on the mono-synaptic ventral root responses and corresponding focal potentials produced by stimulation of the lateral columns in the isolated spinal cord of the frog (Fig. 11).

The increase of the first monosynaptic response in the train after 4-AP can be well explained by an increased probability of transmitter release, as has been already documented in the neuro-muscular junction (75). The question is if this also explains the reduction (or abolition) of the response decrement, or if it is necessary to postulate a concomitant increase in the amount of transmitter available for release. In this context it will be useful to examine the problem using the two pool model of transmitter storage developed by Capek and Esplin (11). This model is based on the assumption that each incoming volley releases instantaneously a constant fraction (p) of the transmitter available for release (Q)

and that the store is being replenished during the time between the
successive impulses by a constant fraction (r) of the depleted part
of the store. With this model, the replenishment would be considered
as a first order process with a time constant 1/r. To have some idea
on the relative importance of r and p in the abolition of the
response decrement produced by 4-AP or by TEA we solved the equation
describing the discontinuous release model (see 11) for several values
of r, p and Q, using values in the range of those obtained
experimentally (see Fig. 11 and also ref. 11). As shown in Fig. 12A,
an increase in p keeping r and Q constant increases the size of the
monosynaptic responses but does not reduce the decrement of the
responses produced by successive stimuli. With this same model we
now increased r, keeping Q and p constant (Fig. 12B). This procedure
reduced the decrement of the reflex responses without affecting the
size of the first response. Finally, we kept p and r constant and
increased Q (Fig. 12C). This increased the reflex size but had small
effects on the response decrement. This result suggests that dishabituation produced by 4-AP may result, at least in part, from the
increased rate of replenishment of available transmitter. This is
an interesting possibility because it implies that increased
transmitter synthesis, increased transmitter mobilization within the
presynaptic terminals, or increased uptake of previously released
transmitter, may change the frequency behaviour of the system.

On the other hand, it is also possible that homosynaptic
depression results from action potential propagation failure at sites
of axon branching, even at such low stimulation frequencies. In
peripheral branching axons conduction blockade has been attributed
to increased extracellular K^+ concentration (52,53,105). 4-AP, by
blocking the K^+ channels could also prevent accumulation of this ion
in the extracellular space, and thus propagation failure. Obviously,
more work is needed to elucidate the mechanisms involved in homosynaptic depression. What is interesting to point out is the
possibility that in vertebrates, as in Aplysia (62), the frequency
behaviour of the pathway may be controlled by activation of specific
neuronal pathways.

CONCLUDING REMARKS

This presentation has been concerned with the analysis of some
of the mechanisms involved in the regulation of impulse transmission
in monosynaptic pathways in the vertebrate spinal cord. Particular
emphasis has been given to the presynaptic mechanisms of control.
This is because there are still many controversies about their mode
of operation and also because it is felt that these mechanisms are,
at least in principle, the basis of a highly specific mechanism for
controlling information transfer in monosynaptic pathways: with
presynaptic inhibition the responses of the second order cells can

be reduced to some inputs but not to others, whereas with post-synaptic inhibition the responses to any input may be equally affected. To further know how <u>specific</u> can be the mechanisms of pre-synaptic depolarization we investigated the effects of sensory nerve and intraspinal stimulation on the excitability of descending and of group I fibers ending in overlapping regions in the motor and intermediate nuclei of the spinal cord. The results obtained show that intraspinal stimulation activates elements which are able to depolarize afferent fibers but not nearby descending fibers. It seems very likely that these elements are not the afferent fibers them-selves, but interneurons interposed in the PAD pathway which synapse specifically with group I fiber terminals.

Further insigth on the functional meaning of presynaptic modulation of synaptic transmission may come from knowledge of fiber systems which are not subjected to a presynaptic control of the type of PAD. We have provided evidence that descending fiber systems such as the vestibulo-spinal and the rubro-spinal are not subjected to specific control mechanisms such as those operating on afferent fibers. Rubro-spinal (and also cortico-spinal) fibers may be depolarized by afferent conditioning stimulation, probably because of accumulation of potassium ions in the extracellular space, but this may not have important consequences on synaptic transmission, particularly during physiological activation of sensory fibers. Although the available information is rather limited, it is tempting to conclude that other descending fiber systems in the spinal cord, and perhaps also propriospinal systems, are devoid of specific mechanisms of presynaptic modulation, but this needs further investigation. Furthermore, within the afferent domain it is not yet fully established which fibers are and which are not subjected to specific PAD mechanisms. We certainly know that group Ia and Ib muscle fibers, as well as large cutaneous fibers, are (see 99,9). However, it is not clear whether these same mechanisms of presynaptic control operate on group II and III muscle afferents or on cutaneous unmyelinated fibers (10,55).

I would like now to further discuss some of the implications of the introduction of correlated fluctuations on afferent fiber terminals (91,98). Information theory states that the information transmitted by a given number of channels will be maximal if each channel transmits independent, that is, uncorrelated, information. Under those conditions the information transmitted through the channel ensemble will be equal to the sum of the information transmitted through each of the channels (90,95,100). The introduction of correlated noise into the system reduces the information transmitted through the ensemble, which will be now less than the sum of the information transmitted through each channel. It then follows that the interneurons in the pathways producing PAD of afferent fibers may be acting as information modulators by introducing variable

amounts of correlated noise into the system. As a consequence, the
information transmitted by the cell ensemble will be an <u>emergent</u>
property of that ensemble, that is, a property which arises from
the mutual interactions among the individual elements belonging to
that population (90,95). This implies that the emergent properties
pertain to the ensemble and may not be predicted from the analysis
of the properties of the individual elements belonging to that
population. Another implication of our findings is that the "degree
of belonging" of an individual element to a given population is not
time invariant but may change with the amount of correlated noise
introduced into the system. This "belonging" can be defined by
measuring the correlation of the individual behaviour relative to
that of the whole population (95). The fact that population behaviour
may not be predicted from the individual behaviour speaks for the
need of using new approaches to study regulation of information
transfer in neuronal networks. This is also necessary in order to
have a more complete understanding of the functional meaning of
homosynaptic depression. From a conceptual point of view the
implications of homosynaptic depression on behavioral habituation are
relatively simple to understand, particularly because the resulting
changes can be directly observed in the effectors. However, homosynaptic depression can also be viewed as a frequency behaviour
limiting mechanism in motor and sensory systems. It is possible that
in control systems such as the Ia fiber-alpha motoneuron loop this
prevents undesired oscillations of the system. Homosynaptic depression
could then serve the purpose of a stabilizing first order filter
inserted into the control loop. In other words, stability or
unstability in a given pathway may depend not only on the balance
between excitation and inhibition (either pre- or postsynaptic),
but also on its frequency following characteristics which appear to
be amenable of control In addition, we should also consider that
homosynaptic depression is a presynaptic event restricted to the
activated fiber terminals and depends on the particular history of
each individual fiber. This means that homosynaptic depression could
probably act as a mechanism opposing to the <u>short</u> <u>term</u> correlated
fluctuations of transmitter release imposed presynaptically to the
ensemble by the PAD pathways. On the other hand, the amount of homosynaptic depression in a given pathway also appears to depend on
metabolic factors which may be externally controlled by mechanisms
affecting many fibers at the same time, such as neurohormones or
neuromodulators. In that sense they could act as correlating
mechanisms, imposing their own (albeit slow) rhythms to the whole
ensemble. This could happen, at least in part, during the cyclic
changes in synaptic efficacy associated with sleep and wakefulness,
or during other kind of circadian rhythms (63).

Acknowledgements: The material presented in this work derives from
research done since 1967 with R.E. Burke, H. Dutton, I. Engberg, S.
Glusman, E. Jankowska, I. Jiménez, J. Madrid and R. Núñez whose

continuous enthusiasm, and patience during long-lasting experiments and their critical views have been encouraging throughout the years. Also I want to thank Mr. A. Rivera for his constant and skillful technical assistance, to G. González and J. Alvarado for the programming, to C. Rivera for drawings and photography and to Drs. O. Calvillo and S. Dueñas for their participation in some of the experiments. This work was supported in part by NIH grant NS 09196 and CONACyT grant 1634.

REFERENCES

1. Altmann, H., Steinberg, R., Ten Bruggencate, G. and Sonhof, U., Actions of amino acids applied iontophoretically in the red nucleus, Pflugers Arch. Ges. Physiol. 335 (1972) Suppl. Abstr. 132.
2. Banna, N.R. and Jabbur, S.J., The effects of depleting GABA on cuneate presynaptic inhibition, Brain Res. 33 (1971) 530-532.
3. Barillot, J.C., Depolarization présynaptique des fibres sensitives vagales et laryngées, J. Physiol. (Paris) 62 (1970) 273-294.
4. Barker, J.L. and Nicoll, R.A., Gamma-aminobutyric acid: role in primary afferent depolarization, Science 176 (1972) 1043-1045.
5. Barker, J.L. and Nicoll, R.A., The pharmacology and ionic dependency of amino acid responses in the frog spinal cord, J. Physiol. (Lond.) 228 (1973) 259-277.
6. Bell, J.A. and Anderson, E.G., The influence of semicarbazide induced depletion of γ-aminobutyric acid on presynaptic inhibition, Brain Res. 43 (1972) 161-169.
7. Brown, D.A. and Marsch, S., Axonal GABA-receptor in mammalian peripheral nerve trunks, Brain Res. 156 (1979) 187-191.
8. Barron, D.H. and Matthews, B.H.C., The Interpretation of potential changes in the spinal cord, J. Physiol. (Lond.) 85 (1939) 73-103.
9. Burke, R.E. and Rudomín, P., Spinal neurons and synapses. In Handbook of Physiology Section 1. The Nervous System. Vol. 1. Part 2 (Ed. E.R. Kandel), Am. Physiol. Soc., Bethesda, 1977, pp 877-944.
10. Calvillo, O., Primary afferent depolarization of C fibres in the spinal cord of the cat, Can. J. Physiol. Pharmacol. 56 (1978) 154-157.
11. Capek, R. and Esplin, B., Homosynaptic depression and transmitter turnover in spinal monosynaptic pathway, J. Neurophysiol. 40 (1977) 95-105.
12. Carlen, P.L. Yaari, Y. and Werman, R., Postsynaptic conductance increase associated with presynaptic inhibition in cat lumbar motoneurones, J. Physiol. (Lond.) 298 (1980) 539-556.
13. Carpenter, D.O. Engberg, I. and Lundberg, A., Primary afferent depolarization evoked from the brain stem and the cerebellum,

Arch. Ital. Biol. 104 (1966) 73–85.

14. Carpenter, D.O., Lundberg, A. and Norrsell, U., Primary afferent depolarization evoked from the sensory-motor cortex, Acta Physiol. Scand. 59 (1963) 126–142.

15. Carpenter, D.O. and Rudomín, P., The organization of primary afferent depolarization in the isolated spinal cord of the frog, J. Physiol. (Lond.) 229 (1973) 471–493.

16. Coehn, L.B. and de Weer, P., Structural and metabolic processes directly related to action potential propagation. In Handbook of Physiology. Section 1. The Nervous System, Vol. 1., Part 1. (Ed. E.R. Kandel), Am. Physiol. Soc., Bethesda, 1977, pp 137–159.

17. Conradi, S., On motoneuron synaptology in adult cats, Acta Physiol. Scand. (1969) Suppl. 332.

18. Cook, W.A. and Cangiano, A., Presynaptic and postsynaptic inhibition of spinal neurons, J. Neurophysiol. 35 (1972) 389–403.

19. Curtis, D.R., Duggan, A.W., Felix, D. and Johnston, G.A.R., Bicuculline, and antagonist of GABA and synaptic inhibitions in the spinal cord of the cat, Brain Res. 32 (1971) 69–96.

20. Curtis, D.R. and Johnston G.A.R., Amino acid transmitters. In Handbook of Neurochemistry (Ed. A. Lajtha), Plenum, New York 1970, Vol. 4; 115–135.

21. Curtis, D.R. and Lodge, D., GABA depolarization of spinal group 1 afferent terminals. In Iontophoresis and Transmitter Mechanisms in the Mammalian Central Nervous System (Eds. R.W. Ryall and J.S. Kelly), Elsevier, Amsterdam, 1978, pp 258–260.

22. Curtis, D.R., Lodge, D. and Headley, P.M., Electrical interactions between motoneurons and afferent terminals in cat spinal cord, J. Neurophysiol. 42 (1979) 635–641.

23. Davidoff, R.A., Gamma-aminobutyric acid antagonism and presynaptic inhibition in the frog spinal cord, Science 175 (1972) 331–333.

24. Davidoff, R.A., The effects of bicuculline on the isolated spinal cord of the frog, Exptl. Neurol. 35 (1972) 179–193.

25. Davidoff, R.A., Grayson, V. and Adair R., GABA-transaminase inhibitors and presynaptic inhibition in the amphibian spinal cord, Am. J. Physiol. 224 (1973) 1230–1234.

26. Decima, E.G. and Goldberg, L.J., Centrifugal dorsal root discharges induced by motoneurone activation, J. Physiol. (Lond.) 207 (1970) 103–118.

27. Deschenes, M. and Feltz, P., GABA-induced rise of extracellular potassium in rat dorsal root ganglia. An electrophysiological study in vivo, Brain Res. 118 (1976) 494–499.

28. Dubner, R. and Sessle, B.J., Presynaptic modification of corticofugal fibers participating in a feedback loop between Trigeminal brain-stem nuclei and sensory-motory cortex. In Oral-facial Sensory and Motor Mechanisms, Appleton-Century – Crofts, N.Y. 1971, pp 299–314.

29. Dubner, R. and Sessle, B.J., Presynaptic excitability changes of primary afferent and corticofugal fibers projecting to trigeminal brain stem nuclei, Exp. Neurol. 30 (1971) 223–238.

30. Dunlap, K. and Fischbach, G.D., Neurotransmitters decrease the calcium component of sensory neurone action potential, Nature (Lond.) 276 (1978) 837–839.

31. Dreifuss, J.J. Kelly, J.S. and Krnjevic, K., Cortical inhibition and ɣ-amino-butyric acid, Exp. Brain Res. 9 (1969) 137–154.

32. Eccles, J.C., The Physiology of Synapses. Academic Press, New York, 1964.

33. Eccles, J.C., Eccles, R.M. and Magni, F., Central inhibitory action attributable to presynaptic depolarization produced by muscle afferent volleys, J. Physiol. (Lond.) 159 (1961) 147–166.

34. Eccles, J.C., Magni F. and Willis, W.D., Depolarization of central terminals of group I afferent fibres from muscle, J. Physiol. (Lond.) 160 (1962) 62–93.

35. Eccles, J.C. and Krnjevic K., Presynaptic changes associated with posttetanic potentiation in the spinal cord, J. Physiol. (Lond.) 149 (1959) 274–287.

36. Eccles, J.C., Schmidt, R.F. and Willis, W.D., Pharmacological studies on presynaptic inhibition, J. Physiol. (Lond.) 168 (1963) 500–530.

37. Eide, E., Jurna, I. and Lundberg, A., Conductance measurements from motoneuron during presynaptic inhibition. In Structure and Function of Inhibitory Neuronal Mechanisms, C. Von Euler and V. Soderberg), Pergamon Press, Oxford, 1968, 215–219.

38. Farel, P.B., Glanzman, D.L. and Thompson, R.F., Habituation of a monosynaptic response in vertebrate central nervous system: Lateral column-motoneuron pathway in isolated frog spinal cord, J. Neurophysiol. 36 (1973), 1117–1130.

39. Feltz, P. and Raminsky, M., A model for the action of GABA in primary afferent terminals: Depolarizing effects of GABA applied iontophoretically to neurones of mammalian dorsal root ganglia, Neuropharmacol. 13 (1974) 553–563.

40. Frank, K., Basic mechanisms of synaptic transmission in the central nervous system, Inst. Radio Eng. Trans. Med. Electron. ME-6 (1959) 85–88.

41. Frank, K. and Fourtes, M.G.F., Presynaptic and postsynaptic inhibition of monosynaptic reflexes, Fed. Proc. 16 (1957) 39–40.

42. Galindo, J. and Rudomín, P., The effects of gallamine on field and dorsal root potentials produced by antidromic stimulation of motor fibres in the frog spinal cord, Exp. Brain Res. 32 (1978) 135–150.

43. Galindo, J., Ibarra, B. and Ayma, M.A., Possible mechanisms involved in the interaction between the fastigial nucleus and the carotid sinus nerve and their implications on cardiovascular function, Proc. Mexican Society of Physiol. Sciences XXIII Annual Meeting, 1980, p. 81.

44. Gallagher, J.P., Higashi, H. and Nishi, S., Characterization and ionic basis of GABA-induced depolarizations recorded in vitro from cat primary afferent neurones, J. Physiol. (Lond.) 275 (1978) 263–282.

45. Glusman, S., Correlation between the topographical distribution of (^{3}H) GABA uptake and primary afferent depolarization in the frog spinal cord, Brain Res. 88 (1975) 109–114.

46. Glusman, S. and Rudomín, P., Presynaptic modulation of synaptic effectiveness of afferent and ventrolateral tract fibers in the frog spinal cord. Exp. Neurol. 45 (1974) 474–490.

47. Glusman, S., Vázquez G. and Rudomín, P., Ultrastructural observations in the frog spinal cord in relation to the generation of primary afferent depolarization, Neuroscience Letters 2 (1976) 137–145.

48. Granit, D., Kellerth, J.O. and Williams, T.D., "Adjacent" and "remote" postsynaptic inhibition in motoneurones activated by muscle stretch, J. Physiol. (Lond.) 174 (1964) 453–472.

49. Gray, E.G., A morphological basis for presynaptic inhibition, Nature 193 (1962) 82–83.

50. Grillner, S. Hongo, T. and Lund, S., The vestibulospinal tract. Effects on alpha-motoneurones in the lumbo-sacral cord in the cat, Exp. Brain Res. 10 (1970) 94–120.

51. Grinnell, A.D., Electrical interaction between antidromically stimulated frog motoneurones and dorsal root afferents: enhancement by gallamine and TEA, J. Physiol. (Lond.) 210 (1970) 17–43.

52. Grossman, Y., Parnas, I. and Spira, M.E., Differential conduction block in branches of a bifurcating axon, J. Physiol. (Lond.) 295 (1979) 283–305.

53. Grossman, Y., Parnas, I. and Spira, M.E., Ionic mechanisms involved in differential conduction of action potentials at high frequency in a branching axon, J. Physiol. (Lond.) 295 (1979) 307–322.

54. Gutnik, M., Rudomín, P., Wall, P.D., and Werman, R., Is there electrical interaction between motoneurones and afferent fibers in the spinal cord?, Brain Res. 93 (1975) 507–510.

55. Hentall, I.D. and Fields, H.L., Segmental and descending influences on intraspinal thresholds of single C-fibers, J. Neurophysiol. 42 (1979) 1527–1537.

56. Hongo, T., Jankowska, E. and Lundberg, A., The rubrospinal tract. III Effects on primary afferent terminals, Exp. Brain Res. 15 (1972) 39–53.

57. Hongo, T., Jankowska, E. and Lundberg, A., The rubrospinal tract. IV. Effects on interneurones, Exp. Brain Res. 15 (1972) 54–78.

58. Howland, B., Lettvin, J.Y., McCulloch, W.S., Pitts, W. and Wall, P.D., Reflex inhibition by dorsal root interaction, J. Neurophysiol. 18 (1955) 1–7.

59. Jankowska, E., Lundberg, A., Rudomín, P. and Sykova, E. Effects of 4-aminopyridine on transmission in excitatory and inhibitory synapses in the spinal cord, Brain Res. 136 (1977) 387–392.

60. Jordan, D. and Spyer, K.M., Studies on the excitability of sinus nerve afferent terminals, J. Physiol. (Lond.) 297 (1979) 123–134.

61. Kandel, E.R., Neuronal plasticity and the modification of behavior. In Handbook of Physiology. Section 1. The Nervous

System. Vol. 1, Part 2. (Ed. E.R. Kandel), Am. Physiol. Soc. Bethesda, 1977, 1137-1182.

62. Kandel, E.R., Cellular insights into behavior and learning, The Harvey Lectures, Series 73 (1979) pp 19-92.

63. Kandel, E.R., Krasne, F.B., Strumwasser, F. and Truman J.W., Cellular mechanisms in the selection and modulation of behaviour, Neurosci. Res. Progr. Bull. 17 (1979) 523-710.

64. Klein, M. and Kandel, E.R., Presynaptic modulation of voltage-dependent Ca^{+2} current: mechanism for behavioral sensitization in Aplysia californica, Proc. Natl. Acad. Sci. U.S.A. 75 (1978) 3512-3516.

65. Kriz, N.E., Sykova, E., Ujec, E. and Vyklicky, L. Changes of extracellular potassium concentration induced by neuronal activity in the spinal cord of the cat, J. Physiol. (Lond.) 23 (1974) 1-15.

66. Kriz, N. Sykova, E. and Vyklicky, L., Extracellular potassium changes in the spinal cord of the cat and their relation to slow potentials, active transport and impulse transmission, J. Physiol. (Lond.) 249 (1975) 167-182.

67. Krnjevic, K. and Morris, E., Correlation between slow potentials and changes in extracellular K^+ concentration evoked by primary afferent activity, Fed. Proc. 32 (1973) 444.

68. Krnjevic, K. and Schwartz, S., The action of γ-aminobutyric acid on cortical neurones, Exp. Brain Res. 3 (1967) 320-336.

69. Kuno, M., Mechanism of facilitation and depression of the excitatory synaptic potential in spinal motoneurones, J. Physiol. (Lond.) 175 (1964) 100-112.

70. Levy, R.A. GABA: A direct depolarizing action at the mammalian afferent terminal, Brain Res. 76 (1974) 155-160.

71. Ljungdahl, A. and Hökfelt, T., Autoradiographic uptake patterns of (^{3}H) GABA and (^{3}H) glycine in central nervous tissues with special reference to the cat spinal cord, Brain Res. 62 (1973) 587-595.

72. Llinás, R., Walton, K. and Bohr, V., Synaptic transmission in squid giant synapse after potassium conductance blockage with external 3 and 4-aminopyridine, Biophys. J. 16 (1976) 83-86.

73. Lloyd, D.P.C., Post-tetanic potentiation of response in mono-synaptic reflex pathways of the spinal cord, J. Gen. Physiol. 33 (1949) 147-170.

74. Lothman, E.W. and Somjen, G.G., Extracellular potassium activity, intracellular and extracellular potential responses in the spinal cord, J. Physiol. (Lond.) 252 (1975) 115-136.

75. Lundh, H. and Thesleff, S., The mode of action of 4-aminopyridine and guanidine on transmitter release from motor nerve terminals, Eur. J. Pharm. 42 (1977) 411-412.

76. Madrid, J., Alvarado, J., Dutton, H. and Rudomín, P., A method for the dynamic continuous estimation of excitability changes of single fiber terminals in the central nervous system, Neuroscience Letters 11 (1979) 253-258.

77. Martin, A.F., Junctional transmission. II. Presynaptic mechanisms. In: Handbook of Physiology, Sect. 1 Vol.: The Nervous System:

The Cellular Biology of Neurons, (Ed. E.R. Kandel), Am. Physiol.
Soc., Washington, D.C., 1977, pp 329-335.

78. McLennan, H., Presynaptic inhibition in the cuneate nucleus. In
    Iontophoresis and Transmitter Mechanisms in the Mammalian Central
    Nervous System (Eds. R.W. Ryall and J.S. Kelly), Elsevier,
    Amsterdam, 1978, pp 270-272.

79. Miyata, Y. and Otsuka, M., Distribution of ʃ-aminobutyric acid in
    the cat spinal cord and the alteration produced by local ischemia,
    J. Neurochem. 19 (1972) 1833-1834.

80. Meves, H. and Pichon, Y., The effect of internal and external
    4-aminopyridine on the potassium current in intracellularly
    perfused squid giant axons, J. Physiol. (Lond.) 268 (1977) 511-
    532.

81. Nicoll, R.A. and Alger, B.E., Presynaptic inhibition: transmitter
    and ionic mechanisms, Internat. Rev. Neurobiol. 21 (1979) 217-
    258.

82. Niijima, A., Afferent impulse discharges from glucoreceptors in
    the liver of a guinea pig, Ann. N.Y. Acad. Sci. 157 (1969) 690-
    700.

83. Obata, K., Takeda, K. and Shinozaki, H., Further study of
    pharmacological properties of the cerebellar-induced inhibition
    of Deiters neurones, Exp. Brain Res. 11 (1970) 327-342.

84. Paintal, A.S. Afferent fibers, Ergebn. Physiol. 52 (1963) 74-156.

85. Pelhate, M. and Pichon, Y., Selective inhibition of potassium
    current in the giant axon of the cockroach, J. Physiol. (Lond.)
    242 (1974) 90-91p.

86. Rethelyi, M., Ultrastructural synaptology of Clarkes column, Exp.
    Brain Res. 11 (1970) 159-174.

87. Roberts, E. and Kuriyama, K., Biochemical-physiological correla-
    tions in studies of the ʃ-aminobutyric acid system, Brain Res. 8
    (1968) 1-35.

88. Rudomín, P., Presynaptic inhibition induced by vagal afferent
    volleys, J. Neurophysiol. 30 (1967) 964-981.

89. Rudomín, P., Excitability changes of superior laryngeal, vagal
    and depressor afferent terminals produced by stimulation of the
    solitary tract nucleus, Exp. Brain Res. 6 (1968) 156-170.

90. Rudomín, P., Information processing at synapses in the vertebrate
    spinal cord: Presynaptic control of information transfer in
    monosynaptic pathways. In: Information Processing in the Nervous
    System (Ed. H.M. Pinsker and W.D. Willis), Raven Press, N.Y.,1980.

91. Rudomín, P., Burke, R.E., Núñez, R., Madrid, J. and Dutton, H.,
    Control by presynaptic correlation: a mechanism affecting
    information transmission form Ia fibers to motoneurons, J.
    Neurophysiol. 38 (1975) 267-284.

92. Rudomín, P., Dueñas, S., Jiménez, I. and Jankowska, E., Localiza-
    tion of the last order interneuron in the pathway producing
    depolarization of Ia fiber terminals in the lumbo-sacral spinal
    cord of the cat, Society for Neuroscience Abstracts, 10th Annual
    Meeting, 1980, p. 436.

93. Rudomín, P., Engberg, I., Jankowska, E. and Jiménez, I., Evidence of two different mechanisms involved in the generation of presynaptic depolarization of afferent and rubrospinal fibers in the cat spinal cord, Brain Res. 189 (1980) 256-261.

94. Rudomín, P., Jankowska, E. and Madrid, J., Presynaptic depolarization of cortico-spinal and rubro-spinal terminal arborizations produced by conditioning volleys to cutaneous nerves, Society for Neuroscience Abstracts, 8th Annual Meeting 1978, p. 571.

95. Rudomín, P. and Madrid, J., Changes in correlation between monosynaptic responses of single motoneurons and in information transmission produced by conditioning volleys to cutaneous nerves, J. Neurophysiol. 35 (1972) 44-64.

96. Rudomín, P. and Muñoz-Martínez, J., A tetrodotoxin-resistant primary afferent depolarization, Exp. Neurol. 25 (1969) 106-115.

97. Rudomín, P., Nuñez, R. and Madrid, J., Modulation of synaptic effectiveness of Ia and descending fibers in cat spinal cord, J. Neurophysiol. 38 (1975) 1181-1195.

98. Rudomín, P., Nuñez, R., Madrid, J. and Burke, R.E., Primary afferent hyperpolarization and presynaptic facilitation of Ia afferent terminals induced by large cutaneous fibers, J. Neurophysiol. 37 (1974) 413-429.

99. Schmidt, R.F. Presynaptic inhibition in the vertebrate central nervous system, Ergeb. Physiol. 63 (1971) 20-101.

100. Shannon, C. and Weaver, W., The Mathematical Theory of Communication, University of Illinois Press, Urbana, Ill, 1949.

101. Shapovalov, A.I. and Shiriaev, B.I., Recurrent interactions between individual motoneurones and dorsal root fibres in the frog, Exp. Brain Res. 38 (1980) 115-116.

102. Shik, M.L. and Orlovsky, G.N., Neurophysiology of locomotor automatism, Physiol. Rev. 56 (1976) 465-501.

103. Singer, W. and Lux, H.D., Presynaptic depolarization and extracellular potassium in the cat geniculate nucleus, Brain Res. 64 (1973) 17-33.

104. Skibo, G.G. Neuronal organization of the vental horn's medial part in the cats spinal cord (In Russian), Neurophysiol. 4 (1972) 183-190.

105. Smith, D.O., Mechanisms of action potential propagation failure at sites of axon branching in the crayfish, J. Physiol. (Lond.) 301 (1980) 243-259.

106. Spencer, W.A. and April, R.S., Plastic properties of monosynaptic pathways in mammals. In Short-term Changes in Neural Activity and Behaviour (Ed. G. Horn and R.A. Hinde), Cambridge Univ. Press, Cambridge, 1970, pp 433-474.

107. Takeuchi, A., Junctional transmission. I. Postsynaptic mechanisms. In Handbook of Physiology. Sect. I. Vol. 1. The Nervous System: The Cellular Biology of Neurons (Ed. E.R. Kandel), Am. Physiol. Soc., Washington, D.C. 1977, pp. 295-327.

108. Tebecis, A.K. and Phillis, J.W., The use of convulsants in studying possible functions of amino-acids in the toad spinal cord, Comp. Biochem. Physiol. 28 (1969) 1303-1315.

109. Tebecis, A.K. and Phillis, J.W., The pharmacology of the isolated toad spinal cord. In Experiments in Physiology and Biochemistry (Ed. C.A. Kerkut), Academic Press, 1969, Vol. 2, pp 361-395.

110. Ten Bruggencate, G. and Engberg, I., Iontophoretic studies in Deiter's nucleus of the inhibitory action of GABA and related aminoacids and the interactions of strychnine and picrotoxin, Brain Res. 25 (1971) 431-448.

111. Ten Bruggencate, G., Lux, H.D. and Liebl., L., Possible relationship between extracellular potassium activity and presynaptic inhibition in the spinal cord of the cat, Eur. J. Physiol. 349 (1974) 301-317.

112. Weiss, G.K. and Crill, W.E., Carotid sinus nerve primary afferent depolarization evoked by hypothalamic stimulation, Brain Res. 16 (1969) 269-272.

113. Wetzel, M.C. and Stuart, D.G., Ensemble characteristics of cat locomotion and its neural control, Progr. Neurobiol. 7 (1976) 1-98.

114. Wilson, V.J. and Yoshida, M., Comparison of effects of stimulation of Deiters nucleus and medial longitudinal fasciculus on neck, forelimb and hindlimb motoneurons, J. Neurophysiol. 32 (1969) 743-758.

115. Wood, J.G., McLaughlin, B.J. and Vaughn J.E. Immunocytochemical localization of GAD in electron microscopic preparations of rodent CNS. In GABA in Nervous System Function (Eds. E. Roberts, T.N. Chase and D.B. Tower), Raven Press, N.Y., 1976, pp 133-148.

SYNAPTIC CHANGES INDUCED BY OPTIC CHIASM LOW INTENSITY

REPETITIVE ELECTRICAL STIMULATION (THE KINDLING EFFECT)

A. Fernández-Guardiola, M. Condés-Lara, and J.M. Calvo

Unidad de Investigaciones Cerebrales, Instituto Nacional
de Neurología y Neurocirugía, México 22, D.F. and Facultad
de Psicología, Universidad Nacional Autónoma de México

INTRODUCTION

The kindling effect has been defined as a relatively permanent
alteration in brain function which results from repeated electrical
stimulation and culminates in the appearance of electrographic and
behavioral convulsions (9). Kindling has been considered as a model
for learning, long term memory, and plasticity (16). Many brain
regions are unresponsive to the kindling effect, specifically the
majority of the neocortex, the thalamus and the brain stem. The most
responsive regions are parts of the limbic system. As Goddard et al.
(9) pointed out, kindling can be produced by electrically stimulating
bundles such as the stria terminalis, the fornix, the fimbria, the
corpus callosum and the anterior commissure. Recently, Cain (4)
observed that the repeated stimulation of visual and auditory thalamic
relay nuclei induced partially or fully generalized seizures in some
animals. Later Cain (5) compared the kindling response of hippocampus
and dentate gyrus with the kindling response of lateral and medial
geniculate nuclei. While between 31% and 44% of the animals with
thalamic stimulation developed generalized convulsions, 100% of the
animals displayed seizures with hippocampal kindling.

Previous studies on kindling have been related to structures
which produce convulsions, while kindled structures which do not
develop them are rarely considered (9). We propose that the
mechanisms of kindling need to be extended to encompass permanent
changes in the electrical activity of the stimulated area and
functionally related structures which do not necessarily culminate
with generalized convulsions. Such changes may involve an increase
of inhibitory action or alterations in sensory processing.

We have explored the kindling of non limbic structures (8).
Daily electrical stimulation of raphe nucleus shortened the
first REM latency of sleep, and long periods of atonic and phasic
events (PGO) were observed during slow wave sleep stages. The
animals never exhibited tonic clonic generalized seizures, even
after 200 days of daily stimulation. The raphe nucleus showed
afterdischarges, and the visual cortex (VC) displayed noticeable
interstimulus spiking while the motor cortex did not.

This led us to look for a model with three main features.
First, it should be an anatomically well known system; second, the
kindling effect should be feasible; third, it should be certain that
only fibers have been stimulated. One such a model is the visual
pathway in which the stimulating electrodes are placed in the optic
chiasm.

EXPERIMENTAL PROCEDURES

Fifteen adult cats (2.5-3 kg) curarized and artificially
ventilated were used. Local anesthesia (xylocaine 2%) was infused
at pressure and incision points, including the mandibular nerve
near the tympanic bulla (to insure analgesia of the auricular zone),
the infraorbital nerve and the tissue covering the external auditory
canals. Body temperature was maintained at 37-38°C. Under ether
anesthesia, concentric bipolar stainless steel electrodes were placed
in the optic chiasm (planes 12.5 A; 2.0 R; -6.0 H) and in both lateral
geniculate bodies (planes 6.5 A; 9.5 LR; +3.5 H) following coordinates
of the Snider and Niemer atlas (18). Both visual cortices were
recorded through implanted epidural bone nails. The electrode
placement was confirmed by recording optimal visual averaged evoked
responses (EP) to a series of 8 flashes. The electroretinogram (ERG)
was also recorded in five experiments. The electrical activity was
amplified, integrated in a polygraph, simultaneously taped, and on-
line analyzed for transient signal averaging. The EKG was
continuously monitored tachographically.

Two hours were allowed for the ether effects to disappear.
The pupils were atropinized (10% local instillated atropine
sulfate). Flash control trials with 5 different intensities
(1,2,4,8,16) from a Grass photostimulator were delivered at 0.2 Hz,
and the photic evoked responses of the optic chiasm (OCh), lateral
geniculated body (LGB) and visual cortices (VC) were averaged. The
kindling procedure started 30 minutes later. The optic chiasm
was then stimulated (200 microamps, 100 Hz, 2 msec, during 2 sec)
every 20 or 30 minutes (K_n) until 18 trials were completed. Twelve
seconds after each chiasmatic stimulation, a testing train of 32
flashes (at 0.2 Hz at the lowest intensity, x1) was applied and
the visual responses from the recorded structures were analyzed.

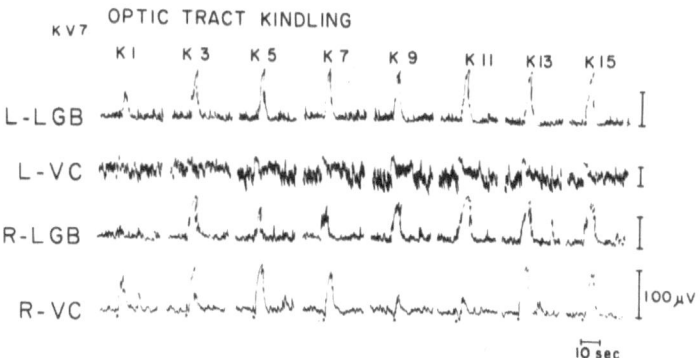

Fig. 1.   Integrated ERG responses to the electrical stimulation
of the right side of the optic chiasm.   The stimulus interval was
20 minutes.   Notice the progressive enhancement in both LGB and
transsynaptic changes in R-VC.   L-LGB, left lateral geniculate
body; 1-VC, left visual cortex; R-LGB, right lateral geniculate
body; R-VC, right visual cortex. Kn, kindling trials.

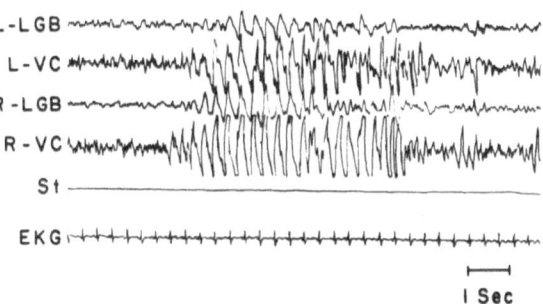

Fig. 2.   Spontaneous paroxystic discharge in visual established
kindling.   Abbreviatures in this and the following figures, as for
Fig. 1.

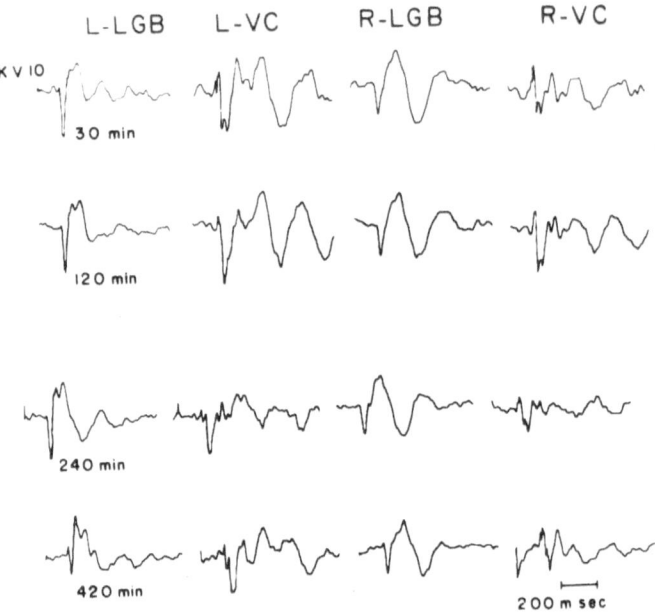

Fig. 3. Photic evoked responses along the visual pathway. Each potential is the average of 32 EPs. In this animal the optic chiasm was not electrically stimulated. Only the visual photic responses were recorded every 30 min. Notice the absence of changes or the slight decrease in amplitude through this control experiment.

Twenty minutes after the last kindling trial, the increasing flash intensity control test was repeated.

The direct responses to the electrical chiasmatic stimulation were recorded and amplitude-frequency integrated. At the end of the experiments, single pulses (0.1 msec., 200 microamps.) were applied to the OCh and both LGB and VC evoked responses were compared in order to detect any lateralization. Histological control was performed by the method of Guzman et al. (11).

EXPERIMENTAL OBSERVATIONS

The electrical chiasmatic stimulation provoked ipsilateral and contralateral responses along the visual pathway. As the stimuli were applied in the right lateral portion of the chiasm (uncrossed fibers), both the synaptic (LGB) and transsynaptic (VC) responses were prominent in the ipsilateral side. These responses progressively increased in amplitude with repetitive

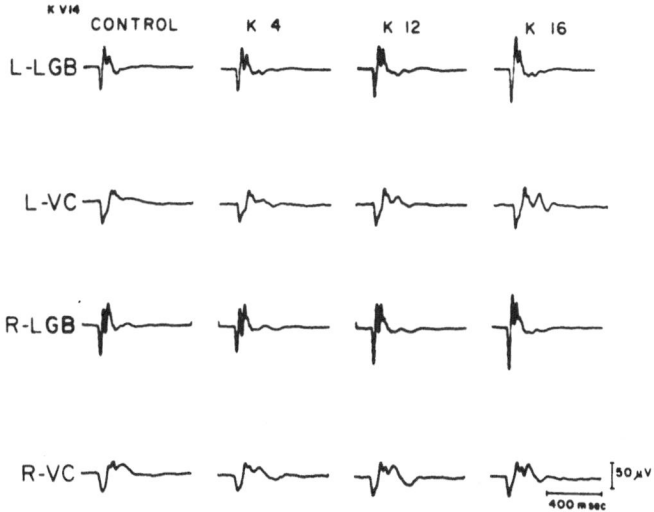

Fig. 4. The kindling effect in the visual pathway. Optic chiasm was stimulated every 20 min and 32 test photic stimuli were delivered after each chiasmatic stimulation. K_4, K_{12}, K_{16}, kindling trials. Notice the enhancement of EPs amplitude in both lateral geniculate bodies.

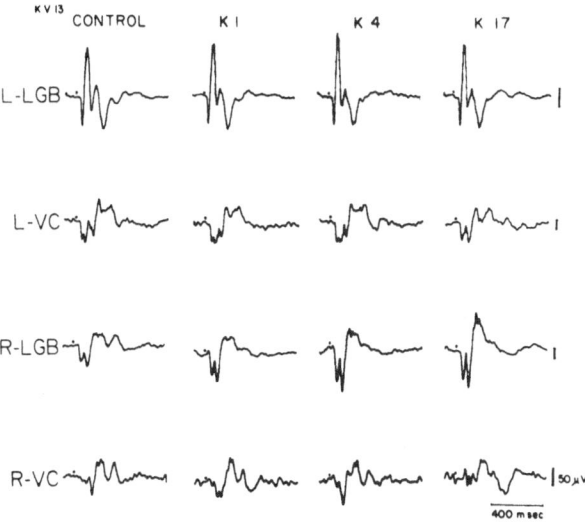

Fig. 5. The kindling effect in the visual pathway. Similar experiment to Fig. 4 showing a noticeable increment in R-LGB, EP amplitude.

electrical stimulation, VC showing a greater variability than LGB (Fig. 1).

The late stages of the kindling procedure (K_5 to K_{18}) were characterized by spontaneous and photically induced paroxystic activity (polyspikes and spike and waves). These bursts of paroxystic activity started in the R-VC (ipsilateral) and then propagated to the contralateral side (Fig. 2).

The kindling evolution was also tested through the changes in visual evoked potentials (EP). For this purpose, control experiments were performed delivering 32 flashes every 20 minutes without optic chiasm electrical stimulation. Figure 3 illustrates these control EP's up to 7 hours, showing almost no changes or a slight amplitude decrement.

When repetitive chiasm stimulation preceded the visual stimuli by 12 seconds, a conspicuous progressive EP's amplitude increase occurred. This enhancement affected the presynaptic as well as the postsynaptic component of the responses of both LGB's. The EPs of VC displayed less (N_1) presynaptic changes, but more evident and long lasting postsynaptic and late components alterations (Figs. 4 and 5).

In the contralateral structures (left side) the EP's underwent slight reduction in the first trials affecting both N_1 and P_1 components. After the 4th or 5th trial, both components increased significantly in LGB ($P < 0.001$) and in VC ($P < 0.01$). In the ipsilateral side the initial depression was not observed and the regression lines were steeper for N_1 and $N_1 - P_1$ amplitudes measured in LGB. In the ipsilateral VC the EPs were augmented exclusively in the postsynaptic component. The presynaptic R-VC component decrease was not significant (Fig. 6).

Optic chiasm and retinal responses (ERG) to the visual stimuli test underwent a striking facilitation following the 4th or 5th kindling trials. The responses were augmented by about 300% (Fig. 7). This enhancement of the EP along the visual pathway from the retina to the visual cortex persisted at least one hour after the last kindling trial.

As described in Experimental Procedures, series of increasing flash intensities were applied before and after the kindling procedure. Figure 8 illustrates the thalamic and cortical EPs facilitation, being the differences between control and kindled responses more important for the lower flash intensities. Similar results for optic chiasm responses and ERG are shown in Figures 9 and 10.

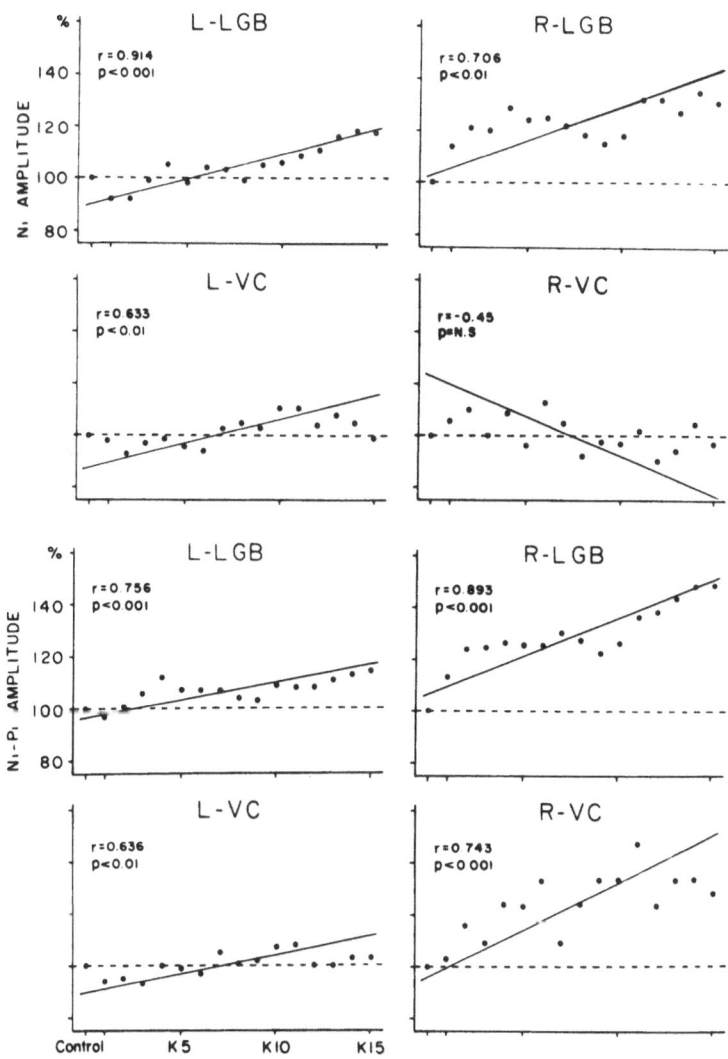

Fig. 6. Graphs of EP amplitude vs kindling trials. The upper four graphs show the amplitude of N_1 component changes and the four lower, the peak to peak amplitude values (measured from N_1 to P_1). Ordinates: values in percent of the EPs amplitude (control considered as 100%). Abscissae: the kindling trials. The regression lines show the most significant changes in R–LGB and also in R–VC for the N_1–P_1 amplitude. Notice the slight decrement of N_1 amplitude on the R–VC. Each point represents the average of 192 visual EPs from six cats.

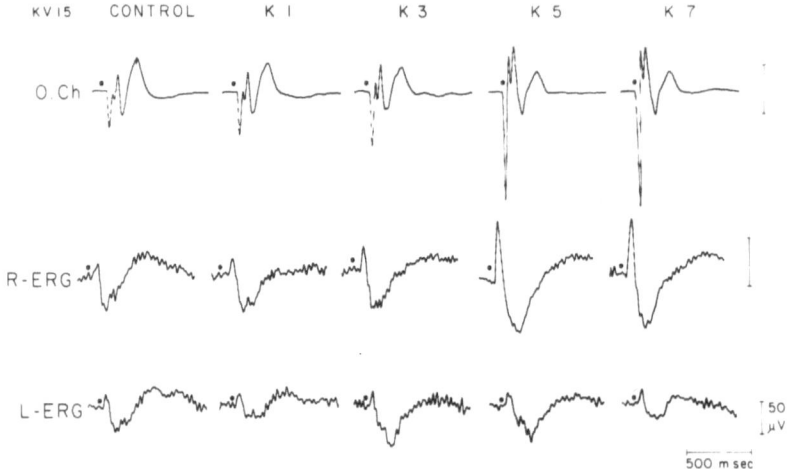

Fig. 7. Optic chiasm recording and electroretinogram (ERG) from
both sides during photic test stimulation after each kindling trial.
These potentials represent the average of 32 EPs. The chiasmatic
EP shows an important increase in both N_1 and N_1-P_1 amplitude. The
right ERG (R-ERG) shows a progressive increment in wave "b" and the
appearance of wave "a" most noticeable between the K5 and K7. Notice
that these changes are ipsilateral (right side) to the electrical
stimulation, and there is no changes in left ERG (L-ERG).

DISCUSSION

It has been reported (12) that kindling can take place in
paralyzed animals with curarizing agents or with a spinal cord
section, as long as the electrical stimulation induces enhancement
in the frequency and duration of the afterdischarges (AD). Goddard
et al. (9), on the other hand, have reported that the interval
between stimulations through the kindling process is optimal within
20 min as the minimum and 7 days as maximum. Intervals shorter
than that provoke adaptation before convulsions can develop. Thus
we can be certain that under the experimental conditions we used
in our studies, kindling is possible.

Kindling studies have dealt mainly with stimulation of
cellular bodies, specially of limbic structures such as the
amygdala. However, Goddard et al. reported that convulsions were
produced by stimulating the white matter (stria terminalis,
anterior commisure). Similar results were observed by Bliss and
Lomo (1) from the perforant path stimulation. Other investigators
have previously reported kindling produced by the stimulation of
sensory structures. Relevant to our experiments is Cain's work (5)

in which he produced convulsions by the stimulation of diverse
auditory and visual thalamic relays.

The optic track or chiasm electrical stimulation described in
the present paper offers the advantage that it specifically
activates a large number of fibers well known as to their origin
(ganlionar cells) and termination (LGB glomeruli). Moreover, we
think that the activation of a synpatic group through an afferent
pathway allows for a better analysis of the synaptic effects as
opposed to those obtained by a stimulation of a dense neuronal
group.

Our experimental model provides a good analysis of the
synaptic transmission in a visual thalamic relay under
physiological conditions, as well as the cortical transsynaptic
changes. Since the animals were curarized, they could not
manifest motor convulsive activity. However, the electrical
recordings of both cortices revealed the progressive installation

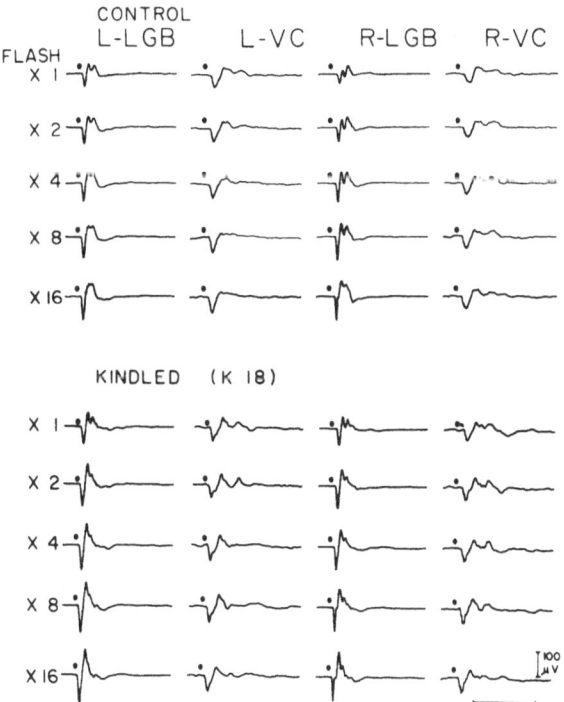

Fig. 8. Increasing flash intensity effects along the visual
pathway in control and after kindling ($K18$). Both LGB show
amplitude enhancements in pre and postsynaptic components. In the
visual cortices the changes are predominantely postsynaptic.

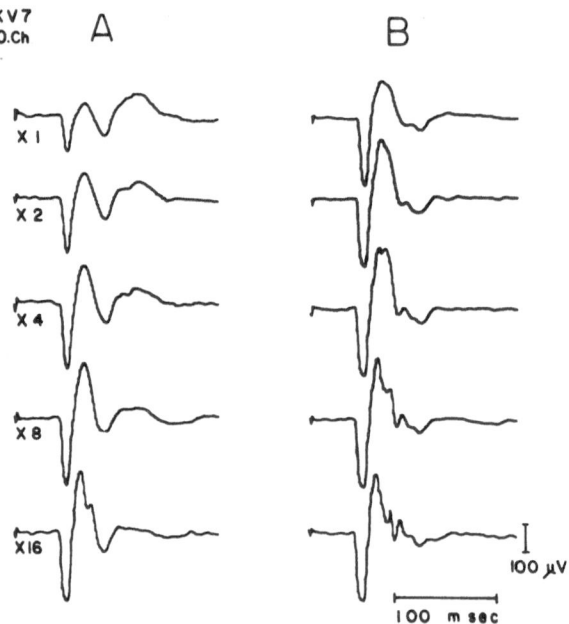

Fig. 9. Optic chiasm visual responses to increasing flash
intensities in control (A) and after kindling (B). Late EPs
components for the higher flash intensities appear and peak to peak
amplitude enhancements for the lower intensities occur as a result
of kindling (K15).

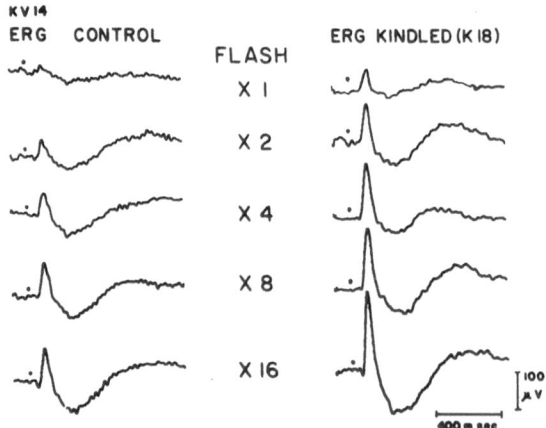

Fig. 10. Effect of increasing flash intensities in control
situation (visual test stimuli alone) and after the kindling
procedure (K18). All flash intensities provoked augmented
responses as a result of kindling. Notice the appearance in the
ERG of wave "a". After kindling lower intensities are needed for
its elicitation.

of paroxystic discharges. These were triggered at first by the
test photic stimuli but later they appeared spontaneously.

It has been known that the evoked sensory potentials (EP)
may be modified by the presence of convulsive activity. Fernández-
Guardiola et al (7) found increased EPs from the specific visual
path and mesencephalic reticular formation (MRF) under a
metrazolic activation. EPs are also modified by kindling. Racine
(15) and Bliss and Lomo (1) demonstrated an increase in the EPs
recorded at a secondary focus, elicited by a test-stimulus applied
to the primary kindled area.

Our results confirmed the above observations, since we found
a progressive increase in the visual EPs, although in our
experiments the test-pulses were not electrical but photic, thus
being a more physiological activation of the visual receptor. We
can conclude that these changes were due to progressive activation
of the optic chiasm since the control photic test stimuli evoked
constant responses which even tended to decrease at the end of the
test (Fig. 3).

The EPs increase were both in pre and postsynaptic
components, being maximal at the ipsilateral structures. Ipsilateral
LGB was expected to receive a bigger volley of impulses than the
contralateral since we were stimulating the lateral portion of the
tract which corresponds to 30% of the direct uncrossed fibers (14).
Nevertheless, a direct stimulation of crossed fibers as responsible
for the contralateral effects cannot be discarded.

The sequence of events in the EPs was as follows: initially
an increase in the presynaptic components (N_1) of the LGB-EPs was
observed (approximately at minute 90), and then the EPs of both
visual cortices increased, most prominently in the late and
postsynaptic components. This indicates that LGB constitutes the
primary and VC the secondary kindled foci, not taking into account
the striking increase in the ganglionic cells components of the
ERG. The antidromic activation of the retina and the induced long
lasting facilitation may account for an increased excitability
primarily of the retina and secondarily of the LGB. This structure
is thus activated twice, first by the direct stimulation
(orthodromic) and second by the activation of ganglion cells
stimulated antidromically. This intense activation makes the LGB
constitute a kindled secondary site sui-generis. The retina
effects viewed as antidromic could also be due to the activation
of centrifugal fibers to the retina (3,10). Odgen (13) described
an intraretinal slow potential (P wave) evoked by optic tract or
lateral geniculate stimulation in primates. He suggested that the
P wave is postsynaptic in nature and could result from the
activation of amacrine cells. It may also represent extracellularly summed IPSPs. There is still a controversy which will

be partially solved by the elucidation of the participation of the
afferent or efferent fibers in $\underline{P}$ wave generation. Centrifugal
fibers have only been demonstrated in birds but they have been
suggested in other species (2). The EP evoked by the optic chiasm
stimulation recorded in the retina is in essence inhibitory, but
the increase we found in our ERG potential could be the result of
a facilitation provoked by the antidromic inhibition of the tonic
centripetal inhibitory discharge in the retina. Similar processes
have been described by Chang (6). The transynaptic cortical effects
were manifested by the increase in the postsynaptic and late
components of the EPs, as in other kindling models (17).

    We conclude that this model provides a good means for
studying the excitability of the receptors, the thalamic relays
and the cortical areas during kindling development.

Acknowledgements. This work was supported in part by Grants
03/03/80 and 03/04/80 from the Instituto Mexicano de Psiquiatría.
We wish to thank Miss T. González-Estrada and Mr. R. Budelli for
correcting the manuscript.

REFERENCES

1. Bliss, T.V.P., and Lømo, T., Long lasting potentiation of
    synaptic transmission in the dentate area of the anaesthetized
    rabbit following stimulation of the perforant path, J. Physiol.
    (Lond.), 232 (1973) 311-356.
2. Branston, N.M. and Fleming, D.G., Efferent fibers in the frog
    optic nerve, Expl. Neurol., 20 (1968) 611-623.
3. Cajal, R.S., Histologie du Systéme Nerveux de l'Homme et
    des Vertébrés. Maloine, París. 1911.
4. Cain, D.P., Seizures development following repeated electrical
    stimulation of central olfatory structures, Ann. N.Y. Acad.
    Sci., 290 (1977) 200-216.
5. Cain, D.P., Kindling in sensory systems: Thalamus, Exp. Neurol.,
    66 (1979) 319-329.
6. Chang, H.T., Cortical responses to stimulation of lateral
    geniculated body and the potentiation thereof by continuous
    illumination of the retina, J. Neurophysiol., 15 (1952) 5-26.
7. Fernández-Guardiola, A., Roldan, E., and Guzmán, C., Activación
    por metrazol de los potenciales evocados en las vías sensoriales
    específicas e inespecíficas. Bol. Inst. Est. Méd. Biol. (Méx.),
    15 (1957) 37-47.
8. Fernández-Guardiola, A., Jurado, J.L., and Calvo, J.M. Repetitive
    low intensity electrical stimulation of cat's non limbic brain
    structures: dorsal raphe nucleus. In: J. Wada (Ed.), 2nd
    Kindling Symposium, Raven Press, New York, (1980) (In Press).
9. Goddard, G.V., Mc Intyre, C.D. and Leech, C.K., A permanent
    change in brain function resulting from daily electrical

stimulation, Exp. Neurol., 25 (1969) 295-330.
10. Granit, R. Centrifugal and antidromic effects of ganglion cells of retina. J. Neurophysiol., 18 (1955) 388-411.
11. Guzmán, F.C., Alcaraz, M., and Fernández-Guardiola, A. Rapid procedure to localize electrodes in experimental neuro-physiology, Bol. Inst. Estud. Méd. Biol. (Méx.), 16 (1958) 29-31.
12. Morell, F., Tsuru, N., Hoeppner, T.J., Morgan, D. and Harrison, W.H. Secondary epileptogenesis in frog forebrain: effect of inhibition of protein synthesis. In: J. Wada (Ed.) Kindling. Raven Press, New York, (1976) pp. 41-61.
13. Ogden, T.E., Intraretinal slow potentials evoked by brain stimulation in the primate, J. Neurophysiol., 29 (1966) 898-909.
14. Polyak, S.L., The Vertebrate Visual System, Univ. Chicago Press, Chicago, 1957.
15. Racine, R.J., Modification of seizure activity by electrical stimulation: I. Afterdischarge threshold, Electroenceph. clin. Neurophysiol., 32 (1972) 281-294.
16. Racine, R., Gartner, J., and Burnham, W., Epileptiform activity and neurol plasticity in limbic structures, Brain Res., 47 (1972) 262-268.
17. Racine, R. and Zaide, J., A further investigation into the mechanisms underlying the kindling phenomenon. In: K.E. Livingston and O. Kornykrewicz (Eds) Limbic Mechanisms, Plenum, New York (1978) pp. 457-493.
18. Snider, R.S. and Niemer, W.T., A Stereotaxic Atlas of the Cat Brain. Univ. Chicago Press, Chicago, 1961.

CALCIUM CURRENT MODULATION AS A MECHANISM IN THE SYNAPTIC

PLASTICITY UNDERLYING HABITUATION AND SENSITIZATION IN APLYSIA

Mark Klein

Department of Physiology and Division of Neurobiology
and Behavior, College of Physicians & Surgeons
Columbia University, New York, N.Y. 10032

INTRODUCTION

Habituation and sensitization are two simple forms of learning which occur widely throughout the animal kingdom (26). Habituation is the decline in reflex responsiveness which occurs when a stimulus is presented repeatedly, while sensitization is the increase in responsiveness which results from the transient presentation of a novel stimulus. Although this definition does not exclude the possibility that what happens in sensitization might simply be the opposite of what happens during habituation, earlier work has shown that sensitization is in actuality a distinct process, superimposable on habituation and having its own discernible properties and time course (3,9). In this report, I present experiments which address the question of the cellular mechanisms of both habituation and sensitization, and which confirm, at the level of the biophysics of single neurons, the distinctiveness of sensitization as a process in its own right, independent of habituation. To be more specific, the results of these experiments imply that both habituation and sensitization result from changes in transmitter release consequent to modulation of the influx of calcium ions into the terminals of the presynaptic element in the reflex pathway. They also imply, however, that habituation results from a decrease in calcium influx due to the properties of the calcium channels themselves, whereas sensitization occurs because of an increased calcium influx which results from a change in the properties of a different set of ionic channels in the presynaptic membrane.

Our experimental preparation consisted of identified neurons of the gill-withdrawal reflex pathway in the marine snail Aplysia

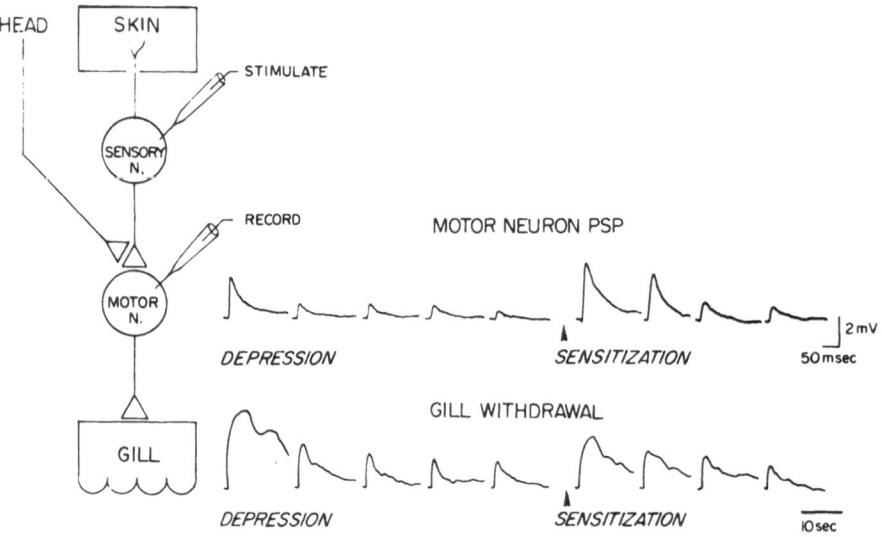

Fig. 1.   The gill-withdrawal reflex pathway and the behavioral and
synaptic aspects of habituation and sensitization.   Tactile
stimulation of the siphon skin activates sensory neurons which
connect to gill motor neurons.   When the skin is stimulated at low
frequency, the gill-withdrawal response decrements, or habituates,
while head shock causes the response to increase, or sensitize
(lower traces).   In the reduced preparation, intracellular
stimulation of the sensory neurons replaces tactile stimulation,
intracellular recording of motor-neuron EPSPs replaces measurement
of gill withdrawal, and stimulation of a nerve from the head
replaces head shock.   Motor-neuron EPSPs show both decrement to
repeated sensory-neuron stimulation and enhancement after nerve
stimulation (upper traces), corresponding to the behavioral
plasticity shown by the gill withdrawal itself. (Data from 5 and 21.)

californica (Fig. 1). In a large body of earlier work, Kandel and
his colleagues have amply documented the characteristics of the
behavioral changes associated with the habituation and the
sensitization of this reflex, and have also pinpointed the cellular
locus of the underlying synaptic processes to the terminals of the
presynaptic mechanoreceptor sensory neurons in the abdominal ganglion
(5,11,21).   Their work showed that habituation results from
decreased release of transmitter from the sensory neuron terminals
with repeated homosynaptic activation, while sensitization results
from presynaptic facilitation (12) of synaptic transmission caused
by activation of another input to the ganglion (Fig. 1).   Their
work also suggested that the presynaptic facilitatory transmitter
substance might be serotonin, and that this transmitter might act

by bringing about an increase in the concentration of cyclic AMP in the presynaptic terminals of the sensory neurons (2,6,7). The increase in cyclic AMP was postulated to increase transmitter release in some way, perhaps by increasing the intraterminal free calcium concentration (2). (For the relationship between calcium and transmitter release, see 13,16 and Llinás and Walton, this volume).

PRELIMINARY CONSIDERATIONS

In all of our experiments, we recorded from sensory neuron cell bodies under conditions of either current or voltage clamp. In some experiments, we also recorded from neurons postsynaptic to the sensory neurons in order to correlate transmitter release with the changes in presynaptic properties we recorded from the cell bodies. In other experiments, we isolated sensory neurons by tying them off from the ganglion in order to maximize our voltage-clamp control. It should be noted that although we could control the duration of transmitter release from the terminals by voltage clamping the soma, most of the data obtained with simultaneous presynaptic and postsynaptic recording is strictly correlational in the sense that we could not control all aspects of release from the terminals. Our experiments were made possible by two, perhaps not unrelated, characteristics of the sensory neurons: first, that the electrical distance between the soma and the terminals is short enough to allow partial voltage control of the terminals from the cell body (18,23, 25), as well as to permit recording in the cell body of some of the phenomena occurring in the synaptic region; and second, that the soma membrane apparently has properties similar to those of the transmitter releasing membrane, enabling us to use isolated cell bodies in some of the experiments.

PRESYNAPTIC FACILITATION IS ASSOCIATED WITH A DECREASE IN MEMBRANE CONDUCTANCE

Our first clue that we might be able to examine changes in the sensory neuron cell body which are correlated with the changes in release from the terminals was our observation that the latency for action potential initiation by a short pulse of constant depolarizing current was decreased after the heterosynaptic stimulation which causes presynaptic facilitation and sensitization. When we examined the resting membrane properties during the synaptic equivalents of habituation and sensitization, we found that, although membrane properties did not change during homosynaptic depression, there appeared to be a small depolarization associated with a decrease in membrane conductance after heterosynaptic stimulation (15) (Fig. 2A). A decrease in membrane conductance would cause a larger voltage change with a pulse of a

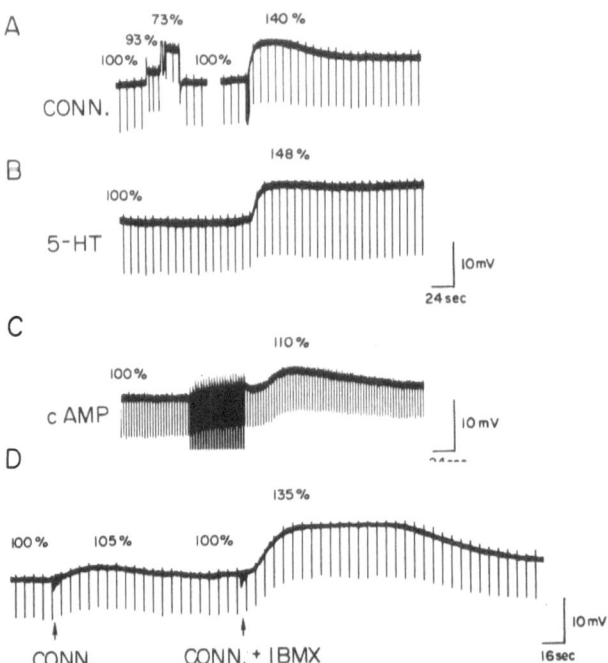

Fig. 2. Effects of nerve stimulation and drugs on presynaptic
membrane potential and resistance. A. Stimulation of the left
connective (a nerve pathway from the head that mediates
sensitization) caused a depolarization and an apparent increase
(40%) in membrane resistance as measured with electrotonic
potentials elicited by constant-current pulses delivered through
the second barrel of a double-barreled microelectrode. The membrane
was depolarized to different levels before connective stimulation
in order to show that depolarization alone caused a decrease in the
size of the electrotonic potentials. B. Serotonin (5-HT) applied
extracellularly. C. Cyclic AMP injected iontophoretically into
a sensory neuron. D. Application of the phosphodiesterase
inhibitor isobutylmethylxanthine (IBMX) prior to connective
stimulation potentiated its effects. (From 15)

constant current and would explain the decrease in spike latency
we had observed. Several characteristics of this decreased-
conductance postsynaptic potential (PSP) were worthy of note. First
of all, it could be extremely long-lasting as far as conventional
synaptic potentials go, lasting anywhere from a few minutes to half
an hour or more. Secondly, extracellular serotonin or intracellular
cyclic AMP gave rise to a similar decrease in membrane conductance,
while a phosphodiesterase inhibitor potentiated the effects of
nerve stimulation (15) (Fig. 2B-D). Thus, not only did

heterosynaptic stimulation, serotonin, and cyclic AMP all facilitate
synaptic transmission at these synapses, they also caused a decrease
in sensory neuron membrane conductance. And the third notable
characteristic of this decreased-conductance PSP was that it appeared
to be voltage-sensitive, so that as the sensory neuron was held more
and more depolarized, the conductance change was seen to become
greater and greater. One possible interpretation of this voltage-
sensitivity is that the ionic channels which are closed by the
neurotransmitter are themselves voltage-sensitive, being opened
by depolarization and closed by hyperpolarization. Thus, when the
membrane is held depolarized, more of the channels are open and
closing them by the action of the heterosynaptic transmitter would
give a greater conductance decrease than when few of the channels
were open to begin with. In this way too, this PSP differed from
conventional PSPs, since in most cases synaptic channels are not
affected significantly by membrane voltage.

PRESYNAPTIC FACILITATION IS ASSOCIATED WITH BROADENING OF THE PRESYNAPTIC ACTION POTENTIAL

The function of synaptic potentials has, until recently, been
considered to be to bring the membrane closer to or further from
threshold for firing an all-or-none action potential. If, however,
a PSP were to affect voltage sensitive channels, it might also control
the configuration and transmitter releasing capabilities of the
action potential, since these voltage-sensitive channels would be
activated during the action potential itself. A voltage-sensitive,
decreased-conductance PSP of the kind we observed in the sensory
neurons would also affect firing threshold, as well as electrotonic
propagation of subthreshold signals in areas of the neuron where
the threshold for action potential generation is high. Although
neither of these possibilities should be discounted, the major part
of this discussion will be confined to what happens to the action
potential itself once it has been generated, since we have at
present no data relating to action potential generating mechanisms
in the terminal arborization of the sensory neurons.

In any case, however, if the synaptic action we observed played
a role in controlling the ability of the action potential to cause
transmitter release, it could do so in one of two, non-mutually
exclusive, ways. If the locus of action of the PSP were the calcium
channel itself, then an action potential could admit more calcium
ions with or without any concomitant change in configuration, since
the spike configuration might be determined by sodium and potassium
currents of much greater magnitude than the concurrently activated
calcium current. On the other hand, the PSP might change the height
or duration of the action potential by acting on one of the other
voltage-sensitive conductances activated in the action potential,
and thereby control the magnitude or duration of the calcium influx

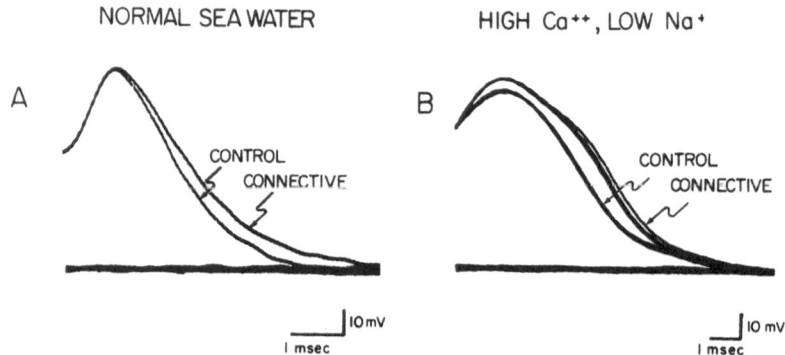

Fig. 3. Presynaptic facilitation produces action potential broadening in drug-free solutions. Each trace represents superimposition of two or three sensory neuron action potentials. A. Normal sea water. B. Sea water containing 0.5X sodium, 5.5X calcium, and 2X magnesium. (Modified from 15).

without changing the properties of the calcium channel itself. The same considerations apply to changes in ionic currents which might underlie habituation: the calcium current might decrease independently of the shape of the action potential, or also it might decrease as a consequence of a decrease in the amplitude or duration of the action potential.

The first course of action that suggested itself, therefore, was to examine closely the shape of the sensory neuron action potential during both synaptic depression and facilitation, bearing in mind that anything we might see would be, at best, a qualitative indication of what was going on at the terminals. When we did this, we found that just as there was no change in the resting membrane properties during habituation, so there was no change in the action potential. However, after heterosynaptic stimulation we did see an increase in the duration of the action potential such as might have been predicted from the voltage-sensitive decreased-conductance PSP we had observed earlier (15) (Fig. 3A).

We also observed a similar, perhaps greater, increase in duration on incubation with serotonin (unpublished). Taking into consideration the likelihood that a larger part of the inward current might be carried by calcium ions relative to sodium at the terminals (14), we bathed the preparation in low sodium/high calcium solution to simulate in a crude way the effect of heterosynaptic stimulation on terminal spikes. We found that the increase in action potential duration was now greater and was accompanied by an increase in amplitude as well (Fig. 3B). We now believe that the action of the PSP is accentuated under the latter

circumstances mainly because the action potential is slower in both
its rising and falling phases, and the presynaptic facilitatory
transmitter has a greater effect later in time than it does within
the first millisecond or so after action potential initiation.

SYNAPTIC DEPRESSION AND FACILITATION INVOLVE CHANGES IN THE RATIO BETWEEN CALCIUM AND POTASSIUM CURRENTS

While our observations up to this point did not tell us
anything about the mechanism of habituation, they did suggest that
our suspicions about the relevance of the decreased-conductance
PSP to presynaptic facilitation might be well-founded, and that
the facilitatory transmitter might indeed cause a decrease in a
voltage-sensitive potassium current which has an important role
in terminating the action potential. By decreasing this current,
the action potential, and therefore the calcium current, would
be prolonged, greater calcium entry would occur, and more
transmitter would be released. At that time, however, we could
not distinguish between this interpretation of presynaptic
facilitation and another one which involved a direct activation
of the voltage-sensitive calcium conductance (20). This second
interpretation could also account for both the apparent decrease
in membrane conductance (19,20) and the broadening of the action
potential. In either case, though, the change in the currents
would involve an increase in the calcium current relative to the
potassium currents, either because of an increase in the calcium
current or because of a diminution of the potassium current.

In order to obtain a more sensitive measure of the balance
between these currents than we had in the rather subtle changes
in input resistance and spike duration in normal sea water, it
turned out to be helpful to decrease the potassium current by
bathing the cells in tetraethylammonium (TEA). This treatment
causes the calcium and potassium currents to be matched more
equally than is normally the case. When this drug is added to the
bathing solution, the action potential develops a plateau on its
falling phase which increases its duration by a factor of fifty or
more. The inward current during the plateau is carried by calcium
ions (10,15), and the plateau is maintained as long as the
outward potassium and leak currents do not outweigh the inward
current. When the outward currents predominate, the action
potential repolarizes. Since the inward and outward currents
balance each other during the plateau phase, a small change in
one or the other of the currents can result in a relatively large
change in action potential duration (14a).

When we repeated our experiments with presynaptic facilitation
in the presence of TEA, we found that the action potential
duration increased dramatically after nerve stimulation (Fig. 4 A).

NORMAL SEA WATER + T.E.A.

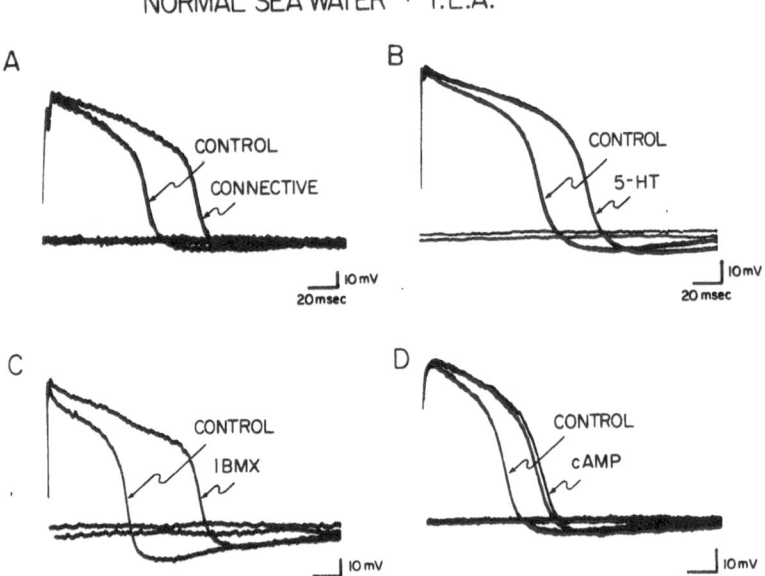

Fig. 4. Effects of connective stimulation and drugs on the calcium plateau of action potentials in TEA. The action potential broadens after connective stimulation (A), application of 0.2mM serotonin (B), application of 0.1mM IBMX (C), or intracellular injection of cyclic AMP (D). Each trace represents the superimposition of two or three action potentials evoked with two-to-five-msec intracellular depolarizations in sea water containing 0.1M TEA. In B and C, a small depolarization occurred after drug application (From 15)

When serotonin or phosphodiesterase inhibitors were added to the bath, or when cyclic AMP was injected into single sensory neurons, we saw similar large increases in spike duration (Fig. 4B-D), again consistent with the increase in synaptic transmission observed in the wake of similar treatments. In control experiments with a variety of other transmitter candidates (T.K. Tomosky-Sykes, unpublished) or with injection of 5'AMP, spike broadening was not observed. We also found that the effects of nerve stimulation and serotonin were not blocked by removing sodium ions (Fig. 5A) or by bathing in tetrodotoxin (TTX), and that sodium spikes were unaffected by nerve stimulation (Fig. 5 C), thus lending further support to our hypothesis that it was the calcium and/or potassium currents which were affected in presynaptic facilitation (15).

Now that we had a way of monitoring the calcium current with relative ease, by simply measuring the duration of the calcium plateau of the TEA action potential, we could also look for a

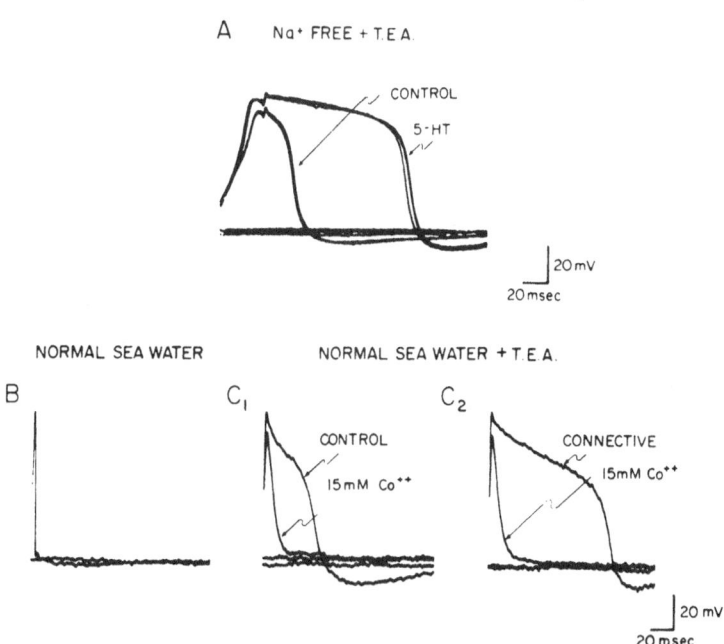

Fig. 5. Spike broadening in TEA solution is independent of sodium current. A. TEA action potentials in sucrose-substituted sodium-free sea water broaden on application of serotonin (two or three sweeps superimposed in each trace). B. Sensory neuron spike before addition of TEA to the bathing solution. C_1. In the same cell, 0.1M TEA was added and action potentials were evoked every ten seconds until they achieved a stable configuration (CONTROL), when cobalt ions were added to the bath, reducing the peak spike amplitude and eliminating the calcium plateau (15mM Co^{++}). Cobalt was then washed out and the action potential recovered. C_2. After recovery, the connective was stimulated and the spike broadened (CONNECTIVE). Cobalt ions were added once more, causing the spike to shrink back to the same configuration it had after cobalt addition before connective stimulation (15mM Co^{++}). Since practically all of the current in cobalt and TEA is carried by sodium ions, the absence of any change with connective stimulation implies that the sodium current is not affected by heterosynaptic stimulation. (Modified from 15)

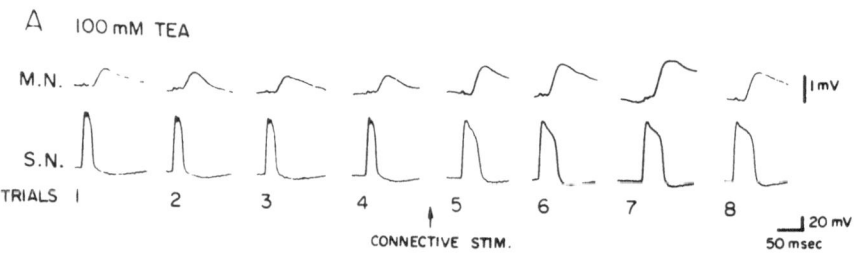

A   100 mM TEA

M.N.

S.N.

TRIALS 1   2   3   4   5   6   7   8

| 1 mV

CONNECTIVE STIM.   20 mV
                                    50 msec

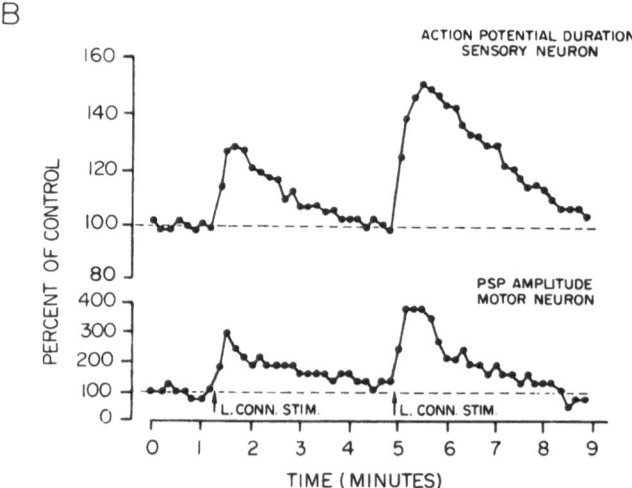

B

ACTION POTENTIAL DURATION
SENSORY NEURON

PERCENT OF CONTROL

160
140
120
100
80

PSP AMPLITUDE
MOTOR NEURON

400
300
200
100
0

L. CONN. STIM.   L. CONN. STIM.

0   1   2   3   4   5   6   7   8   9

TIME (MINUTES)

Fig. 6. Increase in calcium plateau that parallels presynaptic
facilitation. A. Broadening of the action potential in TEA
accompanying presynaptic facilitation of monosynaptic sensory-to-
motor EPSP. B. Graph of EPSP amplitude and presynaptic spike
duration during an experiment in which the facilitating pathway
was stimulated first weakly, and then more strongly. In both A
and B spikes were evoked at 0.1 Hz in 0.1M TEA sea water.
(Modified from 15). M.N., motor neuron; S.N., sensory neuron.

correlation between the presynaptic changes in the sensory neurons
and changes in the monosynaptic excitatory postsynaptic potential
(EPSP) in the interneurons and motor neurons of the reflex pathway.
When we did such experiments, we found a good correlation between
presynaptic action potential duration and EPSP amplitude (15)
(Fig. 6). In addition, we were now in a better position to examine
the presynaptic concomitants of homosynaptic depression, since
while in normal sea water the contribution of the calcium current
to action potential configuration was minimal, in TEA solution
we might be able to see shortening of the calcium plateau. And
indeed, that is what we found: in the case of synaptic depression

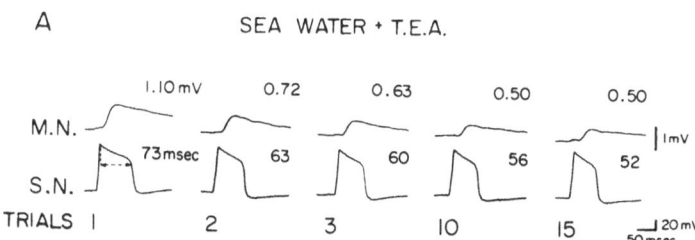

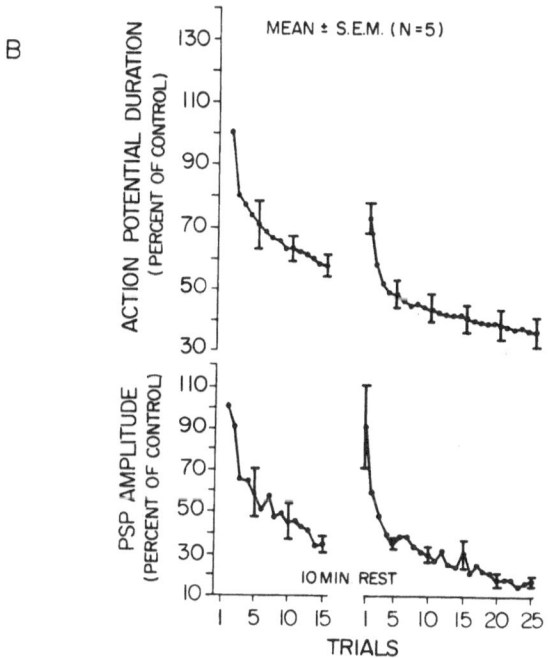

Fig. 7. Change in calcium current during homosynaptic depression.
A. Correlation between sensory neuron action potential in sea water
containing 0.1M TEA and monosynaptic EPSP with repeated stimulation.
Action potentials fired at 0.1 Hz. B. Average action potential
duration and EPSP amplitude (based on five preparations) during a
first habituation run, after a ten-minute rest, and during a
second run. Spikes evoked at 0.1 Hz in sea water containing TEA.
(From 15b).

as well we saw a parallel between the calcium plateau duration and
EPSP amplitude (Fig. 7).

In summary, then, our observations of sensory neuron spikes
and postsynaptic neuron EPSPs in TEA during the synaptic processes
which underlie habituation and sensitization revealed that both
fomrs of synaptic plasticity were accompanied by changes in the
presynaptic calcium current. Thus, we found a decrease in the
current during homosynaptic depression, and an increase with
presynaptic facilitation. However, we could be certain only that
the calcium current was changing relative to the total potassium
current; we could not be sure whether the primary locus of either
presynaptic change was the calcium current or the potassium
current. Nevertheless, the results of the experiments in TEA and
in normal sea water, when taken together, do suggest a particular
interpretation, which is borne out by the results of the voltage-
clamp experiments to be discussed later.

SYNAPTIC DEPRESSION AND FACILITATION HAVE DIFFERENT MECHANISMS

In normal sea water, sensory neuron action potentials do not
change during synaptic depression, while they broaden during
facilitation. This finding already suggests that the presynaptic
processes underlying the two forms of learning are distinct, and
that sensitization is not caused by the reversal of whatever
happens during habituation. On the other hand, when one examines
action potentials in TEA, the two forms of plasticity are
symmetrically related: spikes narrow with synaptic depression,
and broaden with synaptic facilitation. The results in TEA
solution imply that the calcium current is being modulated in
some way during both depression and facilitation. Why then is
there no narrowing of the action potential during depression in
drug-free solution, while presynaptic facilitation is accompanied
by spike broadening? One way to resolve this problem is to suppose
that during habituation the calcium current decreases because of
homosynaptic activation, independent of other currents, while
facilitation is caused by a prolongation of spikes, and of the
calcium current, secondary to a decrease in the potassium current.
In normal sea water, where it is the sodium and potassium currents
which determine the configuration of the action potential, the
decrease in calcium current during habituation would not affect
the action potential, while decreasing the potassium current with
facilitation would reduce its rate of repolarization. In TEA
solution, however, since the duration of the plateau is determined
by the balance between the calcium and the potassium currents, a
change in either one would affect action potential duration.
During habituation, therefore, a progressive decline in the
calcium current would cause the action potential plateau to be
terminated earlier, while facilitation would give rise to a

prolongation of the plateau by decreasing the opposing potassium current.

VOLTAGE-CLAMP ANALYSIS OF SYNAPTIC DEPRESSION: SYNAPTIC DEPRESSION PARALLELS PRESYNAPTIC CALCIUM CURRENT INACTIVATION

The next set of experiments was done using the technique of voltage clamp in order to allow us to visualize the presynaptic currents directly while at the same time we could control the membrane potential changes.When we recorded from postsynaptic cells at the same time in these experiments, we found that we could not control transmitter release from the soma in the sense that we could not achieve graded increases in transmitter release by increasing the amplitude of depolarizing command pulses. Neverthe- less, we could in many cases increase the amplitude of the PSP by increasing the duration of the presynaptic depolarization. In many of the experiments where we show simultaneous presynaptic currents and postsynaptic potentials, therefore, we can only claim to show correlations between pre- and postsynaptic events which are not causally related in a simple way. It should also be noted that most of the current we see in these experiments flows across the cell body membrane, and that only a small portion of the current we record may be involved in transmitter release. Bearing these qualifications in mind, we will proceed to discuss the somewhat preliminary voltage-clamp analysis of the synaptic depression which causes habituation.

When a sensory neuron is voltage clamped in drug-free solution and EPSPs are elicited postsynaptically with short depolarizing command steps every ten seconds in the presynaptic cell, the presynaptic membrane currents do not change, while the EPSP undergoes homosynaptic depression (Fig. 8A). In this case, the inward current during the depolarizing step is carried mainly by sodium ions, and the outward current is carried by potassium ions. The stability of the currents underlies the constancy of the action potential during synaptic depression. On the other hand, when sodium and potassium currents are largely blocked, with TTX and TEA respectively, the inward current seen for the duration of the step is carried by calcium ions. Under these conditions, the inward current decreases progressively as the EPSP decrements (Fig. 8B). If we plot out peak inward current and EPSP amplitude we find that they decrease in parallel during a first habituation run, recover partially after a short rest, and then decrement together once again during a second run. If we plot EPSP amplitude against peak inward current on the same set of coordinates, we find that the two correlate very well with each other, giving a curve which can be fit by a straight line. The interpretation of this correlation is not straightforward, and might reflect either the presence of different currents in different regions of the cell, only some of which cause transmitter release, or

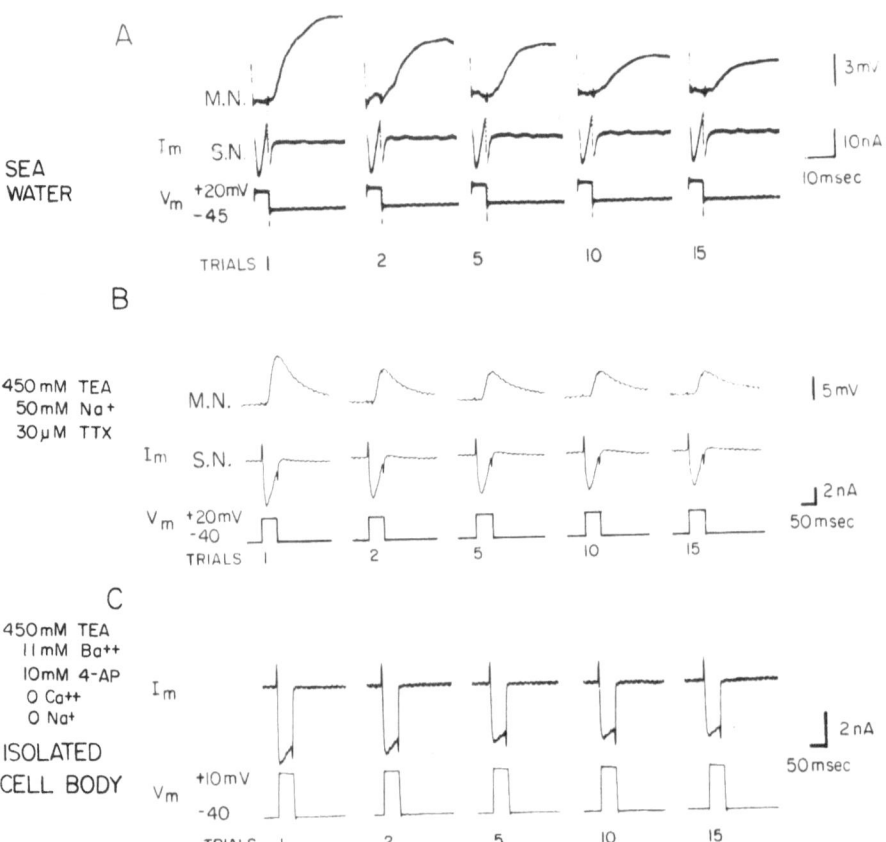

Fig. 8. Homosynaptic depression and decline in calcium current under voltage clamp. A. In drug-free sea water the inward current is largely due to sodium. The presynaptic voltage–clamp pulses elicit EPSPs that undergo depression with repetition of the depolarizing command. The decrease in EPSP is not accompanied by any change in the current. B. With the sodium current blocked with TTX and a large part of the potassium current blocked with TEA presynaptic voltage–clamp pulses elicit EPSPs which undergo depression with repetition of the depolarizing command. The decrease in EPSP is now paralleled by a decrease in the inward current. C. With virtually all the potassium current blocked (by bathing in a high concentration of 4-aminopyridine in addition to TEA and barium) and with voltage–clamp control maximized by ligating the sensory neuron cell body, inward current decreases with 0.1 Hz stimulation, indicating that current moving through the calcium channels undergoes a progressive decline independent of changes in sodium or potassium currents. (Modified from 15b).

else the existence of different components of the inward current
which have different inactivation properties. In addition, the
possibility exists, in the intact neuron, that the inward current
reflects to some extent the presence of calcium spikes in
uncontrolled regions of the cell, and that some of the decrement
we see in the total current is a consequence of failure of
these calcium spikes in some parts of the neuron, rather than
being simply a consequence of the graded process of calcium
current inactivation.

To simplify the analysis of this decline in inward current,
while at the same time reducing any residual potassium current
which might still be present, we isolated sensory neuron somata
by tying them off and bathed them in a solution in which barium
ions were substituted for calcium and 4-aminopyridine was added
in addition to TEA. Although we could not now monitor transmitter
release, we could still follow changes in the inward current in
experiments where we used the same protocol as we did with intact
neurons. Here again we found that the peak inward current appeared
to inactivate at slow rates of stimulation (Fig. 8C). In these
experiments, the likelihood that the inward current decreased
because of an increase in outward current or because of all-or-none
block of uncontrolled active responses is negligible.

To exclude the possibility that the inward current declined
because of intracellular accumulation of divalent cations, and
consequent reduction of the driving force on the charge carrier,
we also interposed smaller depolarizing steps in between the
larger command pulses shown in the figure. The currents elicited
by the interposed steps were smaller inward currents which showed
no decline either during the smaller pulse or between pulses.
If the decline observed across trials for the larger currents were
caused by a decreased driving force for divalent cations, then
there should have been a decrease in the driving force for the
currents in the smaller steps as well. The fact that the latter
currents did not change suggests that the decline in the larger
currents was due to a true inactivation of the current (28),
whether as an effect of voltage per se, or as a result of the entry
of barium or calcium ions (see 1, 27). In any case, however,
the correlations we observed between apparent calcium current
inactivation in presynaptic neurons and the decline in EPSP in
follower cells suggests that this inactivation may play a role in
homosynaptic depression.

VOLTAGE-CLAMP ANALYSIS OF PRESYNAPTIC FACILITATION: FACILITATION
IS ASSOCIATED WITH DECREASED POTASSIUM CURRENT

The same techniques utilized in the analysis of the
presynaptic correlates of synaptic depression can be used to

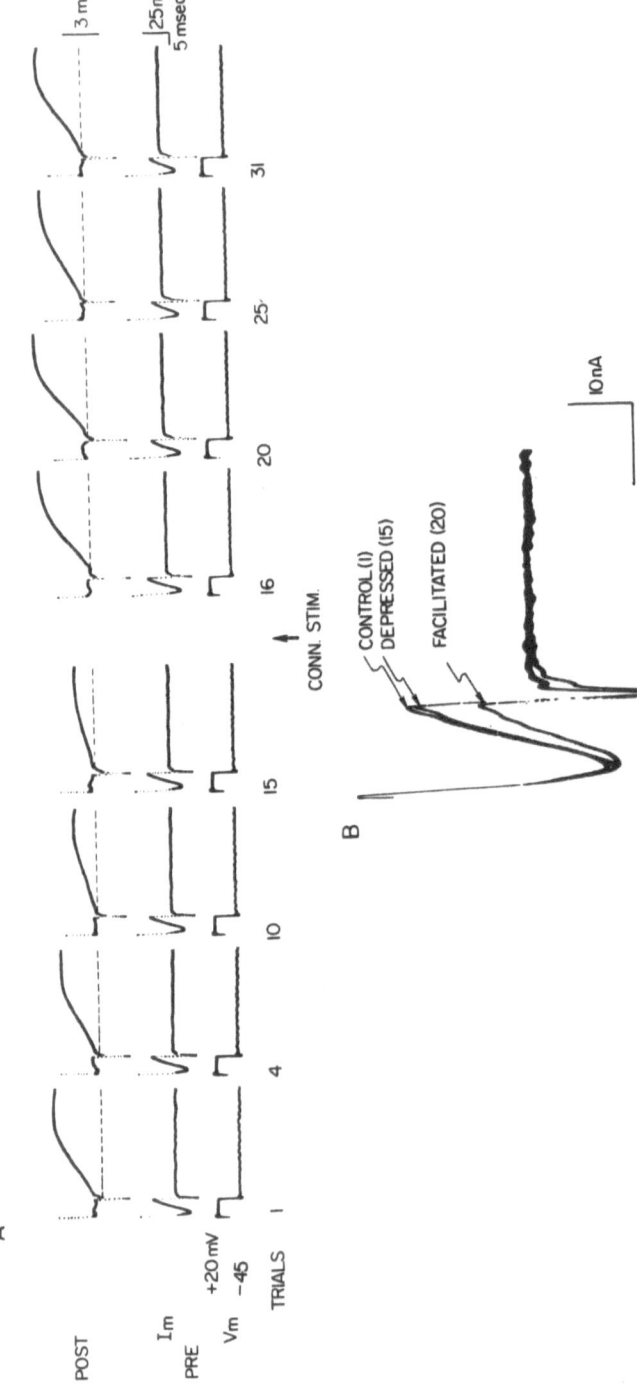

Fig. 9. Presynaptic facilitation and decrease in outward current under voltage clamp. A. Five-millisecond depolarizing commands in a sensory neuron in drug-free solution caused EPSPs in a postsynaptic cell. The EPSPs declined in amplitude with pulses delivered every ten seconds, while they increased in amplitude after connective stimulation. The increase in the EPSPs was accompanied by a decrease in outward current in the sensory neuron. B. Superimposed current traces from part A displayed at higher gain and sweep speed show that while the currents remained stable throughout synaptic depression, the later part of the total current moved inward after connective stimulation. Such an inward shift causes the current to cross the zero-current line at a later time, and would lead to prolongation of an action potential in an unclamped cell (From 15a).

examine presynaptic facilitation as well, and the same
reservations regarding the relationship between the presynaptic
soma currents and the EPSP apply here. When presynaptic facilitation
of synaptic transmission is elicited with the presynaptic neuron
held under voltage clamp in drug-free solution, two things are
seen to happen. First, one observes an inward shift of the
holding current and a decrease in the leak (Fig. $10C_1$),
corresponding to the slow decreased-conductance PSP described in
an earlier section. Second, one also sees a decrease in the
transient outward current evoked by a depolarizing command (Fig. 9),
corresponding to the increase in action potential duration observed
in normal sea water. In order to answer the question as to whether
these changes are the result of increased calcium current or decreased
potassium current, we isolated the cell bodies by ligation to
maximize voltage-clamp control, and also eliminated sodium and
potassium currents by removing sodium ions and adding potassium
current blockers. When serotonin, the putative facilitating
transmitter, was added to the bath, there was no change in the
holding current, the leak current, or the transient calcium current
elicited by step depolarization (Fig. 10A, C_2). However, when
this bathing solution was replaced by normal sea water, application
of serotonin now produced all three of the changes in sensory
neuron currents that are normally observed with heterosynaptic
stimulation (Fig. 10B, C_1). This experiment implies that serotonin
acts to decrease potassium current, rather than to increase the
current through the calcium channel directly. (It should be noted
that the same results are obtained whether the charge carrier
through the calcium channels is calcium or barium). However, since
it is possible that the potassium blockers might also block the
serotonin receptors on the cell membrane, we performed an
additional experiment to strengthen our conclusion that serotonin
acts exclusively on potassium current.

Another way to eliminate potassium current is to remove the
potassium ions from the interior of the cell. This can be
accomplished by making the cell permeable to monovalent cations
using the antibiotic nystatin, and washing out the potassium
ions by replacing potassium with impermeant cesium ions (22, 28).
After nystatin treatment, a solution containing only calcium,
magnesium, Tris, and chloride ions is used to replace the nystatin-
containing solution. When a neuron is voltage clamped using 3M
cesium chloride electrodes and stepped to a depolarized membrane
potential, an inward current, carried by calcium ions, is observed.
When we added serotonin to the bath we saw no change in the
current. When, after washing out the serotonin, we removed one
of the cesium chloride electrodes and reimpaled the cell with a
potassium chloride electrode, we elicited an outward current
carried by potassium ions leaking into the cell from the potassium
chloride electrode in place of the inward current seen with
cesium chloride electrodes. When we now added serotonin to the

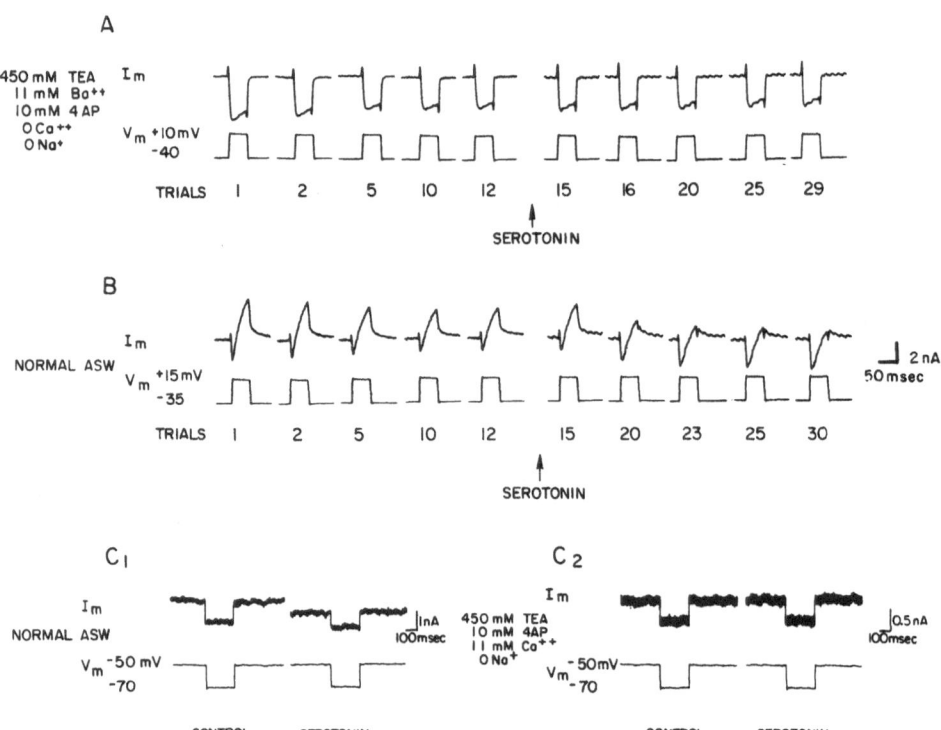

Fig. 10. Elimination of serotonin-induced changes in transient and
steady-state conductances by potassium-current blocking agents. A.
In solution containing potassium-current blockers serotonin failed
to affect calcium-channel currents elicited by fifty-millisecond
commands presented every ten seconds. The preparation was a sensory
neuron cell body isolated by ligation. B. Drug-containing solution
was replaced with normal sea water and experiment shown in part A
was repeated using the same cell. Serotonin now caused inward
movement of both the transient and the holding currents. C. Steady-
state conductance at a holding potential of -50 millivolts was
measured with a step hyperpolarization to -70. In normal sea water
holding current moved inward and steady-state conductance decreased
after serotonin application (C_1). In the presence of potassium-
current blocking agents serotonin had no effect (C_2). The results
shown in this figure imply that serotonin reduces voltage-dependent
potassium current and does not affect current through the calcium
channel in a direct manner. (From 15a).

bath this outward current decreased and the leak current was decreased. This experiment, too, implies that the action of serotonin, and, presumably, nerve stimulation, on both resting and transiently activated currents is to decrease the potassium current.

CHANGES IN THE CONFIGURATION OF MONOSYNAPTIC EPSPs REFLECT THE CHANGES IN THE PRESYNAPTIC CURRENTS

The goal of the final set of experiments we performed was two-fold. We wanted to see whether we could detect anything in the monosynaptic EPSP that might reveal something about changes in the currents of the presynaptic terminals, and we also wanted to specify a little more clearly how a decrease in potassium current leads to synaptic facilitation. We will begin with the second question.

A decrease in the resting potassium current, expressed as an increase in the input resistance of the presynaptic membrane, would cause passively propagated signals to increase in amplitude and to spread further from their site of initiation. Thus, if action potentials were to fail before they reach the presynaptic terminals and transmitter release were caused by an electrotonic potential spreading from the site of failure, increasing the input resistance would cause an increase in the amount of transmitter released as a result of an action potential. A second possibility is that potassium currents might determine the threshold for action potential generation in the fine branches of the presynaptic arborization on which the synaptic terminals are found. Decreasing the potassium conductance would lower the threshold for action potential initiation and allow more presynaptic terminals to be invaded by action potentials, giving rise to increased transmitter release. Lastly, decreasing the potassium current activated in an action potential could cause the action potential to broaden, allowing a greater calcium inflow and giving rise to increased transmitter release. Our observations on cell body spikes led us to suspect that the third possibility, at least, might contribute to synaptic facilitation. Since the first two possibilities are also more difficult to approach directly, we decided to examine the relationship between duration of presynaptic depolarization and EPSP amplitude.

One problem we had with the spike broadening mechanism for synaptic facilitation was a quantitative one: were the modest changes in duration which we could detect in cell body action potentials large enough to account for the large changes in EPSP size which occur in presynaptic facilitation? To answer this question we attempted to control transmitter release by varying the duration of the cell body depolarization under voltage clamp. When we tried such an experiment, we found that changing the pulse duration did in fact cause relatively large changes in EPSP amplitude. For example, we found that changing the duration of the step from twenty to twenty-five milliseconds, doubled the size of the EPSP. The steep dependence of transmitter output on duration of presynaptic depolarization might be a consequence of the slow activation and inactivation of the calcium current. Since the calcium current takes a long time to reach its peak relative to the duration of the presynaptic depolarization, calcium influx is still increasing steeply at the end of the twenty millisecond depolarization. Increasing the duration by five milliseconds allows significantly more calcium to enter the cell, substantially increasing transmitter output. In the case of a sensory neuron action potential which is only two to three milliseconds long at half amplitude, we would expect the relation between action potential duration and EPSP amplitude to be even steeper, since the relative increment in calcium entering the cell would be even greater with the same percent change in action potential duration. On the basis of this semi-quantitative argument, then, it appears that the small changes in spike duration which we observed in drug-free solution could cause significant synaptic facilitation.

The preceding experiment also allowed us to look for a postsynaptic indication of what might have been happening pre-synaptically in presynaptic facilitation. If we increase EPSP amplitude by increasing the duration of the presynaptic depolariza-tion, we might also expect to change the shape of the EPSP, or at least of its initial portion. When we normalized EPSPs elicited with depolarizing commands of different duration for amplitude and superimposed them we found, not surprisingly, that longer depolarizations gave EPSPs with longer rise times and slower initial rates of decay. On the other hand, we knew from earlier work that changing EPSP amplitude by altering the calcium concentration produces no change in the shape of the EPSP (4). The information we could gain from examining the shape of EPSPs of different amplitudes could therefore shed light on the presynaptic mechanisms of habituation and sensitization.

We examined EPSPs elicited by firing single spikes in sensory neurons, in normal sea water, under conditions of both synaptic depression and facilitation. If our interpretation was correct —that depression results from calcium current inactivation, while facilitation results from spike broadening— then we should have seen no change in EPSP shape with homosynaptic activation, and a slowing in the initial portion of the EPSP after heterosynaptic stimulation. This is what we did in fact find (Fig. 11). During habituation, EPSP amplitude decreased to a fraction of the initial value with no change in shape. EPSPs after a facilitated stimulus, on the other hand, had prolonged initial phases even when they were smaller than the EPSPs elicited at the beginning of the experiment.

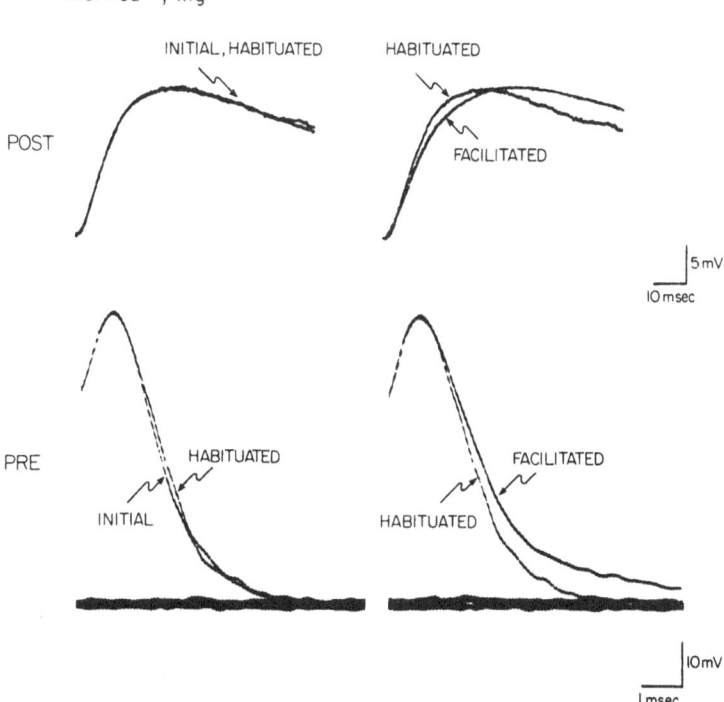

HIGH Ca++, Mg++

Fig. 11. Comparison of EPSP shapes after habituation and sensitization. Monosynaptic EPSPs were evoked with sensory neuron action
potentials fired once every ten seconds in high divalent cation
sea water. Blocks of five EPSPs at the beginning and end of the
habituation run and after connective stimulation were averaged
using a signal averager. The average EPSP at the beginning of the
run is 20 mV in amplitude; the average after habituation is 16 mV,
and after nerve stimulation, 24 mV. The averages were scaled so
that their peak amplitudes were the same and were then superimposed
in pairs (top traces). The traces at the top left show
superimposition of the averages taken from the beginning and end
of the habituation run. Note that despite the fact that the EPSPs
had decremented considerably during the run, their shape has
remained constant. The traces on the top right show the
superimposition of the averages of the habituated EPSPs and the
facilitated EPSPs. The facilitated EPSPs have a longer time-to-
peak and a slower initial decline than the habituated ones. The
traces on the bottom show superimpositions of action potentials from
the same experiment (on a different time scale). Action potentials
remained fairly constant in duration during habituation (left), and
broadened significantly after heterosynaptic stimulation. These
data suggest that habituation occurs through a mechanism which is
independent of action potential duration, while in sensitization
action potential duration is increased (Klein and Kandel, unpublished)

SUMMARY: MODULATION OF CALCIUM ENTRY IN BEHAVIORAL AND
SYNAPTIC PLASTICITY

   Although we have not yet analyzed the changes in presynaptic
membrane properties which accompany habituation as carefully as we
have those which occur during sensitization, the data we do have
suggest that habituation is caused, at least in part, by the
progressive, long-lasting inactivation of presynaptic calcium current,
and not by changes in resting membrane conductances or in action
potential configuration. We will be unable to determine the
quantitative contribution of calcium current inactivation to
habituation until additional experiments have been done, and it is
quite possible that more than one presynaptic process is involved
in the synaptic depression underlying habituation.

   In the case of sensitization, we can assert more confidently
that presynaptic facilitation is caused by increased calcium influx
due to a decrease in presynaptic potassium current. The experiments
in which we controlled transmitter release by varying the duration
of voltage-clamp depolarizations suggest, in a semi-quantitative
way, that the action potential broadening which we actually observe
after heterosynaptic stimulation may be of sufficient magnitude
to account completely for presynaptic facilitation. Our analysis
has shown that this action potential prolongation is the result
of a decrease in voltage-sensitive potassium conductance induced
by the presynaptic facilitatory transmitter.

   The final common pathway for the regulation of synaptic
transmission which is the physiological basis for habituation and
behavioral sensitization appears, from our analysis, to be the
calcium influx during the action potential. Control of the
calcium current is also thought to underlie other forms of short-
term synaptic plasticity such as presynaptic inhibition (8,17,24)
and the control of transmitter release by membrane potential (18,
23,25). In view of the multiplicity of ways in which the calcium
current is regulated in several forms of synaptic plasticity, it
would be of great interest to determine what, if any, role calcium
plays in long-term forms of behavioral plasticity on the one hand,
and in associative learning on the other.

Acknowledgements. This work was done in the laboratory of, and
in collaboration with, Dr. Eric Kandel. The support of NIH
Grants NS12744 and GM 23540 to Dr. Kandel is acknowledged.

REFERENCES

1. Brehm, P. and Eckert, R., Calcium entry leads to inactivation
    of calcium channel in paramecium, Science, 202 (1978) 1203-1206.

2. Brunelli, M., Castellucci, V. and Kandel, E.R., Synaptic facilita-
    tion and behavioral sensitization in Aplysia: Possible role of
    serotonin and cyclic AMP, Science, 194 (1976) 1178-1181.
3. Carew, T. J., Castellucci, F. and Kandel, E. R., An analysis of
    dishabituation and sensitization of the gill-withdrawal reflex
    in Aplysia, Int. J. Neurosci., 2 (1971) 79-98.
4. Castellucci, V. F. and Kandel, E. R., A quantal analysis of the
    synaptic depression underlying habituation of the gill-withdrawal
    reflex in Aplysia, Proc. Nat. Acad. Sci. USA, 71 (1974) 5004-5003.
5. Castellucci, V. F. and Kandel, E. R., Presynaptic facilitation
    as a mechanism for behavioral sensitization in Aplysia,Science,
    194 (1976) 1176-1178.
6. Cedar, H., Kandel, E. R. and Schwartz, J. H., Cyclic adenosine
    monophosphate in the nervous system of Aplysia californica. I.
    Increased synthesis in response to synaptic stimulation, J. Gen.
    Physiol., 60 (1971) 558-569.
7. Cedar, H. and Schwartz, J. M., Cyclic adenosine monophosphate
    in the nervous system of Aplysia californica. II. Effect of
    serotonin and dopamine, J. Gen. Physiol., 60 (1972) 570-587.
8. Dunlap, K. and Fischbach, G. D., Neurotransmitters decrease the
    calcium component of sensory neuron action potentials, Nature,
    276 (1978) 837-839.
9. Groves, P. M. and Thompson, R. F., Habituation: A dual-process
    theory, Psychol. Rev., 77 (1970) 419-450.
10. Horn, R. and Miller, J. J., A prolonged, voltage-dependent cal-
    cium permeability revealed by tetraethylammonium in the soma
    and axon of Aplysia giant neuron, J. Neurobiol., 8 (1977)
    399-415.
11. Kandel, E. R., A Cell-Biological Approach to Learning, Grass
    Lecture Monograph 1, Society for Neuroscience, Bethesda, Md.
    (1978).
12. Kandel, E. R. and Tauc, L., Heterosynaptic facilitation in
    neurones of the abdominal ganglion of Aplysia depilans, J.
    Physiol. (London), 181 (1965) 1-27.
13. Katz, B. and Miledi, R., The timing of calcium action during
    neuromuscular transmission, J. Physiol. (London), 189 (1967)
    535-544.
14. Katz, B. and Miledi, R., Tetrodotoxin-resistant electrical
    activity in presynaptic terminals, J. Physiol. (London), 203
    (1969) 459-487.
14a.Katz, B. and Miledi, R., The effects of prolonged depolarization
    on synaptic transfer in the stellate ganglion of the squid,
    J. Physiol. (London), 216 (1971) 503-512.
15. Klein, M. and Kandel, E. R., Presynaptic modulation of voltage-
    dependent Ca^{++} current: Mechanism for behavioral sensitization
    in Aplysia californica, Proc. Nat. Acad. Sci. USA, 75 (1978)
    3512-3516.
15a.Klein, M. and Kandel, E. R., Mechanism of calcium current
    modulation underlying presynaptic facilitation and behavioral
    sensitization in Aplysia, Proc. Nat. Acad. Sci. U.S.A.,

(1980) in press.

15b. Klein, M., Shapiro, E. and Kandel, E. R., Synaptic plasticity
and the modulation of the Ca^{2+} current, J. Exp. Biol., (1980)
in press.

16. Llinas, R. R., Calcium and transmitter release in squid synapse,
In: Society for Neuroscience Symposia, Vol. 2, M. W. Cowan and
J. A. Ferendelli, eds., Society for Neuroscience, Bethesda, Md.
(1977) 139–160.

17. Mudge, A. W., Leeman, S. E. and Fischbach, G. D., Enkephalin
inhibits release of substance P from sensory neurons in culture
and decreases action potential duration, Proc. Nat. Acad. Sci.
USA, 76 (1979) 526–530.

18. Nicholls, J. and Wallace, B. G., Modulation of transmission at
an inhibitory synapse in the central nervous system of the leech,
J. Physiol. (London), 281 (1978) 157–170.

19. Pellmar, T. C. and Wilson, W. A., Unconventional serotonergic
excitation in Aplysia, Nature, 269 (1977) 76–78.

20. Pellmar, T. C. and Carpenter, D. O., Voltage-dependent calcium
current induced by serotonin, Nature, 277 (1979) 483–484.

21. Pinsker, H., Kupfermann, I., Castellucci, V. and Kandel, E. R.,
Habituation and dishabituation of the gill-withdrawal reflex
in Aplysia, Science, 167 (1970) 1740–1742.

22. Russell, J. M., Eaton, D. C. and Brodwick, M. S., Effects of
nystatin on membrane conductance and internal ion activities
in Aplysia neurons, J. Membrane Biol. 37 (1977) 137–156.

23. Shapiro, E., Castellucci, V. F. and Kandel, E. R., Presynaptic
membrane potential affects transmitter release in an identified
neuron in Aplysia by modulating the Ca^{++} and the K^+ currents,
Proc. Nat. Acad. Sci. USA, 77 (1980) 629–633.

24. Shapiro, E., Castellucci, V. F. and Kandel, E. R., Presynaptic
inhibition in Aplysia involves a decrease in the Ca^{++} current
of the presynaptic neuron, Proc. Nat. Acad. Sci. USA, 77 (1980)
1185–1189.

25. Shimahara, T. and Peretz, B., Soma potential of an interneurone
controls transmitter release in a monosynaptic pathway in
Aplysia, Nature, 273 (1978) 158–160.

26. Thompson, R. F. and Spencer, W. A., Habituation: A model for the
study of neuronal substrates of behavior, Psychol. Rev., 73
(1966) 16–43.

27. Tillotson, D., Inactivation of Ca^{++} conductance dependent on
entry of Ca^{++} ions in molluscan neurons, Proc. Nat. Acad. Sci.
USA, 76 (1979) 1497–1500.

28. Tillotson, D. and Horn, R., Inactivation without facilitation
of calcium conductance in caesium-loaded neurones of Aplysia,
Nature, 273 (1978) 312–314.

MODULATION OF SYNAPTIC PLASTICITY: A THEORETICAL APPROACH

Rolando Lara*

Center for Systems Neuroscience, University of
Massachusetts
Amherst, Massachusetts 01003

INTRODUCTION

The study of the processing of information during learning and memory has been approached from different points of view: behaviorally (25,26,38,42), theoretically (1,2,12,21,26,37,41) and physiologically (6,22,23,24,27,28,35,49). It has been difficult, however, to relate the results obtained by these different approaches. For this reason, the study of simple processes such as habituation, which has been extensively studied in vertebrates and invertebrates behaviorally (15,16,17,26,38,43,44,45,52,53), physiologically (10,13,14,22,23 24,27,28,49,50,51,54,55,57), and theoretically (1,13,14,16,17,43, 44,45,46,52,53), provides the possibility of analyzing the complex processing of information and establishing similarities and differences in the mechanisms. We will briefly review the relevant points in both the behavioral and the physiological studies of habituation, indicating their similarities and differences, and possible ways to relate them.

Habituation has been defined as the most ubiquitous behavioral modification found in animals, including humans. It refers to a decrease in behavioral response that occurs when an initially novel stimulus is repeatedly presented. Habituation is the means whereby animals (including humans) learn to ignore stimuli that have lost novelty or meaning (22).

* On leave from the Departamento de Neurociencias, Centro de Investigaciones en Fisiología Celular, Universidad Nacional Autónoma de México, México 20, D. F.

The general properties of habituation both behaviorally and physiological have been defined as follows (49) (see Hinde, (15) and our own comments for the limitations of these properties): 1) Repeated stimulation leads to a decrease of the response; 2) the more frequent the stimulus presentation, the more rapid and pronounced is the decrease; 3) the weaker the stimulus, the more pronounced is the decreased; strong stimuli may yield insignificant habituation; 4) if the stimulus is witheld, spontaneous recovery of the response occurs; 5) recovery may be further delayed by repeated stimulation, even after depression has been severe enough to abolish the response; 6) habituation occurs more rapidly after repeated periods of habituation trainings and spontaneous recovery; 7) presentation of a strong stimulus to the same or a different pathway produces restoration of response (dishabituation); 8) with repeated presentations, the effectiveness of dishabituation decreases; and 9) habituation of the response to a given stimulus can generalize to other stimuli.

Behavioral Habituation

Probably the first systematic behavioral studies of habituation were carried out on the orienting reflex in humans by Sokolov (43,44, 45). This reflex was first described by Pavlov as a non-specific reflex initiated by an increase, decrease, or qualitative change of a stimulus, independent of the modality of the stimulating agent and subject to extinction or habituation on repeated presentations. Using stimuli such as sound, cold or warm objects, and electrical shock, Sokolov measured the EEG. He used the EEG of the motor region, galvanic skin resistance, muscle tension, eye movements, and respiration to show that the first presentation of the stimulus produced a generalized response which decreased after stimulus repetition, until it completely disappeared, except for the specific modality stimulated.

Sokolov demonstrated that habituation is not a decrease in sensitivity, because a change in intensity or any other parameter of the habituated stimulus restores the orienting response. Based on this, he proposed that the brain creates a model of the stimulus which is compared with the actual object. If a mismatch occurs, the arousal response reappears, whereas if the model and the stimulus match, then the response gradually disappears. This led him to propose that habituation is the elaboration of an inhibitory conditioned reflex from the cortex, regulating the transmission of impulses to the reticular formation, possibly by hyperpolarization of synaptic connections. Later physiological studies on the hippocampus have supported the hypothesis of the creation of the model of the stimulus, because some cells start to respond to the different temporal parameters of the stimulus (50,51). Other behavioral studies (9,15) have also confirmed the hypothesis that habituation is stimulus specific.

Based on his behavioral studies on habituation, Konorski (26), proposed that habituation is produced by an inhibitory perceptive recurrent reflex, which is mediated by a gnostic unit that processes and stores information. When the stimulus presented is already in the gnostic unit, the arousal system is inhibited, and therefore habituation is present. Similarly, Wagner (52,53) proposed that habituation is related to the content of the short term memory and to the presentation of the stimulus. If the stimulus received and the stimulus stored in short term memory are the same, then no response is produced; while if they are different, the retrieval mechanism is activated and a general arousal response is present. In this way, habituation is related to the time necessary to make a model of the stimulus. He explains long-term habituation as the retrieval from long to short-term memory of the stimulus when a cue object is presented. Stein (46) postulated that habituation is simply the classical conditioning of an inhibitory effect that supresses the general response of arousal to an excitatory input. Hernandez Peon (13,14) suggested that habituation is related to an increased inhibitory activity from the reticular formation to the most peripheral sensory nerves, controlling in this way the inflow of sensory information.

Ewert and Ingle (7,20) studying habituation in amphibia have shown that this process is dependent on the general motivational state of the animal. These authors have shown that the rate and level of habituation is greatly reduced in good feeder frogs, in toads in an environment where food odor is present, and in animals with pretectal lession or intoxicated by alcohol, which may reduce the inhibitory effect of the fear drive upon the feeding drive. Conversely, when a threatening stimulus is presented, the rate and level of habituation is greatly increased.

Based on the above conclusions we can propose that behavioral habituation has the following characteristics: 1) habituation is stimulus specific: there is a model of the stimulus in the brain which is continually compared with the present stimulus; the time required to create this model in the brain defines the level of habituation; 2) habituation and dishabituation have a global effect: habituation is the result of a general inhibitory effect upon different behaviors controlled by different brain regions; whereas dishabituation produces a general arousal response when the stimulus is presented; 3) habituation is independent of the site of stimulation: the representation of the stimulus in the brain is processed independently of the sensors excited. This suggests that the model of the stimulus is stored and retrieved depending on the representation of its temporal parameters; and 4) habituation is dependent on the general state of the animal: the animal behavior is controlled by internal variables, the environment and past experience.

Physiological Studies

From a physiological point of view, studies on habituation in the spinal cord, the neuromuscular junction, and invertebrates have shown that habituation is produced by a homosynaptic depression of the amount of transmitter released by the presynaptic terminal. Furthermore, Kandel et al. (22,23) have shown that the reduction in amount of transmitter release is due to a decrease in membrane permeability to calcium ions (see Klein, this volume). The properties of habituation described in the spinal cord are also observed in Aplysia. This suggests that physiological habituation, as a general phenomenon, may be produced by the same mechanisms in different animals. Physiological studies in other preparations have confirmed this hypothesis (27).

The studies of habituation in the spinal cord (11) led to the postulate that habituation is controlled by a dual process: habituation as a homosynaptic depression and sensitization by other cells. This theory proposes that the sensitizing cells are subject to changes depending on the general state of the animal, trying to correlate the behavioral and physiological studies of habituation. The general postulates of the dual process theory are as follows: 1) every stimulus that evokes a response also influences the state of the organism; 2) habituation has all the parametric properties mentioned above; 3) sensitization must be present with all its properties; 4) habituation and sensitization are independent.

On the other hand, studies of simple plastic properties, including habituation, have shown that the intrinsic synaptic plastic properties can be controlled by exogenous factors, which involve mechanisms in addition to those proposed by Groves and Thompson, eg. presynaptic and postsynaptic inhibition and excitation both with short and long term effects (28,47,56). These studies indicate that the state of the animal can modify the plastic properties of cells in different ways. Krasne (28) proposes the following ways by which exogenous control can modify inherent neuronal plasticity: 1) permission for change; 2) suppression of change; 3) erasure of change; 4) extrinsic maintenance of change; 5) stabilization of change; and 6) controlled expression of change. (For more details of how these exogenous control may be applied to behavioral and physiological habituation see below).

Physiological habituation, in contrast to behavioral habituation, and in addition to the already defined properties of this phenomenon, has the following characteristics: 1) habituation is not stimulus specific: a model of the stimulus in the brain is not created, consequently arbitrary changes in the parameters of the stimulus do not produce dishabituation; 2) habituation is a local, spatially located, intrinsic process: the reduced response is specific to a given region, even when generalization of habituation occurs; 3) habituation is dependent on the site of stimulation; and 4) habituation at a cellular level is, similar to behavioral habituation, dependent on the general state of the animal.

We have indicated that behavioral and physiological habituation share common properties but also have great differences. The mechanisms responsible for these changes, however, independent of the type of habituation, are both modulated by the general state of the animal. In the present paper, we propose a model of how both behavioral and physiological habituation can be modulated by the motivational state of the animal. We assume, for simplicity, that behavioral habituation is independent of the stimulus, except when we specifically simulate how a model of the stimulus can be created, compared with the present object, and if a mismatch occurs, produce an arousal response. We also postulate that both types of habituation are produced by an intrinsic process that can be exogenously controlled. Furthermore, we propose that the intrinsic plastic process of behavioral habituation is similar to those found in physiological studies. These assumptions permit us to study the modulation of behavioral habituation by the general state of the animal.

Modulation of Behavioral and Physiological Habituation

In this section we describe some of the possible ways that habituation, both behavioral and physiological, may be modulated based on the results obtained in different animals and following the postulates proposed by Krasne (28) for exogenous control of intrinsic plastic properties, not only for single cells but for a group of neurons controlling a given behavior.

The possible mechanisms for controlling the intrinsic plastic properties of cells are: 1) presynaptic inhibition and excitation; 2) sensitization and heterosynaptic inhibition (also presynaptic); and 3) postsynaptic excitation and inhibition. From a behavioral point of view these mechanisms will be equivalent to short and long term excitatory and inhibitory effects on a given behavior.

Exogenous control of habituation has been described in the following way:

1) Permission for change: Krasne et al. (27,28) have shown that the tail flip response to phasic tactile stimulation of the abdomen of crayfish habituates to a repetitive stimulus. However, if the animal is executing a motor action, habituation is prevented by presynaptic inhibition, as habituation of a given pathway to motion would be a very disadvantageous property of the animal for its adaptation to environmental changes. This evidence and similar results in locust and humans (for example, inhibition of habituation

in the auditory system to one's own voice) suggest that presynaptic inhibition controlling habituation could represent a possible potential gate for the inflow of information. We have seen that this hypothesis has also been made from behavioral studies (13,14).

2) Suppression of change: when habituation has occurred, the animal needs a fast way to suppress the change to confront new stimuli; this suppression of the habituation has been named dishabituation. Dishabituation has been described in humans by Sokolov in his studies of the orienting reflex. If there is a change in any of the parameters of the stimulus (intensity, duration, pattern, etc.) the orienting response reappears. Behavioral dishabituation has also been described in several animals, such as birds (15), toads (9), rabbits (50,51), and other animals (15,28). Physiological dishabituation has been described in spinal cord of cat and in Aplysia, and has been called sensitization. Sensitization occurs when a different pathway is activated or when the intensity of the habituated stimulus is greatly increased. Groves and Thompson (11) have proposed that sensitization is produced by a potentiated excitatory pathway arriving at the same cell as the habituated neuron. Kandel et al. (23) have shown in the abdominal ganglion of Aplysia that sensitization is produced by the presynaptic action of another transmitter in the habituated pathway, increasing the amount of transmitter release. The increase in transmitter release has been associated with an increase in Ca^{2+} membrane permeability mediated by cyclic AMP. Based on points 1 and 2, we conclude that physiological habituation may be modulated by a short term effect, presynaptic inhibition, and a long term effect, sensitization. The behavioral equivalent for these controls are short term inhibition and long term excitation. It is important to notice that sensitization is not an erasure of habituation, but a superimposed plastic process that facilitates habituated and nonhabituated pathways.

3) Erasure of change: the only example of real erasure of habituation, in contrast to sensitization, occurs in the locust, when an excitatory stimulus different from the habituated stimulus is presented. The nature and mechanisms for this change are unknown.

4) Extrinsic maintenance of change: habituation can be maintained if an external input excites the habituated pathway postsynaptically at a frequency that normally does not result in habituation.

5) Stabilization of change: the stabilization of habituation seems to be an intrinsic property of this plastic process, as indicated by repetitive trials of habituation and dishabituation by rest; however it is not known which mechanisms could control the transition from short to long term habituation. In the model proposed in this paper, we postulate that inhibitory and excitatory effects may control this transition; from a physiological point of view,

we postulate, again, that presynaptic inhibition and sensitization may regulate the consolidation of habituation.

6) Controlled expression of change: with this mechanism the phenomenon of habituation can be finely regulated, due to its general response properties for different intensities of stimulation. We have mentioned that habituation is stronger for weaker stimuli. For this reason, if there is a mechanism that control the intensity of the stimulus, then, at the same time, it will regulate the amount and level of habituation. This regulation will be a step further toward the suppression of habituation by presynaptic inhibition, described above, and consequently, a finer mechanism for the control of input information. This fine control of habituation has been described in the peripheral ganglion of Aplysia (31,32,39,40), where different internal conditions of the animal, like age, satiation etc., change the rate and level of habituation. This control has been related to an inhibitory and an excitatory effect of the branchial and ctenidio-genital nerve upon the peripheral cells, respectively. From a behavioral point of view, a similar modulation (or control) has been described in amphibia (7,8,9,18,19,20), where the rate and level of habituation can be controlled by changes in the internal and external variables of the animal and its environment, respectively, such as the ingestion of alcohol, lesions in the pretectum and forebrain, odor related to preys, etc.

In the next sections, we propose a general model of how the state of the animal and its relationship with the environment regulates the phenomenon of habituation as an adaptive mechanism for its different needs. The model is based on the physiological studies of habituation as well as upon the possible exogenous control that can be exerted by the general state of the animal. We define the model in terms of blocks, where in the case of physiological studies each block may represent a single neuron, while for behavioral studies each block may represent brain regions regulating the proper behavior. We simulate the phenomenon of habituation as a homosynaptic depression of the amount of transmitter release that can be modulated by presynaptic inhibition and sensitization. These mechanisms will constitute the means by which the general state of the animal will control the manifestation of habituation. For the simulation of the state of the animal, we have considered the different factors that may change the motivational state of the animal, using a hunger model as a case study to exemplify how these factors may modify the behavior of the animal and its habituation.

MATHEMATICAL MODEL OF HABITUATION

The mathematical model of habituation has been described else-where (29); therefore, just a brief summary of the relevant points will be presented here.

We have postulated that the amount of transmitter release by the presynaptic terminal is given by the following expression:

$$E = n*(I(u)-INH)*F*q \qquad (1)$$

where n represents the number of vesicles, considered constant; q is the average synaptic potential per vesicle, also constant; $I(u)$ is the physiological stimulus; INH is the effect of presynaptic inhibition over the physiological stimulus; F is the membrane permeability to Ca^{++} ions; and E is the amount of transmitter release.

It has been shown that habituation is produced by a reduced membrane permeability to calcium ions (23). For this reason, and taking into consideration that habituation depends on the amplitude and frequency of stimulation, F can be defined as follows:

$$\dot{F}=KO/LT(FO-F)-f(B-E)+SEN \qquad (2)$$

where F is the membrane permeability to Ca^{++}; FO is the equilibrium value of F; KO and B are constants; LT is the variable that regulates the transition from short to long term habituation; and SEN is the sensitizing effect of another synapse, that, as Kandel (23) has suggested is a mirror phenomenon to habituation. The function $f(B-E)$ is defined as follows:

$$f(B-E)= \begin{cases} K1(B-E)+ & IF\ E>0 \\ 0 & ELSE \end{cases}$$

The term $(B-E)+$ is introduced as an indirect measure of the amount of Ca^{++} introduced by the physiological stimulus through the variable E, and permits us to simulate the amplitude and frequency dependence of habituation.

The variable regulating the transition from short to long term habituation, LT, is defined as follows:

$$\dot{LT}=g(F)-(LT)\ (K3) \qquad (3)$$

where K3 is constant; and $g(F)$ is defined as follows:

$$g(F)= \begin{cases} K4 & IF\ K5<F<K6\ and\ \dot{F}>K7 \\ 0 & ELSE \end{cases}$$

This function simulates the fact that long term habituation is dependent on the number of times the pathway has been habituated $(K5<F<K6)$ and recovered spontaneously $(\dot{F}>K7)$.

Based on the mathematical definition of this model, we will

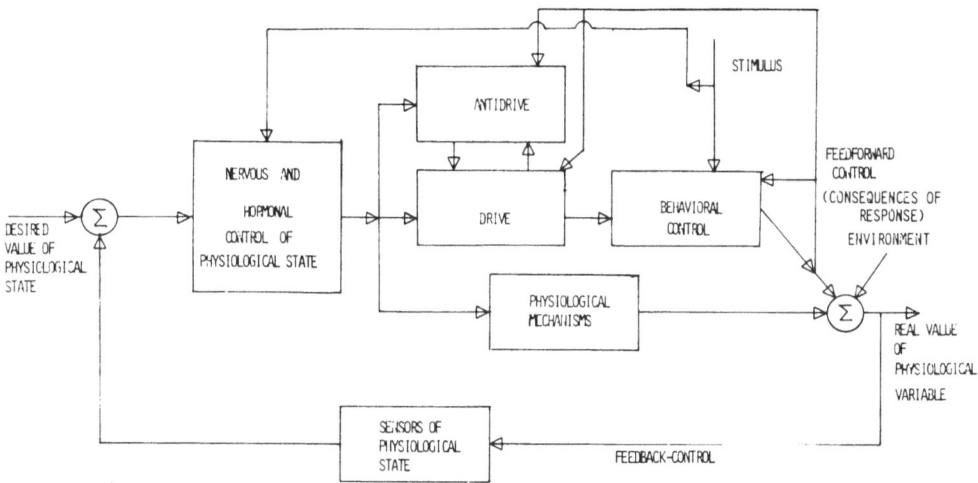

Fig. 1.   Mechanisms that control the state of the drive: i)' feedback
control, through changes in the physiological state as a consequence
of the response; ii) feedforward control, as the immediate effect
of the response; iii) the stimulus, through its association to the
drive; and iv) drive-antidrive interactions, through their mutual
inhibitory effects.

show in the next section how the general state of the animal and the
interaction with its environment can extrinsically control the
phenomenon of habituation.

BEHAVIORAL MODEL OF STATE

We view the animal as having a number of distinct categories of
behavior each of which is controlled by one of a number of separate
systems, called instincts by Tinbergen, and more recently called
motivational systems (drives) (26,33,34). Konorski (26) has
classified the drives as preservatory and protective. He defines as
preservatory drives the following: 1) assimilation of necessary
materials (feeding and drinking); 2) excretion; 3) recuperation
(sleep); and 4) preservation (sex). He considers as protective
drives: 1) withdrawal of the whole body (fear); 2) rejection of
harmful agents; 3) annihilating harmful agents (rage).   In general,
all these drives have mutually inhibitory interactions, the strongest
one controls the behavior of the animal; but this mutually inhibitory
effect is most dramatically observed between protective and
preservative drives: fear against feeding, drinking etc. He postulates
that each drive has its respective antidrive, which controls the
general state of the drive depending on the consequences of the
response.   Examples of drive-antidrives are: hunger-satiation,

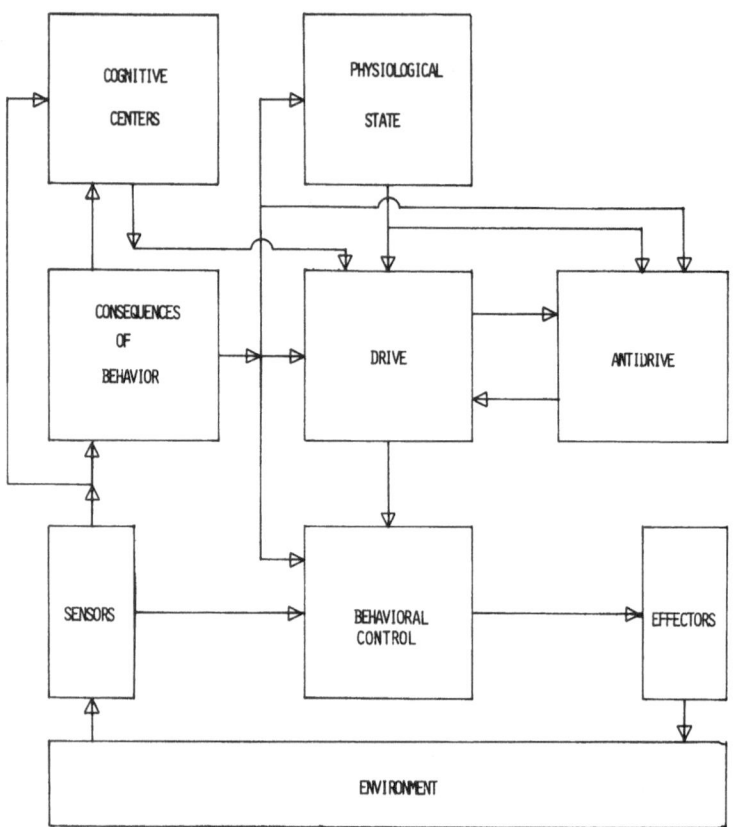

Fig. 2. Block diagram of the factors that control the state of the
drive (see text).

fear-relief, sex-sexual satisfaction, anger- placidity, etc. Moreover,
Konorski postulates that the emotive brain, the brain regions that
regulate drives, can have interactions with the brain regions related
to the processing of information, such as perception, recognition
etc. For example the associative value of a stimulus to a drive is
given genetically or through learned experience.

     The general state of the drive system can be controlled by the
following variables (26,33): 1) the physiological state of the animal,
which through nervous and hormonal centers regulate the drive state
through feedback and feedforward control systems (see Fig. 1); 2) the
stimulus, through its significance for the drive (stored information
in memory, probably genetically given or learned by experience) (Fig.
2); 3) state of other drive centers of the same motivational system,
for example, increased activity of the fear drive reduces the activity
of the feeding and drinking drives and vice versa (Fig. 3);

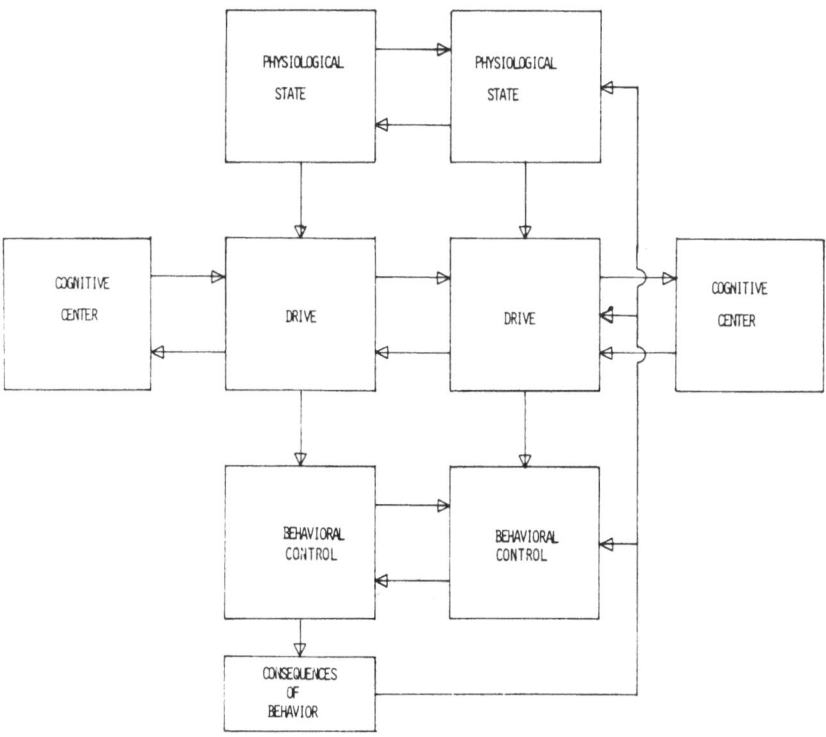

Fig. 3. Block diagram of the interaction between drives. The relationship among different drives and among different behaviors is mutually inhibitory, while the changes in the physiological state of the animal can have facilitatory or inhibitory effects. The consequences of the response produced by one drive may change the factors regulating the state of another drive (see text).

4) behavioral consequences of the response, both with short and long term effects. Among the short term effects, we have the immediate consequence of the response, a feedforward control, on the drive that elicits the behavior in order to continue it (i.e. feeding behavior), on the antidrive for its ending (satiation), and the effect on other drives, as mentioned above. Long term effects are produced through feedback systems controlling the physiological variables of the animal (see Fig. 1,2, and 3).

We have considered that both the drive center and the information processing system control the state of the modulator of the plastic property, which defines the development, expression and maintenance of the plastic phenomenon depending on the different conditions of the animal (Fig. 4). In some cases, the information processing system can act independently from the drive center; for example, a change in any of the parameters of the stimulus produces a general arousal

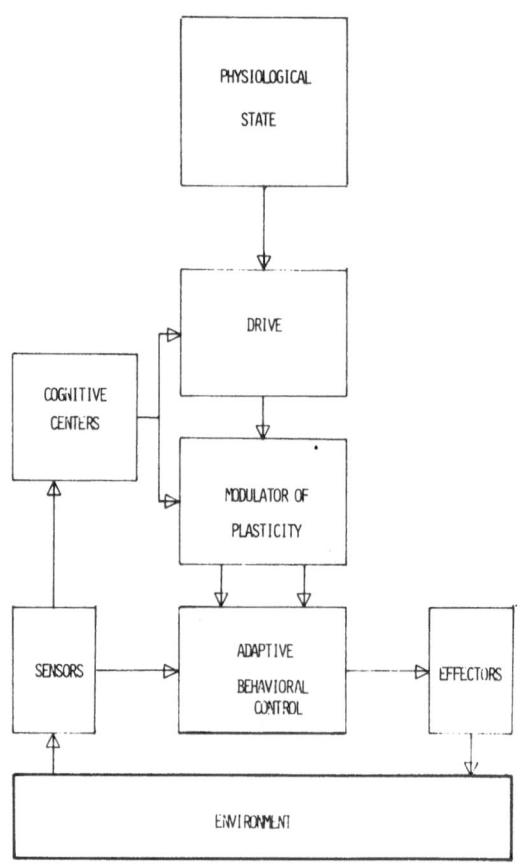

Fig. 4. Block diagram of the process related to the modulation of
a plastic behavior. The modulator of plasticity is controlled by
the state of the drive and by the cognitive centers. The cognitive
centers may act upon the modulator in two ways: indirectly through
the association of the stimulus to the drive; and directly, when
any of the parameters of the stimulus has been modified. The action
of the modulator upon the control of behavior can be both excitatory
and inhibitory.

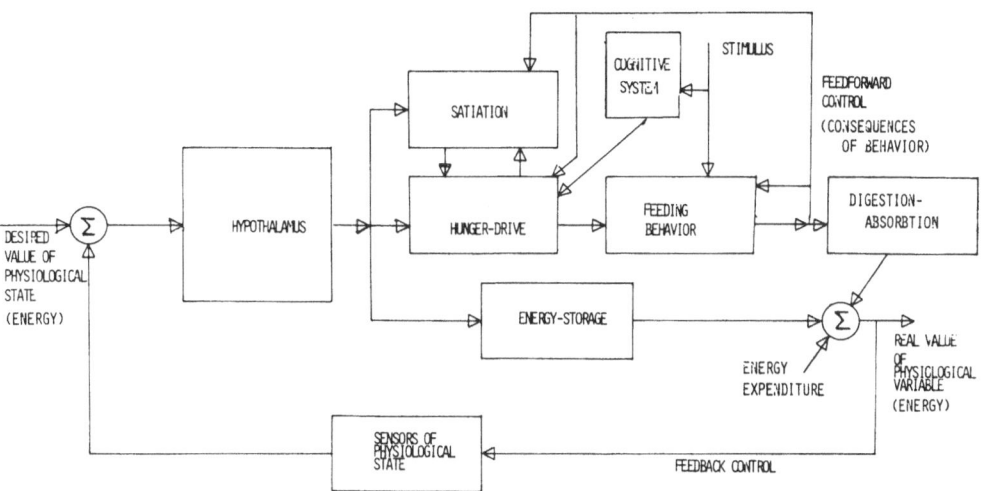

Fig. 5. Block diagram of the control system regulating the amount
of energy available in an animal. The hypothalamus regulates the
state of the drive and antidrive depending on the error signal coming
from the comparator (the sum). The hunger drive controls feeding
behavior, when a stimulus is presented this behavior is released.
The consequences of the response may modify the feeding drive in
two ways: through feedforward effects on both the hunger and satiation
center, and through feedback control through digestion and absorption
of food into the organism. The stimulus can increase the hunger
drive through its association with food which is recognized by the
cognitive centers.

response. In this way, the behavioral response is released when the
stimulus is present depending on the state of the drive center, the
results of the information processed, and the history of the plastic
process.

As we have mentioned above, the modulator controls the expression
of the plastic behavior through inhibition and excitation
(physiologically through presynaptic inhibition and sensitization).
In the next section, we will briefly describe a simple hunger control
system; this model will be used to study the effects on the state of
this drive upon feeding behavior and its habituation under different
circumstances.

Model of a Hunger Control System

The control system regulating the feeding drive is shown in Fig.
5. We have postulated that the amount of energy available for the
animal (RE) depends on the food ingested and absorbed (DA), the
energy released from the energy store (ES), and the energy

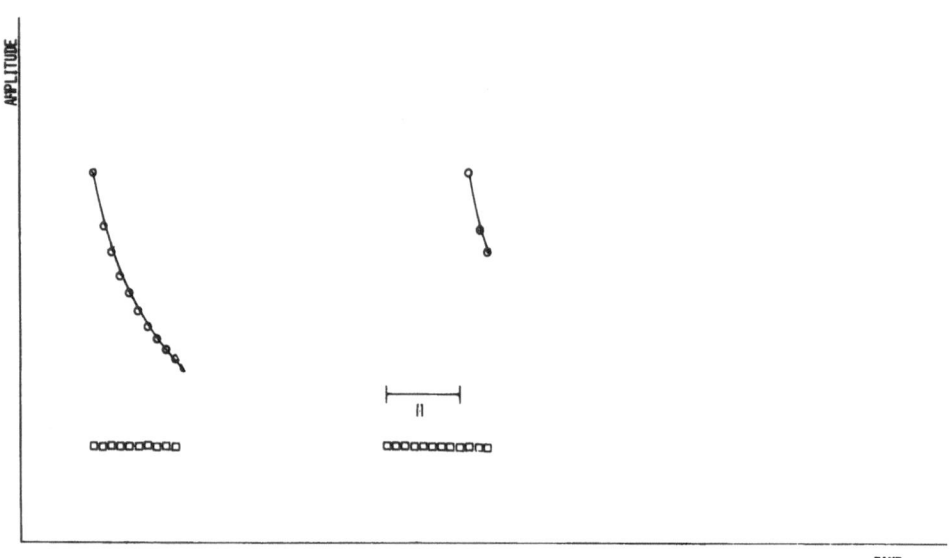

Fig. 6. Computer simulation of control of habituation through
presynaptic inhibition in the tail flip escape response in crayfish.
When the second trial of stimuli is presented (squares), the motor
center is simultaneously active (M), blocking the transmitter release
from the terminal, thus preventing the occurrence of habituation.
When the motor activity is not present, the pathway starts to
habituate in the normal way.

expenditure (EE) of the animal. When (RE) is below the desired normal
energy (NE), an error signal (E) is generated which activates the
feeding drive (Fdd) and releases energy from ES; while if RE is
above NE, then E activates the satiation center (SAT) and stores
energy to ES. In the first case, when E>0, the feeding drive is
increased until stimulus is presented releasing feeding behavior. If
the stimulus is food, then both the feeding drive and the satiation
centers are feedforwardly activated and the physiological variables
are modified by the feedback system through the digestion and
absorption of food (for a more detailed analysis of this model see
36).

     Mathematical model of the hunger control system. The amount of
energy available for the animal (RE) is given by the following
expression:

$$\dot{RE}=(E)^{+}(ES)\ (w1)-EE+DA-(E)^{-}(w2) \qquad (4)$$

where RE is the energy available for the animal; ES is the amount of
energy stores; EE is the energy expenditure, that we have considered
constant; w's are constants; and E, the error signal, is given by:

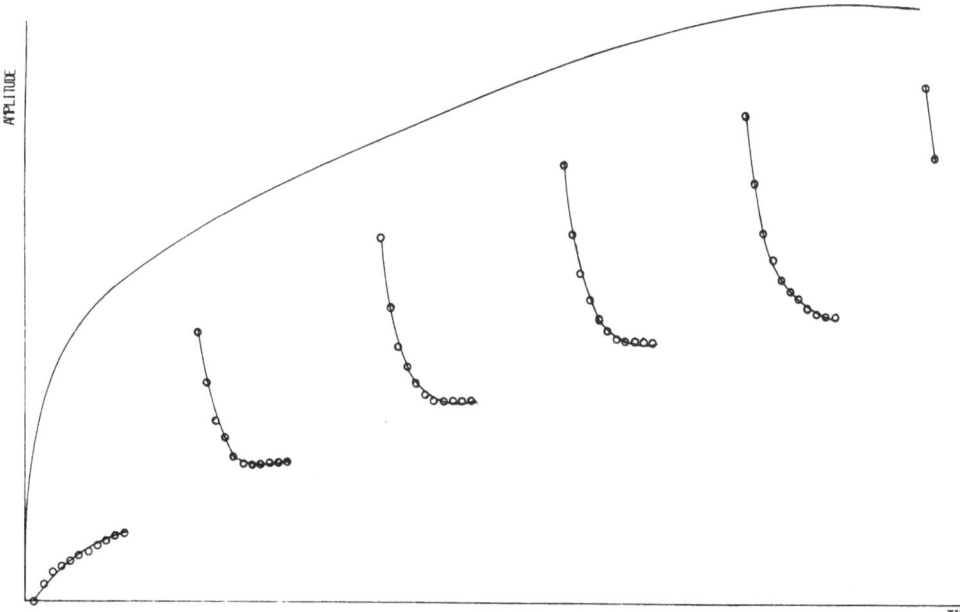

Fig. 7. Computer simulation of the control of habituation of prey orienting behavior by the hunger drive. In this and subsequent figures the continuous line represents the state of the hunger drive, while the circles represent the strength of the response when a stimulus is presented. This figure shows that the strength of the response increases proportionally to the intensity of the hunger drive, reducing the rate and level of habituation.

$$E = NE - RE \qquad (5)$$

where NE is the equilibrium value of the energy of the animal; and RE is the real energy. (E)+ and (E)- are defined by:

$$(E)+ = \begin{cases} E & \text{IF } E > 0 \\ 0 & \text{ELSE} \end{cases}$$

$$(E)-= \begin{cases} -E & \text{IF } E < 0 \\ 0 & \text{ELSE} \end{cases}$$

The factor $(E)^-$ (w2) simulates the storing of energy on ES when RE>NE; while $(E)^+(ES)$(w1) represents the amount of energy released from ES when RE<NE.

The variable defining the digestion and absorption (DA) of

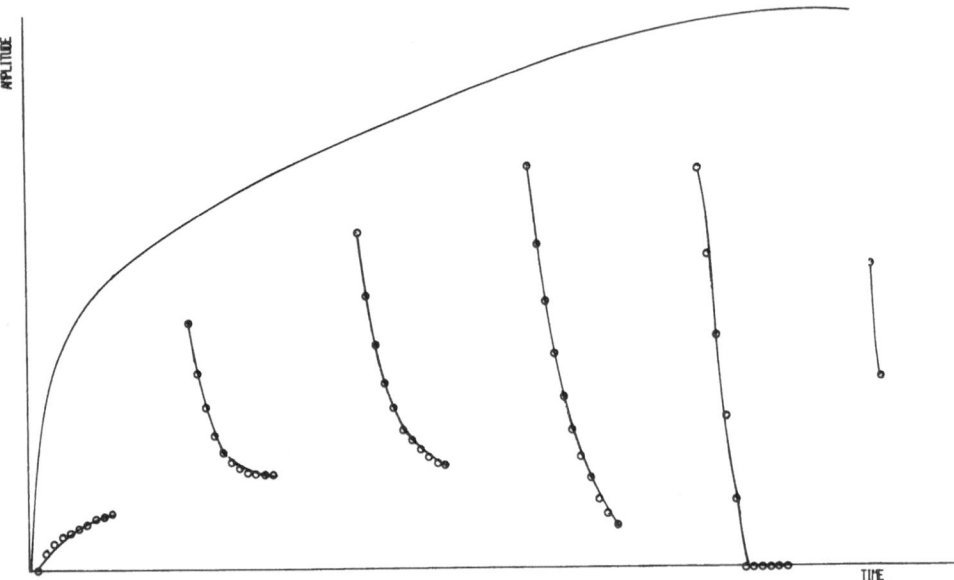

Fig. 8. Computer simulation of the control of long term habituation of the prey orienting behavior by the hunger drive. The transition from short to long term habituation increases the rate and level of habituation dependent upon the number of times the pathway has been habituated and dishabituated by rest.

food is given by:

$$\dot{DA} = (St) - (w3)(DA) \qquad (6)$$

where constant w3 regulates the rate of absorption of food; and St is the stimulus ingested defined by:

$$St = \begin{cases} 1 & \text{If food is ingested} \\ 0 & \text{ELSE} \end{cases}$$

The energy storage (ES) is controlled by the following variables:

$$\dot{ES} = (E)^{-}(w2) - (E)^{+}(ES)(w1) \qquad (7)$$

which are the energy released and stored, respectively, as mentioned above.

The feeding drive Fdd is expressed as follows:

$$\dot{Fdd} = (w4)(E)^{+} - (w4)(SAT) - (w5)(Fed)(Fdd) + (w6)(CS)$$
$$+ (w7)(BC) - (wB)(Fdd) \qquad (8)$$

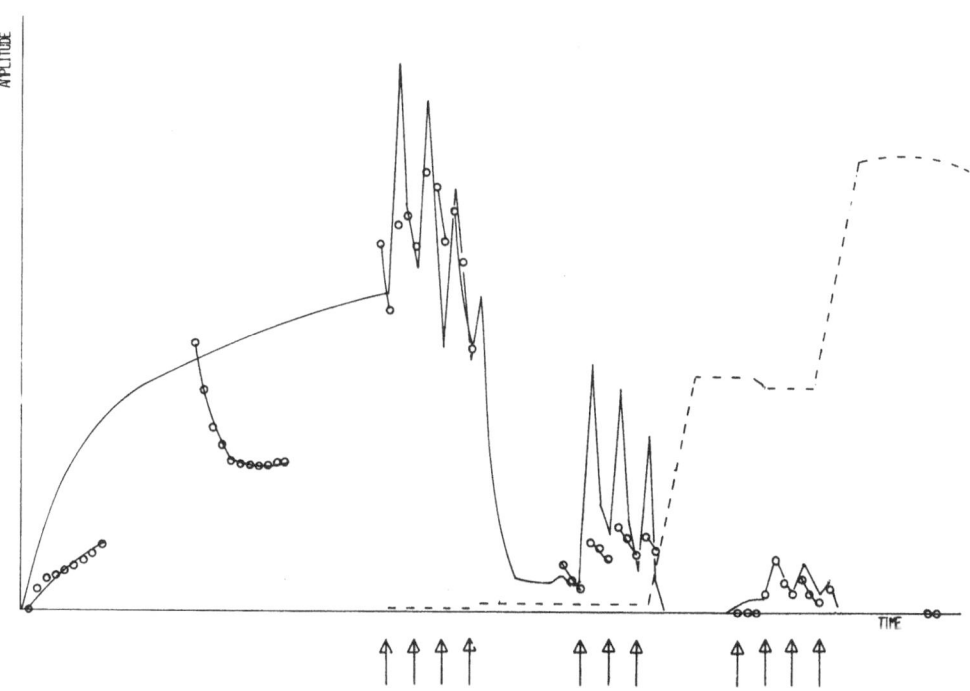

Fig. 9. Computer simulation of the relationship among the hunger
drive, the satiation center, and short term habituation of prey
orienting behavior. In the third trial of stimulus presentation,
a food stimuli (arrows) is introduced every third presentation of
the object. Each time the food stimulus is present, the feeding and
the satiation center are feedforwardly activated (the state of the
satiation center is shown by the dotted line). These effects on the
satiation center are low in comparisson to the feedback action
exerted by the feedback control (see next figure to see clearly the
effects of the feedforward signal). Notice that when the stimulus
is presented and eaten, the feeding drive is stimulated but is
subsequently reduced by the inhibitory effects of the satiation
center. At the end of the fourth trial, after seven food stimuli
have been ingested, the feedback control system is activated, because
part of the food has been absorbed and digested, reducing
dramatically the level of the feeding drive and, consequently, the
level of the response. In this situation the rate and level of
habituation is greatly increased.

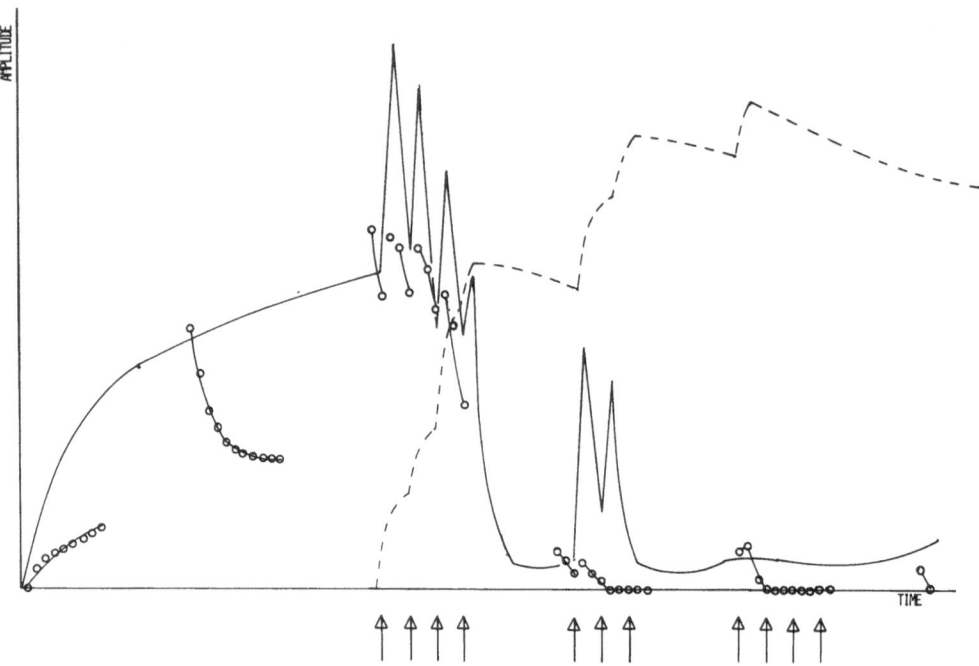

Fig. 10. Computer simulation of the relationship among the hunger
drive, satiation and long term habituation. In this case, it can be
seen that the effects of long term habituation prevents the ingestion
of the seventh food stimulus. For this reason the satiation center
is only affected by the feedforward control, because the feedback
control has been delayed. When the satiation center starts to
decrease after the trial, the feeding drive increases proportionally
producing a new response when the stimulus is presented again. The
intensity of the response, however, is lower and the rate and level
of habituation is faster than in short term habituation.

where w's are constants; E is the error signal; SAT is the state
variable defining the level of satiation; Fed is the effect of the
fear drive over the feeding center (see Fig. 3); CS is the conditioned
stimulus associated with food, i.e. food odor; BC represents the
consequences of the response, i.e. palatability of the stimulus. The
satiation level (SAT) is defined as:

$$\dot{SAT} = (w9)(E) + (St)(w10) - (SAT)(w11) - (Fdd)(w17) \qquad (9)$$

where E is the error signal; St is the ingested stimulus; Fdd is the
inhibitory effect from the feeding center; and w's are constants. The
first factor in this equation represents the feedback control

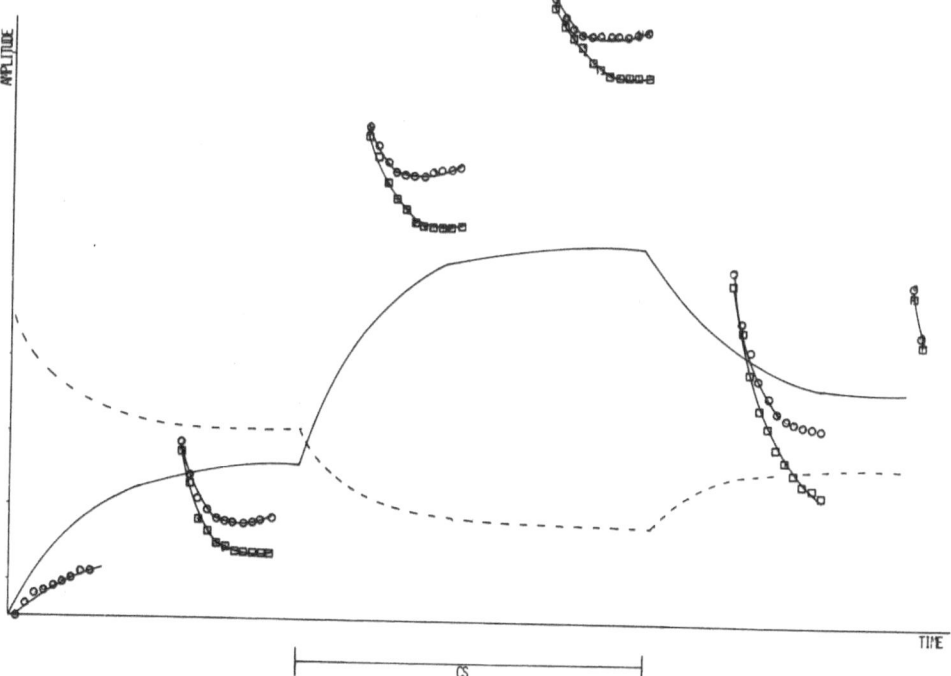

Fig. 11. Computer simulation of the effects upon the hunger drive
and habituation of prey orienting behavior when a stimulus associated
with food is presented. At the end ot the second trial, a stimulus
associated with food is introduced (CS). In this conditions, the
feeding drive is greatly increased and the strength of the response
during the second and third trial of presentation of the stimuli
is greatly enhanced, reducing the rate and level of habituation.
It can also be seen that the transition from short to long term
habituation (squares) is greatly reduced. The dotted line indicates
the state of the fear drive and how is inhibited by the enhancement
of the feeding drive.

produced by the changes in the physiological variables; while the
second one is the feedforward effect on this center.

The variable regulating the fear drive is given by:

$$\dot{Fed}=FCS+FUS-(Fdd)(Fed)+(FEDO-Fed)(w12) \qquad (10)$$

where FCS is the conditioned stimulus associated with fear; FUS is
the strength associated with the threatening stimulus; Fdd is the
inhibitor effect from the feeding center upon the fear drive; and
w12 is a constant.

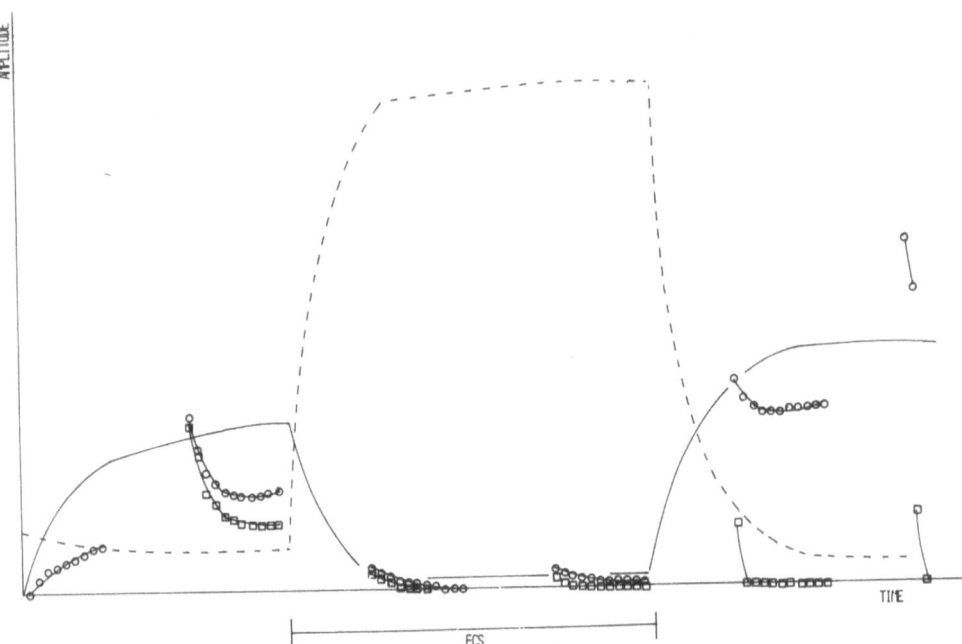

Fig. 12. Computer simulation of the effects upon the hunger drive
and habituation of prey orienting behavior when a threatening
stimulus is introduced. At the end of the second trial, a stimulus
associated with fear is introduced (FCS). In this case the fear
drive (dotted line) is greatly increased, reducing dramatically the
feeding drive; therefore, the response to the presentation of the
stimulus is almost completely abolished. Notice that when the fear
stimulus is withheld, the long term effects of habituation (squares)
do not allow the recovery of the normal response (circles), and the
rate and level of habituation are greatly increased.

Finally, CS, BC, FCS, and FUS are defined as follows:

$$CS = \begin{cases} 1 & \text{If the stimulus is associated with food} \\ 0 & \text{ELSE} \end{cases}$$

$$BC = \begin{cases} 1 & \text{If positively rewarded} \\ 0 & \text{ELSE} \end{cases}$$

$$FCS = \begin{cases} 1 & \text{If stimulus is associated with threat} \\ 0 & \text{ELSE} \end{cases}$$

$$FUS= \begin{cases} 1 & \text{If FUS is a threatening stimulus} \\ 0 & \text{ELSE} \end{cases}$$

Feeding Drive and Habituation

We have indicated that the feeding drive controls feeding behavior and its habituation when the stimulus is presented. This relationship can be seen in Fig. 4. The mathematical definition of the modulator is given by:

$$\dot{MOD}= (w13)(NMOD-MOD)^+(Fdd)(w14) \qquad (11)$$

where MOD is the activity of the modulator; NMOD is the equilibrium value of MOD, considered as its spontaneous activity; Fdd is the feeding drive; and w's are constants.

The modulator exerts the inhibitory effect over the plastic process INH in the following way:

$$INH= (C1-MOD)+(w15) \qquad (12)$$

where MOD is the state of the modulator; C1 is a constant which simulates the tonic inhibitory effect exerted upon the adaptive behavior; and w15 is constant.

The excitatory activity (SEN) exerted by MOD to the plastic center is defined by:

$$SEN= (MOD-NMOD)^+(w16) \qquad (13)$$

where MOD is the activity of the modulator; NMOD is the equilibrium value of MOD; and w16 is constant. This equation shows that SEN is only active if MOD>NMOD.

The block defined as plastic center in Fig. 4, which in this case is habituation, was defined mathematically above, and it can be considered to be applicable to a single cell, in the case of physiological habituation, or to a group of cells, in case of behavioral habituation. INH and SEN are simply what we considered in the mathematical model of habituation as presynaptic inhibition and sensitization.

COMPUTER SIMULATION OF MODULATION OF HABITUATION

Permission for Habituation During Motor Behavior

It has been shown (3,4,27,28) that the tail flip reflex of the crayfish can be protected from habituation when the animal is moving

and that the physiological mechanism for this protection is presynaptic inhibition. Fig. 6 shows the simulation of the protection
of habituation during active motor behavior. In the figure it can
be seen that a pathway that would be habituated by the presentation
of the stimulus gives a normal response when the inhibitory effect
is released. For the simulation of this experiment we simply define
INH as:

$$INH= \begin{cases} 1 & \text{If motor activity is on} \\ 0 & \text{ELSE} \end{cases}$$

For other physiological modulation of habituation see Lara et al.
(29).

Hunger Drive and Habituation

For the purpose of showing how habituation of a given behavior
is modified by the different conditions of the animal and its
interactions with its environment, in the simulation we present a
series of trials consisting of a train of pulses and a period of rest.
Unfortunately, the experimental studies of the relationship between
the hunger drive and habituation have been mostly done from a
qualitative point of view; therefore, a quantitative comparison of
the results of the model with actual data cannot be done. For this
reason only a qualitative comparison will be indicated.

Fig. 7 shows the changes of response to the presentation of the
stimulus, using the trials mentioned above, when the feeding drive
is increasing. In the first trial can be seen a gradual increase of
the response, with subsequent trials starting to show habituation.
This figure shows that the strength of the response, and the rate and
level of habituation, are controlled by the level of the feeding
drive. When the feeding drive increases, the strength of the response
is also increased but the rate and level of habituation are reduced.
This experiment shows that habituation is less pronounced when the
animal is hungry, which could have an adaptive value in terms of
the probability of finding food. This general effect of the hunger
drive upon habituation of prey catching behavior has been observed
in good feeder frogs (20) or in toads whose hunger drive has been
increased by external factors (7) (see below).

If in the above conditions we show the effect of the transition
from short to long term habituation, we see that depending on the
number of trials habituation is greater and faster and dishabituation
is slower, which increases the adaptive behavior to unimportant
stimulus of a hungry animal (Fig. 8).

Hunger Drive, Satiation and Habituation

If in the above paradigm, when the feeding drive has considerably increased, we introduce a food stimulus every third presentation of stimuli, we can see the effect of satiation upon short term habituation (Fig. 9). When the food stimulus is presented (arrows), the feeding drive is sensitized, because BC=1, but the feedforward effect of the stimulus upon the satiation center is also activated, inhibiting, then, the state of activity of the feeding drive; when several food stimuli have been ingested, the feedback control of satiation is activated, inhibiting further feeding behavior. When the satiation center reduces its inhibitory effect over the feeding drive a new response is released but of smaller strength and with faster habituation. In this way we simulate the dynamic changes occurring during feeding behavior and its effects on habituation.

If in the above conditions, long term habituation is presented, we can see (Fig. 10) that the response takes longer to reappear, showing that the transition from short to long term habituation increases the general effects of satiation. As a consequence of this effect the animal does not eat as many preys as in the above case, delaying the effects of the feedback control upon the satiation center. Moreover, in these conditions the transition from short to long term habituation is faster than in the above cases, suggesting a possible way of regulating the consolidation of a plastic process through changes in the internal variables of the animal.

Changes in the Environment, Hunger Drive, and Habituation

When we introduce a stimulus (CS) that has been associated with food, such as food odor or some other cue, then the hunger drive is increased, augmenting the strength of the response and reducing the rate and level of habituation (Fig. 11). These results reproduce, in general terms, the behavior observed in amphibia when worm odor is present in the environment, which increases the strength of the response and reduces the rate, the level of habituation, and the fear drive to threatening stimuli (see below) (7,20). When long-term habituation is present, it can be seen that the transition from short to long term habituation is greatly reduced, representing another example of how the different conditions of the animal-environment complex can regulate the expression and maintenance of change (Fig. 11).

Competition Between Drives and Habituation

Several drives can interact among each other in a given motivational system (20,26,33,34). For example, it has been observed in frogs and toads that feeding and fear drives have mutually inhibitory effects: good feeder frogs, frogs in an environment

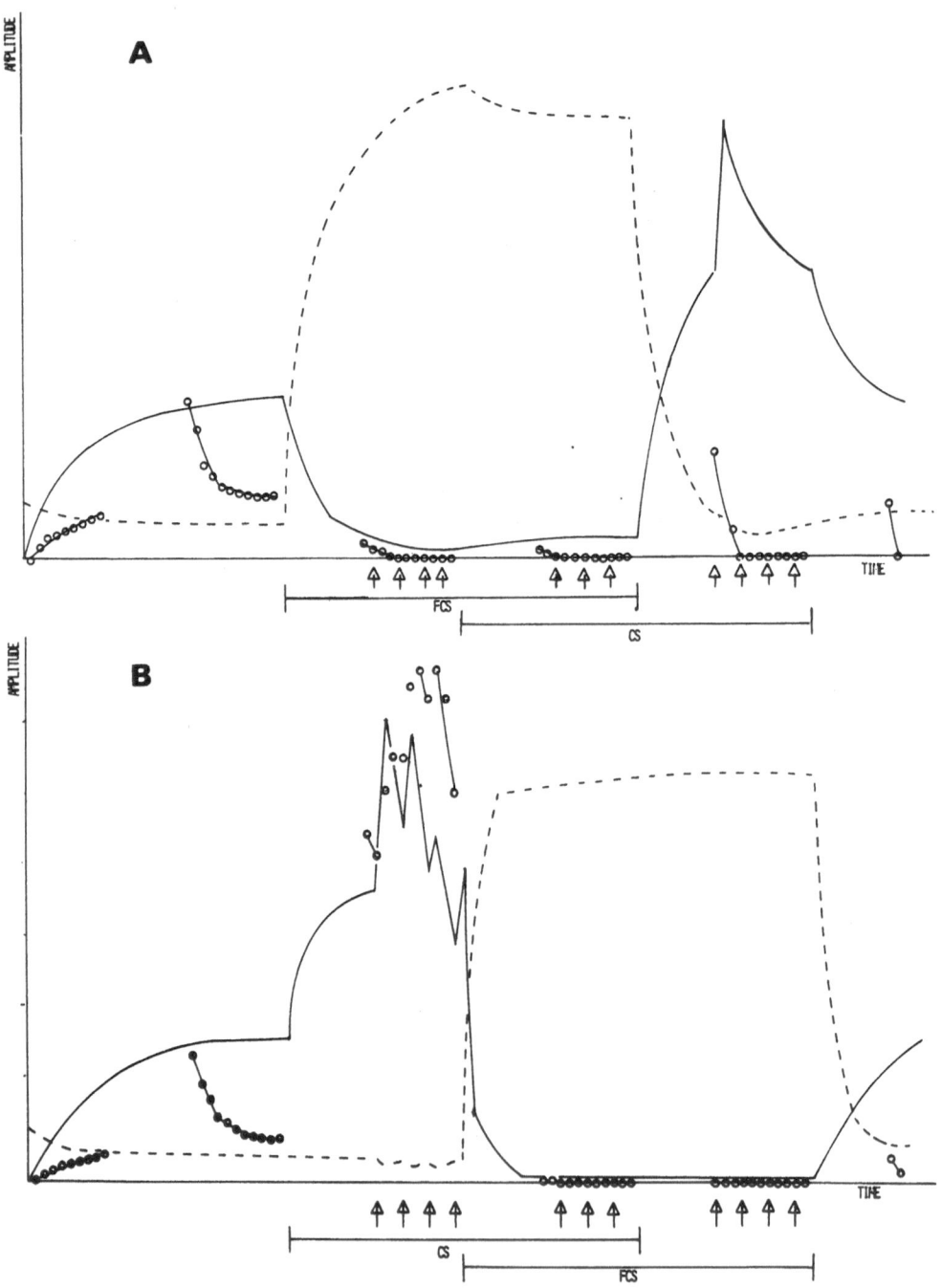

where food odor is present, animals with lesions that reduce the
fear drive and increase feeding drive, and animals intoxicated with
ethanol, which probably increases the feeding drive through an
inhibition of the fear drive, have the following general behavior
(18,19,20): 1) the rate of habituation of prey catching behavior is
greatly reduced; 2) the usual preference of small 7 degree objects
over 16 degree prey stimuli can be reversed; 3) avoidance tendencies
to threatening large objects are reduced.

Fig. 12 shows the effect of introducing a stimulus associated
with a predator FUS during the usual paradigm. It can be seen that
the strength of the response is greatly reduced, because the
feeding drive has been inhibited, and the rate and level of
habituation greatly increased. Conversely (Fig. 11), when a stimulus
associated with food is presented (CS), the fear drive is greatly
reduced, the strength of the response is increased and the rate and
level of habituation are reduced. The effect over the fear drive
increases the threshold for avoidance behavior as mentioned above.
In the first experiment (presentation of predator) the transition
from short to long term memory is greatly increased; while the
opposite effect occurred in the second simulation.

Figs. 13A and B show the behavior of the model when alternate
feeding and fear stimuli are introduced. Fig. 13A shows the effect
of introducing first a threatening stimulus: reducing the strength of

---

Fig. 13. Computer simulation of the effects upon the hunger drive
and habituation of prey orienting behavior when first a threatening
stimulus is introduced (FCS) and then a prey object is presented
(CS) (Fig. 13A) and the opposite order of presentation, first the
prey and then the fear stimulus (Fig. 13B). In the first case it
can be seen that the fear drive is greatly increased, reducing the
feeding drive and increasing the transition from short to long term
habituation. When the food stimulus is presented, simultaneously
with the threatening one, no response is activated because of the
strong inhibitory effect of the fear drive over the feeding center;
when the fear drive is witheld, but the prey stimulus is still there,
the feeding drive starts to increase, but the recovery of the
response is retarded because the long term effects of habituation.
As a consequence, the response is low and habituation is very fast.
In the second situation, Fig. 13B, the introduction of the feeding
stimulus enhances the feeding drive, while the fear drive is
inhibited; but when the threatening stimulus appears, simultaneously
with the food object, the feeding drive is completely suppressed by
the inhibitory effects both of the satiation center and the fear
drive. When the fear drive is witheld, the response recovers very
slowly.

the response, increasing the rate, the level, and the consolidation
of habituation. When the stimulus associated with food is presented,
the response is weak, and the rate and level of habituation are fast,
which means that the past experience of the animal modifies
dramatically the behavior when a new condition is presented (see
Fig. 11 for comparison of behavior). In the opposite experiment
(Fig. 13B), presentation of the stimulus associated with food first
and then the threatening one, the strength of the response is
initially increased and the rate and level of habituation is reduced;
when the threatening stimulus appears, although the feeding response
is strongly reduced, its extinction is less rapid than if only the
fear stimulus would be presented. These results are also in
accordance with the above behavioral experiments.

Stimulus Specific Habituation

In our introduction we stressed that one of the most important
differences between physiological and behavioral habituation is that
behavioral habituation is stimulus specific. This implies that a
trace of the stimulus has to be stored in the brain of the animal
and there is a continuous comparison between the trace and the present
stimulus. When the trace is equal to the stimulus, habituation is
present; while if there is a mismatch between model and object, then
the arousal response is activated.

In this model we will only show how a one-feature stimulus can
be stored, compared and, if a mismatch occurs, generate an arousal
response. A generalization to feature vectors, however, is
straightforward. We have considered that the creation of the model of
the stimulus and its comparison with the actual object is
accomplished by cognitive centers of the brain; these centers send
the final output signal to the modulator of the plastic property
(Fig. 4).

We simply propose that the creation of the stimulus is obtained
by a control system whose set point is the feature of the stimulus.
When the stimulus is presented, the variable will tend to acquire
the value of the feature. If the feature is changed, the system
immediately will tend to reach the new value of the stimulus (Fig. 14,
inset). It is interesting that this system has similar behavior to
some cells considered as novelty detectors, which have been
extensively described in physiological studies of different animals
(19,28,30,49). This system can be defined mathematically as follows:

$$\dot{MO} = (L1)*h(FT-MO) \qquad (14)$$

where FT is the parametric representation of the feature of the
stimulus; MO is the modeled feature by the nervous structure,
(perhaps a cell?); and the function h is defined as follows:

$$h(FT-MO) = \begin{cases} FT-MO & \text{If } FT>0 \\ 0 & \text{ELSE} \end{cases}$$

The comparison between model and stimulus is simply the absolute value of the difference in the following way:

$$EM = |FT-MO| \qquad (15)$$

This signal is sent to the modulator, therefore equation 11 is redefined as follows:

$$\dot{MOD} = (w13)(NMOD-MOD) + (Fdd)(w14) + (EM)(w15) \qquad (16)$$

This equation shows that when a difference between model and stimulus exists (EM>0) the arousal response is activated.

The computer simulation of stimulus specific habituation is shown in Fig. 14. In this experiment we have changed the general

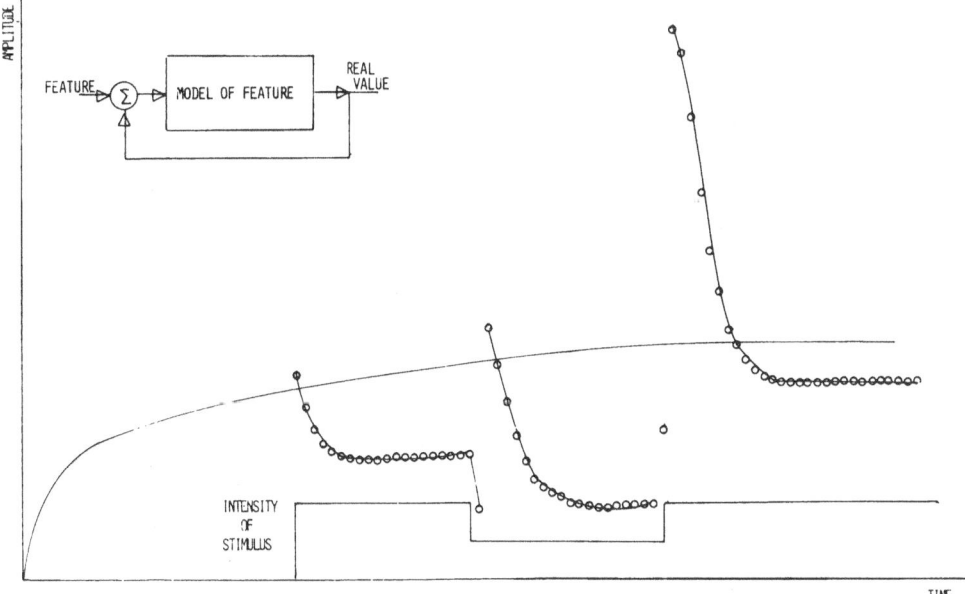

Fig. 14. Computer simulation of stimulus specific habituation. When the stimulus is presented repetitively with a given intensity (shown as a continuous line at the bottom of the graph) the response starts to habituate. When the intensity is lowered, the pathway is dishabituated. A new change in the intensity of stimulation to the original intensity, produces, again, dishabituation of the response. Notice that the strength of the response, the rate and level of habituation are different with different intensities. The inset shows the control system that makes a model of the feature stimulus (FT).

paradigm used in the above experiments to show how a change in any
of the parameters of the stimulus induces an arousal response. For
this reason, we present repetitively a stimulus until the response
is habituated; then, we change its intensity (to a lower and then to
the same value) and we see that the behavior is dishabituated by the
arousal response. Notice that the level and rate of habituation to
a lowering of intensity is greater than in the original strength of
stimulation; when we reestablish the initial value of the stimulus,
the arousal response is reactivated and the rate and level of
habituation are reduced. With this experiment we simulate the general
results obtained in the dishabituation of the orienting response when
any of the parameters of the stimulus are changed (9,42,43,44). We
have not considered, however, how this stimulus could be processed
independently of the region stimulated nor how its temporal
parameters could be stored.

DISCUSSION

In general, physiological studies of habituation have neglected
the importance of state (5), while behavioral theories have been
mostly directed to the complex processing of information required
to store the model of the stimulus, without considering in their
theories the physiological mechanisms needed for this process.
Theories of habituation have, therefore, the tendency to ignore the
differences between physiological and behavioral habituation trying
to explain one in terms of the other, which, naturally, instead of
narrowing the gap between these studies, makes it more divergent.

We presented a model of how the general state of the animal
could control the development, expression, and maintenance of
habituation. The model is based on the physiological mechanisms
postulated for this process to occur and on the hypothesis that
the behavior is controlled by the interaction of motivational and
cognitive centers. In the simulation of motivational systems, we
studied the different factors that may regulate their expression,
their interaction with other motivational centers, and their
relationship with cognitive regions. We have also postulated a
mechanism by which cognitive centers may create a model of the
stimulus, compare it with the present object, and produce an arousal
response if a mismatch occurs. We have not, however, simulated how
habituation can occur independently of the site of stimulation, nor
how the temporal representation of the stimulus is stored or
retrieved. From a physiological point of view, we have simulated
habituation as a homosynaptic depression of the amount of transmitter
release, through changes in membrane permeability to Ca^{++} ions, its
dependence on the amplitude and frequency of stimulation, its
transition from short to long term habituation, and the modulation of
this plastic process by presynaptic inhibition and sensitization.

This model has permitted us to simulate the following physiological results in relation to the modulation of habituation: 1) modulation of habituation, in the gill withdrawal reflex behavior in Aplysia (30,31,38,39); 2) blocking of habituation during locomotion in crayfish (3,4); 3) dishabituation in Aplysia and spinal cord of vertebrates (22,23,49).

From a behavioral point of view, we have simulated the following phenomena: 1) modulation of habituation of the orienting behavior by the feeding drive; good feeder animals (18,19,20), animals in an environment where there is a stimulus associated with food (7,8,9) give a stronger response to the presentation of the stimulus, show less habituation, and avoidance behavior is reduced; 2) interaction between different drives in the regulation of habituation: stimulus associated with predator increases the fear drive, decreases the feeding drive, reduces the strength of the response, and increases the rate and level of habituation (20); 3) dishabituation when any of the parameters of the stimulus changes (9,15,42,43, 44); and 4) the interaction between drive and antidrive in the regulation of habituation.

We consider that this model gives a good idea of the complexity implied in the control exerted on a plastic process such as habituation by the general state of the animal. We consider that this model narrows the gap between behavioral and physiological studies of habituation.

Acknowledgement. I thank Dr. Michael Arbib for his helpful comments and criticisms on the manuscript. This work has been supported in part by grant number RO1NS14971-02 from NIH to Dr. Arbib.

REFERENCES

1. Bartlett, F.C., Remembering, Cambridge University Press, 1967.
2. Bergson, H., Matiere et memoire, Oeuvres, Presses Universitaires de France, Paris, 1963.
3. Bryan, J.S. and Krasne, F.B., Protection from habituation of the crayfish lateral giant fibre escape response, J. Physiol. (Lond.), 271 (1977) 351-368.
4. Bryan, J.S., and Krasne, F.B., Presynaptic inhibition: the mechanism of protection from habituation of the crayfish lateral giant escape response, J. Physiol. (Lond.), 271 (1977) 369-390.
5. Clifton, R.K., and Nelson, M.N., Developmental study of habituation in infants: the importance of paradigm, response system, and state, In: Habituation, Perspectives from Child Development, Animal Behavior, and Neurophysiology, (Tighe, T.J. and Leaton, R.N., eds.), LEA Publishers, New Jersey, 1976.
6. Davis, W.S., and Guilette, R., Neural correlates of behavioral plasticity in command neurons of Pleurobranchea, Science

199 (1978) 301-804.

7. Ewert, P., The visual system of the toad: behavioral and physiological studies on a pattern recognition system, In: The Amphibian Visual System. A Multidisciplinary Approach, (Fite, K., ed.), Academic Press, New York, 1976.

8. Ewert, P., and Ingle, D., Excitatory effects following habituation of prey catching activity in frogs and toads, J. Comp. Physiol. Psychol. 3 (1971) 369-374.

9. Ewert, P. and Kehl, N., Configurational prey selection by individual experience in the toad Bufo bufo, J. Comp. Physiol. 126 (1978) 105-114.

10. Frank, K., and Fuortes, M.G., Presynaptic and postsynaptic inhibition of monosynaptic reflexes, Fed. Proc., 16 (1957) 39-40.

11. Groves, P.M., and Thompson, R.F., Habituation: a dual process theory, Psychol. Rev., 77 (1970) 419-450.

12. Hebb, D.O., The Organization of Behavior: a Neurophysiological Theory, John Wiley and Sons, New York, 1949.

13. Hernández Peon, R., Scherrer, H., and Velasco, M., Central influences on afferent conduction in the somatic and visual pathways, In: Brain and Behavior, Perception and Action. (K. H. Pribram, ed.), Penguin Books, London, 1969, pp. 321-338.

14. Hernandez Peon, R. and Brust, H., Functional role of subcortical structures in habituation and conditioning, In: Brain Mechanisms and Learning, Oxford Press, 1961, pp. 392-412.

15. Hinde, R.A., Behavioral habituation, In: Short Term Changes in Neural Activity and Behavior (Horn, G. and Hinde, R.A., eds.), Cambridge, 1970, pp. 3-40.

16. Horn, G., Neuronal mechanisms of habituation, Nature, 215 (1967) 707-711.

17. Horn, G., Changes in neuronal activity and their relationship to behavior, In: Short Term Changes in Neural Activity and Behavior, (Horn, G. and Hinde, R.A., eds.), Cambridge, 1970, pp. 567-606.

18. Ingle, D., Reduction of habituation of prey catching activity by alcohol intoxication in the frog, Behav. Biol., 8 (1973) 123-129.

19. Ingle, D., Evolutionary perspectives on the function of the optic tectum, Brain Behav. Evol., 8 (1973) 211-237.

20. Ingle, D., Motivation and prey selection by frogs and toads: a neuro-ethological model, In: Handbook of Motivation, (Tetelbaum and Satinoff, eds.), in press.

21. John, E.R., Mechanisms of Memory, Academic Press, New York, 1967.

22. Kandel, E., Cellular Basis of Behavior: An Introduction to Behavioral Neurobiology, W. H. Freeman, 1976.

23. Kandel, E., A cell biological approach to learning, Grass Lecture No. 1, Society of Neuroscience, 1978.

24. Kennedy, Inhibition in the center and the periphery, In: Identified Neurons and Behavior of Arthropods, (Hoyle, G., ed.) Plenum press, New York, 1977.

25. Kohler, W., The Selected Papers of Wolfgang Kohler, Leveright,

New York, 1971.

26. Konorski, J., _Integrative Activity of the Brain_, University of Chicago Press, Chicago, 1967.

27. Krasne, F., Invertebrate systems as a means of gaining insight into the nature of learning and memory, In: _Invertebrate Systems in the Study of Learning and Memory_. (Rosenzweig, R. and Bennet, F.L., eds.), MIT Press, 1976, pp. 401-429.

28. Krasne, F., Extrinsic control of intrinsic neuronal plasticity: an hypothesis framework on simple systems, _Brain Res._, 140 (1978) 197-215.

29. Lara, R., Tapia, R., Cervantes, F., Moreno, A., and Trujillo, H., Mathematical models of synaptic plasticity: II Habituation, _Neurol. Res._, in press.

30. Lettvin, J.Y., Maturana, H.R., McCulloch, W.S., and Pitts, W.H., What the frog's eye tell the frog's brain, _Proc. Inst. Radio Eng._, 47 (1959) 1940-1951.

31. Lukowiak, K., Stimulation of the brachial nerve evokes suppression of the gill withdrawal reflex in young Aplysia, _Brain Res._, 132 (1978) 553-557.

32. Lukowiak, K., and Peretz, B., The interaction between the central and peripheral nervous system in the modulation of gill withdrawal reflex behavior in _Aplysia, J. Comp. Physiol._, 117 (1977) 219-244.

33. McFarland, D.J., _Feedback Mechanisms in Animal Behavior_, Academic Press, New York, 1971.

34. McFarland, D.J., Hunger in interaction with other aspects of motivation, In: _Hunger Models, Computable Theory of Feeding Control_ (Booth, D.A., ed.), Academic Press, New York, 1978, pp. 315-406.

35. Mpistos, G.J., Collins, S.D. and McClellan, A.D., Learning: a model system for physiological studies, _Science_ 188 (1978) 954-957.

36. Mogenson, G.J., and Calaresu, F.R., Food intake considered from a viewpoint of system analysis, In: _Hunger Models, Computable Theory of Feeding Control_ (Booth, D.A., ed.), Academic Press, New York, 1978, pp. 1-24.

37. Neisser, U., _Cognitive Psychology_, Prentice Hall, New Jersey, 1967.

38. Pavlov, I.P., _Conditioned Reflexes_: An Investigation of the _Physiological Activity of the Cerebral Cortex_, Oxford University press, London, 1927.

39. Peretz, B., and Lukowiak, K., Age dependent CNS control of habituation gill withdrawal reflex and correlated activity in identified neurons in Aplysia, _J. Comp. Physiol._, 103 (1975) 1-77

40. Peretz, B., and Howieson, P.B., Central influence on peripheral nervous system on the mediation of gill withdrawal reflex behavior in _Aplysia, J. Comp. Physiol._, 84 (1973) 1-18.

41. Piaget, J., _Biology and Knowledge_, The University of Chicago Press, 1971.

42. Skinner, B.F., Science and Human Behavior, McMillan, New York, 1953.
43. Sokolov, E.N., Neuronal mechanisms of the orienting reflex, In: Neuronal Mechanisms of the Orienting Reflex (Sokolov, E.N., and Vinogradova, O.N., eds.), 1975, pp. 217-238.
44. Sokolov, E.N., Perception and the Conditioned Reflex, Pergamon, New York, 1963.
45. Sokolov, E.N., Neuronal models and the orienting reflex, In: The Central Nervous System and Behavior (Brazier, M.A., ed.), 1961, pp. 187-276.
46. Stein, J., Habituation and stimulus novelty. A model based on classical conditioning, Psychol. Rev., 73 (1966) 352-356.
47. Tauc, L., Presynaptic inhibition in the abdominal ganglion of Aplysia, J. Physiol. (Lond.), 181 (1963) 283-307.
48. Thatcher, R.W., and John, E.R., Foundations of Cognitive Processes, Functional Neuroscience, Vol. 1, LEA Publishers, New Jersey, 1977.
49. Thompson, R.F., and Spencer, W.A., Habituation a model phenomena for the study of neuronal substrate of behavior, Psychol. Rev., 73 (1966) 16-43.
50. Vinogradova, O., Registration of information and the limbic system, In: Short Term Changes in Neural Activity and Behavior (Horn, G. and Hinde, R.A., eds.), Cambridge, 1970.
51. Vinogradova, O., Hippocampus and the orienting reflex, In: Neuronal Mechanisms of the Orienting Reflex (Sokolov, E.N. and Vinogradova, O.S., eds.), LEA Publishers, New Jersey, 1975.
52. Wagner, A.R., Priming in STM: an information processing mechanism for self generated or retrieval generated depression performance, In: Habituation, Perspectives from Child Development, Animal Behavior and Neurophysiology (Tighe, T.J. and Leaton, R.N. eds.), LEA Publishers, New Jersey, 1976, pp. 95-128.
53. Wagner, A.R., Habituation and memory, In: Mechanisms of Learning and Motivation (Dickinson, A. and Boakes, P.A., eds.), LEA Publishers, New Jersey, 1979, pp. 58-82.
54. Wickelgren, B.G., Habituation of spinal motoneurons, J. Neurophysiol., 30 (1967) 1424-1438.
55. Wickelgren, B.G., Habituation of spinal interneurons, J. Neurophysiol., 30 (1967) 1404-1423.
56. Woodson, P.B.J., Tremblay, J.P., Sclapfer, W.T. and Barondes, S.H., Heterosynaptic inhibition modifies the presynaptic plasticities of the transmission process at a synapse in Aplysia californica, Brain Res., 109 (1976) 83-95.
57. Zucker, R.J., Physiological mechanisms underlying habituation, J. Neurophysiol., 35 (1972) 599-651.

REGULATORY MECHANISMS OF REM SLEEP IN THE CAT

René Drucker-Colín, Lourdes Lugo and Raúl Aguilar

Departamento de Neurociencias,
Centro de Investigaciones en Fisiología Celular
Universidad Nacional Autónoma de México, México 20, D.F.

INTRODUCTION

Today it is universally accepted that mammals and primates present at least two basic stages of sleep. The state of sleep is first characterized electroencephalographically by the appearance of 14-18 Hz cortical spindles, which are subsequently replaced by 2-4 Hz slow waves. At the same time, high-voltage (500-800μV) sharp waves are recorded from the hippocampus, while the electromyogram (EMG) decreases slightly. Usually after some 30-40 minutes the electro-physiological signs of slow wave sleep (SWS) are replaced by low-voltage fast cortical EEG activity, regular hippocampal theta rhythm (5-6 Hz), isoelectric EMG, burst of rapid eye movements (REM), and pontogeniculooccipital (PGO) waves. PGO waves appear in SWS approximately one minute before all REM sleep periods. These PGO spikes exhibit a fairly constant daily rate of about 14,000 in the cat (18), may exist in man (30) and are made up of simple Type I and complex Type II spikes (24).

The events that occur within REM sleep have been classified as tonic and phasic. The tonic events refer to those characteristics which more or less define the REM sleep period, such as EEG activation and EMG suppression. The phasic events are short-lasting but recurrent activities, including, in addition to eye movements and PGO spikes, cardiovascular irregularities (32), respiratory changes (2), changes in pupil diameter (4), fluctuations in penile tumescence (16), muscular twitchings (3), middle ear muscle contractions (6), brain stem unit (28) and multiple unit bursts of activity (12).

MACROMOLECULES AND REM SLEEP

Some ten years ago, Oswald (25) suggested that the rebound of
REM sleep that occurs during drug withdrawal, reflects a phase of
neuronal repair indicated by increased protein synthesis, and further
suggested that situations in which protein synthesis increases should
lead to REM sleep. It has been reported (38) and now confirmed many
times (30), that phases 3 and 4 of sleep in man are associated with a
rise in plasma growth hormone (GH). Since in man, the sleep related
surge of plasma GH occurs in the early phase of the night, before
the late appearing REM sleep stage, it has been argued (35) that GH
may play a role in triggering REM sleep. This hypothesis has been
tested by injecting GH in cats (36) and rats (11). Both studies
showed that growth hormone induced a dose-dependent increase of REM
sleep, about 3 hours following the injection.

Since GH is an anabolic hormone which exerts important actions
on protein synthesis (21), it could be argued that the increase of
REM sleep following GH is indirectly produced by an enhanced formation
of macromolecules. However, there are scanty data on the relationship
between protein synthesis and sleep. Stern et al. (38) reported an
increase of REM sleep for 7 days after the intraventricular
administration of cycloheximide in cats. Since protein synthesis
inhibition was at 75% on the first day and 50% by the fourth day,
they suggested that the increase of REM sleep was due to protein
synthesis patterns in the brain which were returning towards a normal
state. In this same study, cycloheximide had no effect on sleep when
injected intraperitoneally. However, Pegram et al. (26) reported that
such injections of cycloheximide in mice produced a specific decrease
of REM sleep. Recently, we also observed a specific decrease of REM
sleep in rats (29) and cats (13) following the administration of the
protein synthesis inhibitors, anisomycin and chloramphenicol. Similar
observations with administration of chloramphenicol have been made
by Kitahama and Valatx (20) in rats, and by Petitjean (27) in cats.
It is important that in all these studies, SWS was unaffected except
when drugs were administered at very high doses (27). It is
interesting to note that the decrease of REM is due to a reduction in
frequency, since mean duration is unaffected (13,26,29). These
results may suggest that protein synthesis could be involved in the
mechanisms that trigger REM sleep.

This possibility was tested by determining the effects of
chloramphenicol on REM sleep rebound, induced either by REM sleep
deprivation or by withdrawal of chronic amphetamine administration.
In one such study, cats were REM sleep-deprived for 72 hours, and the
period of recovery recorded for 12 hours on each of two consecutive
days with or without chloramphenicol. Chloramphenicol blocked the
REM sleep rebound that normally occurs following sleep deprivation,
and again it was frequency and not duration of REM which was affected.

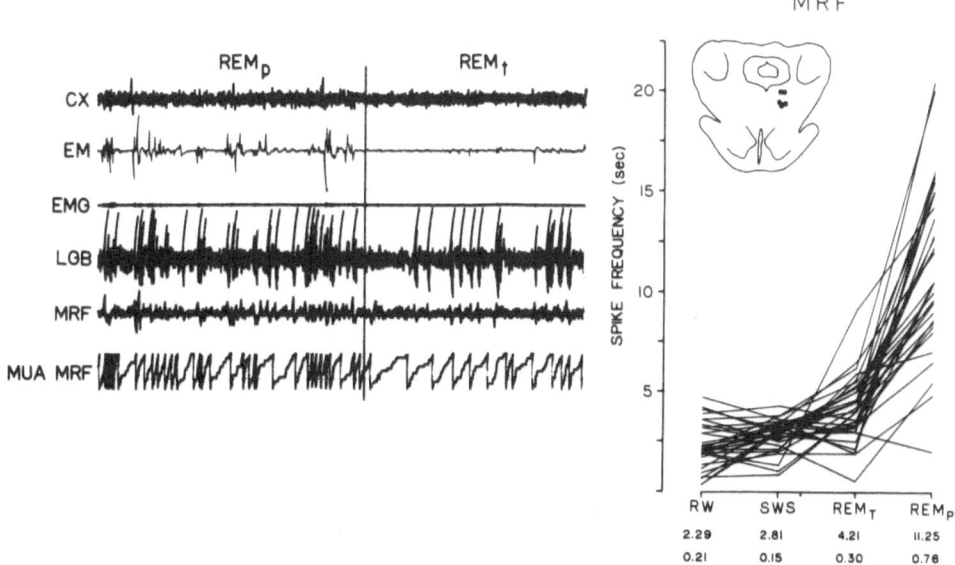

Fig. 1. The left hand side of the figure represents a portion of a
REM sleep period in a cat, where it can be clearly seen that there
is a phasic period (REM_p) characterized by intense eye movements
(EM), high frequency of PGO spikes in the lateral geniculate body
(LGB) and high frequency of multiple unit activity (MUA) in the
midbrain reticular formation (MRF), and a tonic phase (REM_t)
characterized by a decrease in all the above. The graph on the right
hand side represents the spike frequency discharge during several
periods of relaxed wakefulness (RW), slow wave sleep (SWS) and REM_t
and REM_p. One can clearly observe that the cause of overall REM
sleep to be characterized by high frequencies of spike discharge is
mainly due to REM_p. Cortex (CX), Electromyogram (EMG).

In another experiment (9), 12 rats were recorded for one baseline
period. They were subsequently administered 10 mg/kg of amphetamine
for 15 consecutive days. During this treatment their sleep-wake cycle
was recorded on days 1, 7 and 15. At the end of the 15th day
(withdrawal period), the animals were divided into two groups of 6,
and amphetamine administration was substituted in one group by saline
and in the other by 100 mg/kg of chloramphenicol. The results showed
that during the withdrawal of amphetamine, the saline group had a
significant rebound of both SWS and REM sleep, while in the
chloramphenicol group the rebound was restricted to SWS only.

Additional experiments have shown that protein synthesis
inhibitors also affect some of the components of REM sleep.

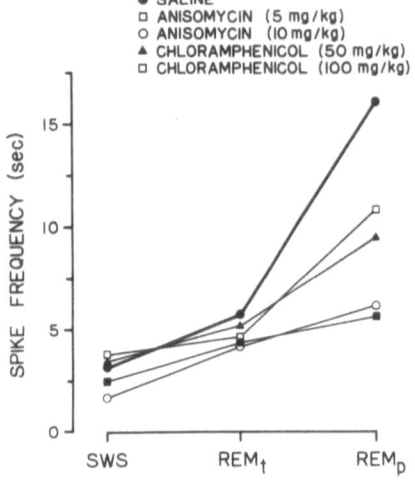

Fig. 2. This graph shows how chloramphenicol and anisomycin produce a dose dependent decrease in spike frequency discharge rate during REM_p. What this graph indicates is that these protein synthesis inhibitors induce a significant decrease in phasic discharge rate.

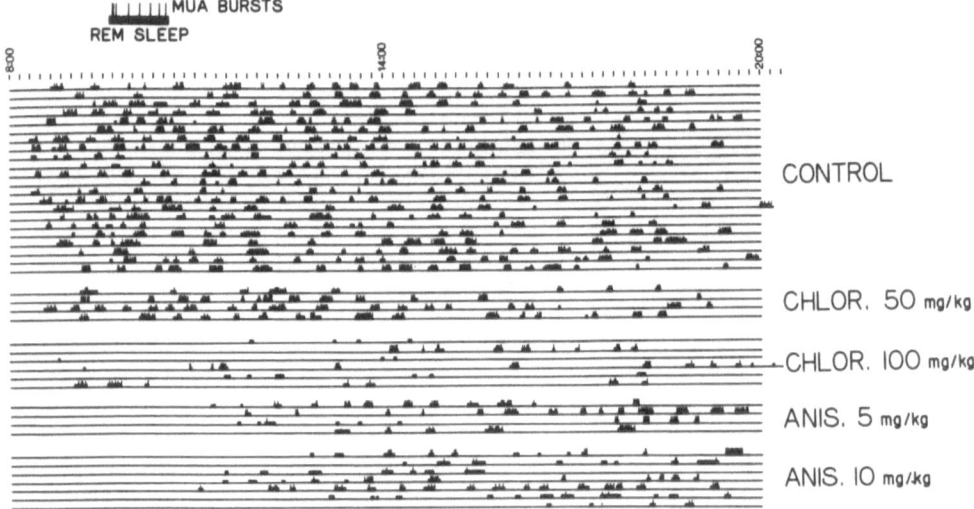

Fig. 3. Graph showing the distribution of REM sleep through 12 h recording period during control and drug administrations. The thick black lines indicate the presence of a REM period and their width gives the approximate duration. The vertical thin line on top of REM sleep represents the appearance of a burst of MUA in the MRF. Note the change in REM frequency and of MUA bursts. (From ref. 13).

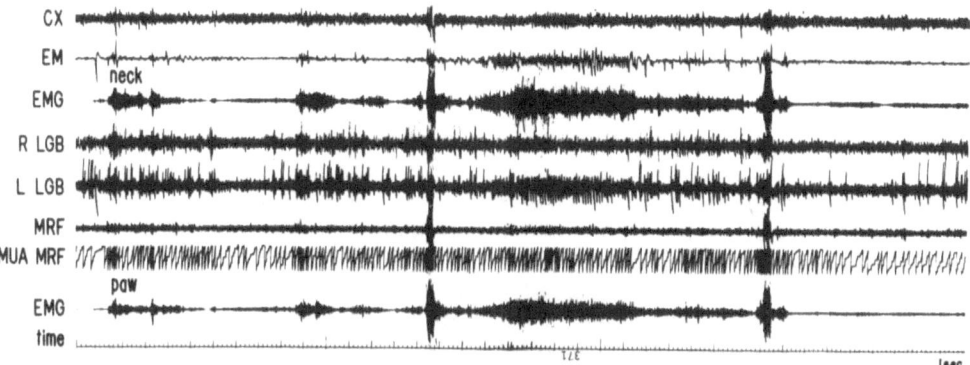

Fig. 4. Portion of a recording during a REM sleep period in a cat
with bilateral Locus Coeruleus α lesion. Note the neck and paw EMG
which show absence of muscle atonia. Towards the middle of the
recording the increase in muscle tone is due to licking movements.
Despite all the changes in muscle tone the animal remains in REM
sleep and presents all the remaining components of this phase such
as eye movements, PGO spikes in LGB, MUA bursts in MRF and cortical
desynchronization.

Observation of any REM sleep period clearly indicates that it is not
a homogeneous period. This means that eye movements, myoclonic
twitches and high frequency bursts of unit activity constantly
oscillate with periods in which these events are absent. Although it
has often been reported that the frequency of unit activity (which
is a good indicator of phasic events) during REM sleep is higher than
seen in any other phase of the sleep-wake cycle (35), such increase
is due solely to bursting periods, since in their absence, unit
activity is as low as that seen in the other stage (12). For this
reason, REM sleep can be (and should be) analyzed in two distinct
periods: phasic REM (REM_p) and tonic REM (REM_t).

A sample recording of REM sleep (Fig. 1) shows that it can be
clearly divided into these two periods. On the right hand graph of
this figure, multiple unit activity (MUA) frequency is much higher
during REM_p, than during any other period. It is very interesting
to note that protein synthesis inhibitors have a very clear and
potent effect on MUA frequency during REM_p. In Figure 2, we see that
the significant increase in MUA frequency during REM_p is completely
abolished by chloramphenicol and anisomycin. A dose-response effect
can also be observed.

We have recently calculated that protein synthesis inhibitors
decrease to about 50% the amount of phasic REM (13). A summary of the
effects of anisomycin and chloramphenicol on REM latency, REM

frequency, and multiple unit activity (MUA) bursts is illustrated
in Figure 3. It is interesting to note in this figure, that short
REM sleep periods, seem to be characterized by an absence of phasic
bursts of MUA.

REM SLEEP WITHOUT ATONIA AND PHASIC ACTIVITY

In 1965 Jouvet (19) reported that lesions in the region of the
pons produced a peculiar motor behavior in cats during REM sleep.
More recently this has been studied more extensively by Morrison (23)
and Sastre and Jouvet (32), and they have termed it REM sleep without
atonia or oneiric behavior respectively. In these cats small bilateral
lesion of the locus coeruleus and its vicinity do not prevent the
animal from going into REM sleep but prevent the muscular atonia
characteristic of this state. Moreover in such preparations the
animals not only do not lose their muscle tone but have several
stereotyped patterns of movement such as licking, grooming, treading,
lunging attack-like behavior and even standing and walking. Many of
these movements seen to be phasic and follow in our experience some
of the phasic activity described above (Fig. 4). It is thus not un-
reasonable to propose that the phasic bursts of unit activity from
the MRF probably activate cortical and pyramidal cells, which in turn
activate spinal motoneurons which are no longer inhibited by the
reticulo-special inhibitory pathway of Magoun and Rhines (22), in
view of the absence of excitatory LC cells. On the other hand it is
also possible that the PGO spikes participate in the genesis of these
phasic motor behaviors.

In our laboratory we have been successful in dissociating phasic
activities by pharmacological means. For example atropine
significantly diminishes the amount of PGO spikes, while
chloramphenicol decreases the multiple unit burst of activity in the
MRF. We have thus lesioned the LC α in a few cats and obtained some
with REM sleep without atonia. Most of our cats present several bouts
of phasic motor behaviors, such as lifting the head, licking and
attack-like leg movements. We administered either 0.8 mg/kg of
atropine or 75 mg/kg of chloramphenicol to LC α lesioned animals and
observed that a decrease in PGO density did not alter neither the
intensity nor the frequency of motor behaviors (Fig. 5). On the other
hand chloramphenicol while eliminating completely the bursts of MUA,
also produced a disappearance of the motor patterns of behavior and
significantly increased the time spent in atonia (Fig. 5). Thus, it
appears therefore that chloramphenicol and therefore protein
synthesis play an important role in regulating phasic activity. An
important question is what role does phasic activity play in REM
sleep regulation.

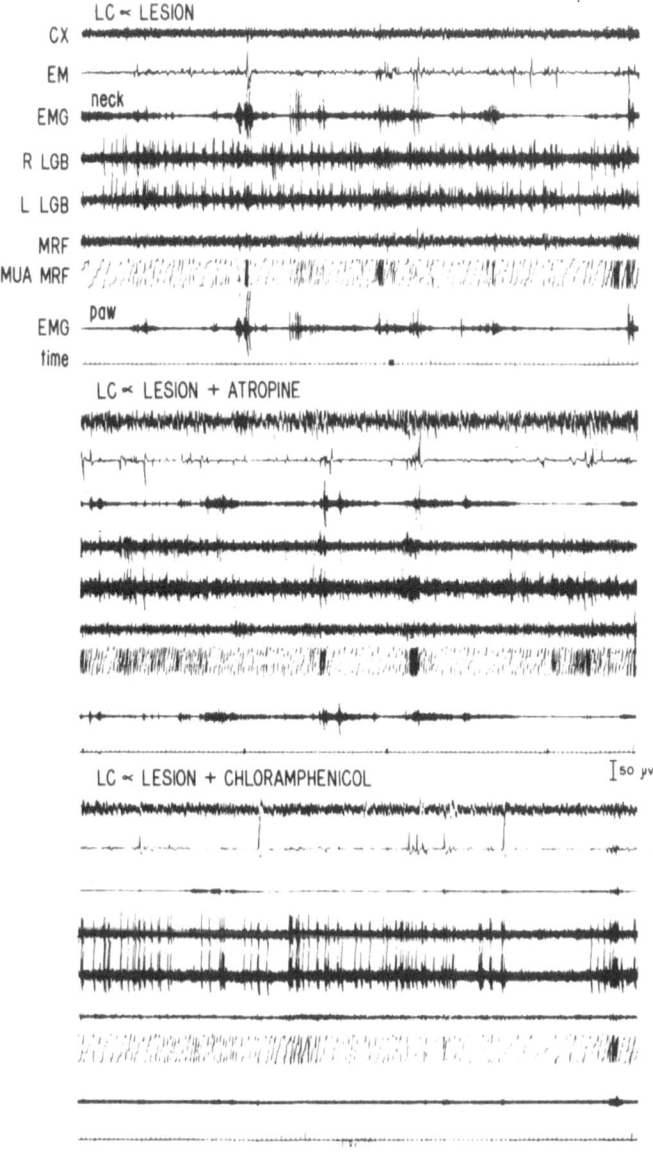

Fig. 5. This graph shows the effects of atropine and chloramphenicol in Locus Coeruleus α lesioned animals during REM sleep. Note in the upper trace the paw and neck muscle tone, which include phasic and jerky movements. In the middle trace following atropine, PGO spike density is reduced but there is no reduction in muscular tone. However following chloramphenicol the phasic bursts of muscle activity are gone and so are the phasic bursts of MUA activity.

THE ROLE OF PHASIC ACTIVITY

Although it is not clear how the decrease of REM_p affects the normal occurrence of REM sleep periods, there is some indication that it may affect its triggering and its duration. Utilizing MUA from MRF neurons as an indicator of phasic activity, the discharge frequency in 30s periods one minute prior to onset of REM and one minute during REM was calculated. In addition REM periods of over two minutes duration were compared with those of less than two minutes. An interesting observation emerged. When REM sleep periods were longer than two minutes in duration, spike frequency increased just prior to onset of REM and showed a further increase during the first minute of that state. Similar events have been recorded in pontine reticular structures (17). On the other hand, when REM sleep duration lasted less than two minutes no such positively inflected rise in discharge frequency was observed (10).

Since, as mentioned previously, it is the phasic discharge of unit activity which contributes to the rise in frequency observed during REM sleep, these results suggest that in the absence of phasic discharges, REM sleep will be short or abortive. Most interesting, is the observation that in cats given protein synthesis inhibitors, the discharge rate of MRF neurons is practically identical to that which occurs in short duration REM sleep periods (10). This could indicate that by preventing phasic activity, these drugs reduce the probability of REM sleep periods occurring, and shorten them when they occur, because of the absence of phasic activity. Unfortunately, the absence of phasic activities with protein synthesis inhibitors is not always accompanied by an absence of long REM sleep periods. In fact, as we mentioned before, these drugs do not seem to alter REM sleep duration. Although these results give evidence indicating that protein synthesis inhibitors have a particular effect on REM_p, they are presently difficult to interpret, since there is little information on the meaning of REM sleep periods with little or no phasic activities. However, in certain pathological states, such as mental retardation, the amount of REM sleep and associated phasic activities appears to be significantly reduced (5,15).

Although it is conceivable that certain macromolecules may participate in the mechanisms which trigger REM sleep, the evidence so far has been provided by indirect neuropharmacological experiments. We have, therefore, attempted to obtain support for this hypothesis through more direct experimental approaches. In one such series of experiments cats implanted with electrodes for recording the sleep-wake cycle and a push-pull cannula in the MRF were continuously perfused and recorded for 12-21 hours. At every hour the protein content of the perfusate, and the time spent in wakefulness, SWS and REM sleep was determined. We found that protein concentrations in the perfusate varied in a cyclic fashion, and that the peaks of protein corresponded to those hours in which REM sleep occupied the greatest

proportion of time (11). Moreover, when insomnia was produced by
bilateral lesions of the preoptic area, the cyclic release of proteins
disappeared (7). Protein levels also lost their cyclicity following
the administration of chloramphenicol, which specifically decreased
REM sleep. When waking perfusates were compared to those from REM
sleep, the latter always contained nearly twice as much protein (11).
In further studies, separation of released proteins by polyacrylamide
SDS gel electrophoresis revealed the existence of some large molecular
weight proteins in REM sleep perfusates, not present in waking
perfusates (34). However, recently we have been unable to confirm
this finding. The differences between the two states, therefore, seem
so far, to be merely that of concentration although the possibility
that with more sensitive methods, specificity of the proteins could
be detected, should not be ignored.

Although these experiments are highly suggestive that proteins
may be involved in the regulation of REM sleep, the evidence is merely
correlative. Moreover, it is impossible to determine whether the
protein release is the cause or the effect of REM sleep. We have
therefore tested an approach which in a more direct form explores
whether specific proteins are indeed involved in the regulation of
sleep. This approach consists in studying the effects of antibodies
against proteins which are obtained from the MRF of cats during REM
sleep.

We were led to this approach for three good reasons. Firstly,
antibodies can be obtained in relatively large quantities by
immunization with small amounts of perfusate proteins. Secondly, if
a protein is indeed involved in the regulation of REM sleep,
antibodies can amplify the effects on sleep. Thirdly, antibodies may
provide the only available probe for determining whether proteins
really participate in sleep regulation, and for determining receptor
regions.

IMMUNOLOGICAL APPROACH TO THE STUDY OF SLEEP

In these experiments some 40 cats were implanted with electrodes
to record the sleep-wake cycle and with a push-pull cannula in the
MRF. For a six month period, perfusates were collected during every
REM sleep period, at a flow rate of 20 μl/min. Some 70 ml of
perfusate was collected, which yielded 1.25 mg of protein. This
material was used to obtain antibodies against proteins that appeared
in REM sleep. Of the 12 bands of proteins of the perfusate, only
four were antigenic. Upon purification of the antibodies on a DEAE-
cellulose column, they were tested for their effects on the sleep-
wake cycle following injection into the MRF of cats who were also
implanted with microelectrodes for recording MUA. The effects of
the antibodies were compared to γ-globulins from pre-immune serum,
antibodies to cat serum proteins, antialbumin antibodies, Ringer and

sham injections.

The results showed that only the anti-MRF antibodies significantly and specifically decreased REM sleep, and most interestingly, produced effects almost identical to those seen following injections of protein synthesis inhibitors. First of all the antibodies increased the latency of REM sleep to about 5 hours, in comparison to controls where mean latency was in order of 50-60 minutes. Secondly, antibodies decreased the frequency but not the duration of REM sleep, and only at high doses (500-1000 µg) did the antibodies have any effect on SWS. This effect on SWS, which manifested itself as an increase in the latency, was only transient, since total SWS time was unchanged in the 12 hours recorded. In addition to the effects on the sleep-wake cycle, bursts of eye movements and MUA were absent during the first few REM sleep periods, but reappeared as the effects of the antibodies started to dissipate.

These significant effects of antibodies, provide the most direct evidence for a role of macromolecules in the regulation of sleep. However, a word of caution should be advanced. Firstly, it should be emphasized that these are preliminary experiments, and the number of observations is relatively small. Secondly, preliminary controls with antibodies from proteins obtained from another area of the brain and antibodies obtained during wakefulness, appear to have similar effects. It could therefore be that the antibodies produce nonspecific reactions which manifest themselves in REM sleep. However, the fact that both antialbumin antibodies and antibodies to cat serum proteins failed to have any effects on the sleep-wake cycle argues against this. The results suggest that a nonspecific antigen-antibody reaction is not responsible for the effects observed, that a polypeptide is indeed involved in the regulation of REM sleep, and that this molecule is in all probability always present in extra-cellular fluid. It could then be suggested that this polypeptide is only activated during REM sleep, or else reaches adequate levels only at the particular time. Some support for the latter possibility comes from the fact that during REM sleep, there is always at least twice the amount of protein present than during wakefulness (11). It is conceivable therefore that the occurrence of REM sleep depends on a "gating" signal of polypeptide levels, which then initiates all the events which produce REM sleep as we normally see it, i.e. atonia, PGO waves, eye movements and phasic unit activity.

On the other hand, even assuming that some specific macromolecule is not involved in the regulation of sleep, the protein-REM sleep relationship is nonetheless interesting. In a recent article, Adam and Oswald (25) suggest that sleep may serve the function of tissue restoration of brain, following waking activity, or perhaps, protein synthesis during REM sleep aids processes of brain development during the neonatal period, and of information processing in later stages of life. Although there is no hard evidence supporting these ideas, they may be worth investigating.

Acknowledgements. Part of this work was supported by Grant No. 1651 of the Consejo Nacional de Ciencia y Tecnología, México, D.F.

REFERENCES

1. Adam, K. and Oswald, I., Sleep is for tissue restoration, J. Roy. Coll. Phycns., 11 (1977) 376-388.
2. Aserinsky, E., Periodic respiratory pattern occurring in conjunction with eye movements during sleep, Science 150 (1965) 763-766.
3. Baldrige, B.J., Whitman, R. and Kramer, M., The concurrence of fine muscle activity and rapid eye movements during sleep, Psychsom. Med., 27 (1965) 19-26.
4. Berlucchi, G., Moruzzi, G., Salvi, G. and Strata, P., Pupil behavior and ocular movements during synchronized and desynchronized sleep, Arch. Ital. Biol., 102 (1965) 230-244.
5. Clausen, J., Sersen, E.A. and Lidsky, A., Sleep patterns in mental retardation: Down's syndrome, EEG Clin. Neurophysiol., 43 (1977) 183-191.
6. Dewson, J., Dement, W.C. and Simmons, F., Middle ear muscle activity in cats during sleep. Exp. Neurol. 12 (1965) 1-8.
7. Drucker-Colín, R.R. and Gutiérrez, M.C., Effects of forebrain lesions on release of proteins from the midbrain reticular formation during the sleep-wake cycle, Exp. Neurol., 52 (1976) 339-344.
8. Drucker-Colín, R.R. and Spanis, C.W., Is there a sleep transmitter? Progr. Neurobiol., 6 (1976) 1-22.
9. Drucker-Colín, R.R. and Benitez, J., REM sleep rebound during withdrawal from chronic amphetamine administration is blocked by chloramphenicol, Neurosci. Lett., 6 (1977) 267-271.
10. Drucker-Colín, R.R., Dreyfus-Cortés, G. and Bernal-Pedraza, J.G., Differences in multiple unit activity discharge frequency during short and long REM sleep periods: Effects of protein synthesis inhibition, Behav. Neural Biol., 26 (1979) 123-127.
11. Drucker-Colín, R.R., Spanis, C.W., Cotman, C.W. and McGaugh, J.L., Changes in protein in perfusates of freely moving cats: relation to behavioral state, Science, 187 (1975) 963-965.
12. Drucker-Colín, R.R., Bernal-Pedraza, J.G., Díaz-Mitoma, F. and Zamora-Quezada, J., Oscillatory changes in multiple unit activity during rapid eye movement sleep, Exp. Neurol., 57 (1977) 331-341.
13. Drucker-Colín, R.R., Zamora, J., Bernal-Pedraza, J. and Sosa, B., Modification of REM sleep and associated phasic activities by protein synthesis inhibitors, Exp. Neurol., 63 (1979) 458-467.
14. Drucker-Colín, R.R., Spanis, C.W., Hunyadi, J., Sassin, J.F. and McGaugh, J.L., Growth hormone effects on sleep and wakefulness in the rat, Neuroendocrinol., 18 (1975) 1-8.

15. Feinberg, I., Eye movement activity during sleep and intellectual function in mental retardation, Science, 159 (1968) 1256.
16. Fisher, C., Gross, J. and Zuch, J., Cycle of penile erection synchronous with dreaming (REM) sleep, Arch. Gen. Psychiat., 12 (1965) 29–45.
17. Hobson, J.A., McCarley, R.W., Freedman, R. and Pivik, R.T., Time course of discharge rate changes by cat pontine brain stem neurons during sleep cycle, J. Neurophysiol., 37 (1974) 1297–1309.
18. Jouvet, M., Biogenic amines and the states of sleep, Science, 163 (1969) 32–41.
19. Jouvet, M. and Delorme, J.F., Locus coeruleus et sommeil paradoxal, C.R. Soc. Biol. París, 159 (1965) 895–899.
20. Kitahama, K. and Valatx, J.L., Effect du chloramphenicol et du thiamphenicol sur le someil de la souris, C.R. Soc. Biol. París 169 (1975) 1522–1525.
21. Korner, A., Growth hormone control of biosynthesis of protein and ribonucleic acid, Recent Prog. Horm. Res., 21 (1965) 205–238.
22. Magoun, H.W. and Rhines, R., An inhibitory mechanism in the bulbar reticular formation, J. Neurophysiol., 9 (1946) 165–171.
23. Morrison, A., Brain stem regulation of behavior during sleep and wakefulness. Prog. Psychobiol., 8 (1979) 91–131.
24. Morrison, A.R. and Pompeino, O., Vestibular influences during sleep IV. Functional relations between vestibular nuclei and lateral geniculate nucleus during desynchronized sleep, Arch. Ital. Biol. 104 (1966) 425–458.
25. Oswald, I., Human brain protein, drugs and dreams, Nature (Lond.) 223 (1969) 893–897.
26. Pegram, V., Hammond, D. and Bridgers, W., The effects of protein synthesis inhibition on sleep in mice, Behav. Biol. 9 (1973) 377–382.
27. Petitjean, F., Antibiotiques et sommeil, These de Doctorat, Lyon, France, 1977.
28. Pivik, R.T., Hobson, J.A. and McCarley, R.W., Eye movement associated rate changes in neuronal activity during desynchronized sleep: a comparison of brain stem regions, Sleep Res., 2 (1973) 35.
29. Rojas-Ramírez, J.A., Aguilar-Jiménez, E., Posadas-A., A., Bernal-Pedroza, J. and Drucker-Colín, R.R., The effects of various protein synthesis inhibitors on the sleep wake cycle of rats, Psychopharmacol., 53 (1977) 147–150.
30. Salzarulo, P., Lairy, G.C., Bancaud, J. and Munari, C., Direct depth recording of the striate cortex curing REM sleep in man: Are there PGO potentials? EEG Clin. Neurophysiol., 38 (1975) 199–202.
31. Sassin, J. Sleep related hormones. In Neurobiology of Sleep and Memory (R.R. Drucker-Colín and J.L. McGaugh, eds.), Academic Press, New York, 1977, pp. 361–372.
32. Sastre, J.P. and Jouvet, M. Le comportement onirique du chat, Physiol. Behav., 22 (1979) 979–989.

33. Snyder, F., Hobson, J., Morrison, D. and Goldfrank, F., Changes in respiration, heart rate and systolic blood pressure in human sleep, J. Appl. Physiol., 19 (1964) 417-422.

34. Spanis, C.W., Gutiérrez, M.C. and Drucker-Colín, R.R., Neuro-humoral correlates of sleep: further biochemical and physiological characterization of sleep perfusates, Pharmacol. Biochem. Behav., 5 (1976) 165-173.

35. Steriade, M. and Hobson, J.A., Neuronal activity during the sleep-waking cycle, Prog. Neurobiol., 6 (1976) 155-376.

36. Stern, W.C. and Morgane, P.J., Sleep and memory: Effects of growth hormone on sleep, brain biochemistry and behavior, in Neurobiology of Sleep and Memory, (R.R. Drucker-Colín and J.L. McGaugh, eds.), Academic Press, New York, 1977 pp. 373-410.

37. Stern, W.C. Jalowiec, E., Shabshalowtiz, H. and Morgane, P.J., Effects of growth hormone on sleep-waking patterns in cats, Horm. Behav., 6 (1975) 189-196.

38. Stern, W.C., Morgane, P.J., Panksepp, J., Solovick, A.J. and Jalowiec, J.E., Elevation of REM sleep following inhibition of protein synthesis, Brain Res., 47 (1972) 254-258.

39. Takahashi, Y., Kipnis, D.M. and Daughaday, W.H., Growth hormone secretion during sleep, J. Clin. Invest., 47 (1968) 2079-2090.

CONTRIBUTORS

RAUL AGUILAR
Departamento de Neurociencias
Centro de Investigaciones en
    Fisiología Celular
Universidad Nacional Autónoma
    de México
México 20, D.F.

K. ANDERSSON
Department of Histology
Karolinska Institute, S-104 01
Stockholm, Sweden

CLORINDA ARIAS
Departamento de Neurociencias
Centro de Investigaciones en
    Fisiología Celular
Universidad Nacional Autónoma
    de México
México 20, D.F.

R. AZAD
Arthur V. Davis Center for
    Behavioral Neurobiology
The Salk Institute
La Jolla CA 92037

ALEJANDRO BAYON
Departamento de Neurociencias
Centro de Investigaciones en
    Fisiología Celular
Universidad Nacional Autónoma
    de México
México 20, D.F.

FLOYD E. BLOOM
Arthur V. Davis Center for
    Behavioral Neurobiology
The Salk Institute
La Jolla CA 92037

HARRY BRADFORD
Department of Biochemistry
Imperial College
London S.W.7 2AZ, England

PIETRO CALISSANO
Laboratorio di Biologia
    Cellulare, C.N.R.
Via Romagnosi 18A, and
Instituto di Anatomia Comparata
Universita di Roma
Via Borelli 50, Roma

JOSE MARIA CALVO
Unidad de Investigaciones
    Cerebrales
Instituto Nacional de Neurología
    y Neurocirugía
México 22, D.F.

L. CASTELLANI
Laboratorio di Biologia
    Cellulare, C.N.R.
Via Romagnosi 18A, and
Instituto di Anatomia Comparata
Universita di Roma
Via Borelli 50, Roma

M. CONDES LARA
Unidad de Investigaciones
    Cerebrales
Instituto Nacional de Neurología
    y Neurocirugía
México 22, D.F.

CARL W. COTMAN
University of California, Irvine
Irvine, California 92717

ROBERT J. DeLORENZO
Yale University School of
    Medicine
Department of Neurology
333 Cedar Street, New Haven,
    CT 06510

RENE DRUCKER-COLIN
Departamento de Neurociencias
Centro de Investigaciones en
    Fisiología Celular
Universidad Nacional Autónoma
    de México
México 20, D.F.

AUGUSTO FERNANDEZ GUARDIOLA
Unidad de Investigaciones
    Cerebrales
Instituto Nacional de Neurología
    y Neurocirugía
México 22, D.F.

FRODE FONNUM
Norwegian Defence Research
    Establishment
Division for Toxicology
N-2007 Kjeller, Norway

KJELL FUXE
Department of Histology
Karolinska Institute, S-104 01
Stockholm, Sweden

KENAN HAVER
Department of Pathology
Mount Sinai School of Medicine
City University of New York
New York, NY 10029

T. HOKFELT
Department of Histology
Karolinska Institute, S-104 01
Stockholm, Sweden

MARK KLEIN
Department of Physiology and
    Division of Neurobiology and
    Behavior
College of Physicians &
    Surgeons
Columbia University
New York, N.Y. 10032

I. KVALE
Norwegian Defence Research
    Establishment
Division for Toxicology
N-2007 Kjeller, Norway

ROLANDO LARA
Departamento de Neurociencias
Centro de Investigaciones en
    Fisiología Celular
Universidad Nacional Autónoma
    de México
México 20, D.F.

RODOLFO LLINAS
Department of Physiology and
    Biophysics
New York University Medical Center
550 First Avenue
New York, N.Y. 10016

LOURDES LUGO
Departamento de Neurociencias
Centro de Investigaciones en
    Fisiología Celular
Universidad Nacional Autónoma
    de México
México 20, D.F.

D. MALTHE-SØRENSSEN
Norwegian Defence Research
    Establishment
Division for Toxicology
N-2007 Kjeller, Norway

D. MERCANTI
Laboratorio di Biologia
    Cellulare C.N.R.
Via Romagnosi 18A, and
Instituto di Anatomia Comparata
Universita di Roma
Via Borelli 50, Roma

G. MONACO
Laboratorio di Biología
    Cellulare C.N.R.
Via Romagnosi 18A, and
Instituto di Anatomia Comparata
Universita di Roma
Via Borelli 50, Roma

JULIO MORAN
Departamento de Neurociencias
Centro de Investigaciones en
    Fisiología Celular
Universidad Nacional Autónoma
    de México
México 20, D.F.

HERMINIA PASANTES-MORALES
Departamento de Neurociencias
Centro de Investigaciones en
    Fisiología Celular
Universidad Nacional Autónoma
    de México
México 20, D.F.

MIGUEL PEREZ DE LA MORA
Departamento de Neurociencias
Centro de Investigaciones en
    Fisiología Celular
Universidad Nacional Autónoma
    de México
México 20, D.F.

LOURIVAL POSSANI
Departamento de Neurociencias
Centro de Investigaciones en
    Fisiología Celular
Universidad Nacional Autónoma
    de México
México 20, D.F.

SAUL PUSZKIN
Department of Pathology
Mount Sinai School of Medicine
City University of New York
New York, NY 10029

PABLO RUDOMIN
Department of Physiology and
    Biophysics
Centro de Investigación y de
    Estudios Avanzados
Instituto Politécnico Nacional
México 14, D.F.

MARIA ELENA SANDOVAL
Departamento de Neurociencias
Centro de Investigaciones en
    Fisiología Celular
Universidad Nacional Autónoma
    de México
México 20, D.F.

WILLIAM SCHOOK
Department of Pathology
Mount Sinai School of Medicine
City University of New York
New York, NY 10029

W.J. SHOEMAKER
Arthur V. Davis Center for
    Behavioral Neurobiology
The Salk Institute
La Jolla CA 92037

GEORGE R. SIGGINS
Arthur V. Davis Center for
    Behavioral Neurobiology
The Salk Institute
La Jolla CA 92037

K.K. SKREDE
Norwegian Defence Research
    Establishment
Division for Toxicology
N-2007 Kjeller, Norway

A. SØREIDE
Norwegian Defence Research
    Establishment
Division for Toxicology
N-2007 Kjeller, Norway

ENRIQUE STEFANI
Department of Physiology and
    Biophysics
Centro de Investigación y
    Estudios Avanzados
Instituto Politécnico Nacional
México 14, D.F.

RICARDO TAPIA
Departamento de Neurociencias
Centro de Investigaciones en
    Fisiología Celular
Universidad Nacional Autónoma
    de México
México 20, D.F.

I. WALAAS
Norwegian Defence Research
    Establishment
Division for Toxicology
N-2007 Kjeller Norway

K. WALTON
Department of Physiology and
    Biophysics
New York University Medical
    Center
550 First Avenue
New York, N.Y. 10016